# MULTIVARIABLE COMPUTER CONTROL

# MULTIVARIABLE COMPUTER CONTROL

## A Case Study

D. Grant Fisher
*and*
Dale E. Seborg

*University of Alberta*
*Department of Chemical Engineering*
*Edmonton, Alberta, Canada*

1976

NORTH-HOLLAND PUBLISHING COMPANY - AMSTERDAM · OXFORD
AMERICAN ELSEVIER PUBLISHING COMPANY, INC. - NEW YORK

North-Holland ISBN: 0 7204 0356 1
American Elsevier ISBN: 0 444 11039 9

Published by:

North-Holland Publishing Company - Amsterdam
North-Holland Publishing Company, Ltd. - Oxford

Sole distributors for the U.S.A. and Canada:

American Elsevier Publishing Company, Inc.
52 Vanderbilt Avenue
New York, N.Y. 10017

PRINTED IN THE NETHERLANDS

## PREFACE

During the past decade there has been a tremendous growth in the number of analytical and design techniques available to the control engineer. Some of these are still in the initial state of development but others have been adapted and extended to the point where their tremendous potential can be exploited in industrial applications. Fortunately, this growth in available techniques has been accompanied, in fact preceded, by the availability of computing hardware that has the capability of implementing almost any control technique an engineer would like to apply to a real process. The rapid growth during the 1960's of real-time computer applications in industrial plants did not level out as many predicted, but due to the introduction of minicomputers and improved interface hardware, has actually expanded. The novelty of real-time computers has been largely overcome and applications of Direct Digital Control (DDC) and many management functions have become quite routine. There have been a few misguided computer applications, some of them highly publicized, that must be classified as failures. However, there have been many more applications that have been outstanding successes, as evidenced by repeated duplication of the original installations rather than by detailed public announcements. In fact the cumulative experience to date has led many knowledgeable people to believe that the future promises more reliable, more technically sophisticated, and more highly integrated process control and management systems.

At this point in time it would appear that the primary needs in the area of computer control are:

(a) the availability of people with the technical training and vision to develop and implement the most promising of the presently available modern control techniques,

(b) engineering evaluation of the currently available control techniques and the needs of industry,

(c) experimental implementation and development of the most promising techniques.

Continuing research into new areas is also required and it was this need, plus considerations such as those listed above, that led to a series of computer-control projects in the Department of Chemical Engineering at the University of Alberta. The objective was to examine promising modern control techniques and with due consideration of the practical constraints of the personnel and equipment available, to evaluate the techniques by application to pilot plant units. This Case Study is concerned with the implementation of computer control on a pilot plant size, double effect evaporator and the subsequent evaluation of multivariable design, analysis and control techniques. The Case Study is compiled with the hope that a presentation of the results of applying several techniques, to a single experimental unit, will enable the reader to see the relative advantages, disadvantages, potential and practicality of these methods.

It can be argued that an evaporator is not the most appropriate piece of equipment for evaluating techniques for industrial applications. However, its basic principles of operation are easily understood by anyone in the process industry, it is simple enough to be modelled from first principles yet complicated enough to require parameter estimation and process identification, and it includes an interacting set of flow, pressure, temperature and concentration variables. The evaporator itself is simpler and more highly instrumented than most industrial units. However, the experimental results should still give a good indication of how the new multivariable methods would work in actual applications. It is hoped that despite the shortcomings of this Case Study, the reader will develop insight into the methods presented and obtain a better basis from which to solve the problems of particular interest to him or her.

Since this Case Study consists of a collection of papers written over a period of several years, the presentation is not as smooth, continuous or complete as a textbook nor is the Case Study intended to be one. The introduction for each section should, however, help bridge the gaps and point out the interrelations between the different parts.

The reader will note from the co-authors for the individual articles that many different individuals have been involved in this work. We gratefully acknowledge the contributions made by our students as part of their thesis projects, by our colleagues, and by the staff in the department's Data Acquisition, Control and Simulation Centre. Without the financial support from the University, The National Research Council of Canada and from individual companies, this work would not have been possible. For any errors, omissions or shortcomings in this case study, we accept full responsibility.

D. GRANT FISHER

DALE E. SEBORG

Edmonton, Canada

TABLE OF CONTENTS

## Section 1: INTRODUCTION



## Section 2: MODELLING, SIMULATION AND DESIGN



## Section 3: CONVENTIONAL CONTROL



## Section 4: MULTIVARIABLE FEEDBACK CONTROL

## Section 5: MULTIVARIABLE SERVO CONTROL



## Section 6: OTHER MULTIVARIABLE CONTROL TECHNIQUES



## Section 7: ON-LINE ESTIMATION AND FILTERING



## Section 8: EDUCATIONAL ASPECTS OF COMPUTER CONTROL



## APPENDIX

# Section 1: INTRODUCTION

CONTENTS:



COMMENTS:

This section presents a review of some of the process dynamics and control projects undertaken in the Department of Chemical Engineering at the University of Alberta. There is no technical detail or theory that is not included in later sections of the Case Study but this section provides a good introduction and overview of the research activities. A description of the computer hardware and software in the Data Acquisition, Control and Simulation (DACS) Centre that was used in this work is available in the reference cited below.

REFERENCE:

[1] "Data Acquisition, Control and Simulation Centre", a booklet available from the Department of Chemical Engineering, University of Alberta, Edmonton, Canada (1974).

# Advanced Computer Control Improves Process Performance

D. G. FISHER and D. E. SEBORG, Univ. of Alberta

Multivariable and other modern control techniques can produce significantly better results than conventional methods and have tremendous potential for industrial application. This article presents specific results that have been obtained on a pilot plant evaporator, and reviews basic concepts and practical aspects of applying various techniques in advanced control.

THE PRIMARY CHALLENGE facing people in the control field today is to formulate their problems more definitely, clarify their objectives and glean from the wealth of available alternatives the practical approaches that can be applied to their specific problems. This challenge led to a number of continuing projects in the Department of Chemical Engineering at the University of Alberta, to investigate and evaluate some of the most promising techniques of modern control theory by applying them to computer-controlled pilot plant units.

This article deals specifically with the application of several techniques to a pilot plant evaporator. The objective is not an all-encompassing review of available techniques but rather a review and discussion of results obtained in specific projects. The discussion does not include rigorous definitions or derivations but instead concentrates on concepts, comparisons and practical aspects of the various techniques. Details, derivations, related work, and additional data are available in the references.

### Modeling multivariable systems

People in industry frequently dismiss modern control techniques because they think a suitable mathematical model cannot be developed. It is true that a model of the process is needed at some stage in most control system design techniques. However, powerful methods are available to derive suitable models through empirical testing or theoretical analysis or a combination of the two. Some control systems can even be designed to learn or adapt, so that they will compensate for unknown or changing factors after they are connected to the process.

The pertinent questions today are "What type of model is required? How complicated? How exact?" The general answer is that development of a suitable model is simply part of the control system design procedure. In practical applications, this might require several iterations rather than a single definitive step.

A schematic diagram of the pilot plant evaporator used for most of the investigations reported here is illustrated in Figure 1. It has a complex feed system which permits operation of the equipment in a cyclic fashion, as well as the introduction of load changes and disturbances in the feed conditions. Controlled flows of concentrated triethylene glycol solution and water are temperature-controlled by means of steam heaters, and then mixed in the proportions necessary to produce a feed stream having the desired concentration and flow rate. This blending equipment is not included in the model, however, because it has a relatively fast dynamic response and does not interact with the state variables of the main evaporator.

The first effect is a short-tube vertical calandria-type unit with natural circulation. The 9-in. diameter unit has an operating holdup of 2 to 4 gallons, and its 32 stainless steel tubes, 3/4-in. o.d. by 18 in. long, provide approximately 10 square feet of heat transfer surface altogether.

The second stage is a long-tube vertical effect set up for either natural or forced circulation. It has a heat

Instrumentation Technology, Vol. 20, No. 9, pp. 71-77 (1973).

transfer area of 5 square feet and is made up of three 6-ft long, 1-in. o.d. tubes. Capacity of the circulating system is about 3 gallons.

The evaporator is fully instrumented with commercial electronic controllers. Solution concentrations are measured by in-line refractometers similar to those described by Stackhouse (Ref. 1).

Experimental studies on the evaporator have clearly shown that different models are desirable for different applications. For example, a steady state model yielded the gains necessary to obtain significant improvement with static feedforward control (Ref. 2). Simple transfer function models, relating specific pairs of input/output variables, have proven adequate for dynamic feedforward compensation (Ref. 2,3). Third-order state-space models have worked as the basis for the design of optimal feedback systems (Ref. 4,5). A fifth-order linear model appears to be reasonable for most multivariable control techniques (Ref. 5,6). A nonlinear dynamic model proved best for off-line simulation purposes (Ref. 5,7). For optimal servo control, parameter estimation (curve fitting) techniques proved necessary to modify some of the theoretically derived constants in a fifth-order model and obtain better agreement with experimental evaporator data (Ref. 7,8,9,10). In other words, a single model is seldom satisfactory for all stages in the design procedure or for use in all parts of the final control system.

The pilot plant evaporator discussed above is obviously simpler than a typical industrial unit as shown in Figure 2. Most industrial installations will also differ from one another, because of various possible combinations of forward, backward and parallel flow systems for both feed and heating medium; and because of design variations among the several evaporator effects. The question therefore arises, "Is there a rational general approach to modeling such systems, or must each one be approached on an individual basis?"

A generalized approach to the modeling of evaporators has been presented by Newell and Fisher (Ref. 11). Using this four-step procedure to develop a model of the pilot plant evaporator of Figure 1 led to a system of 10 nonlinear, first-order ordinary differential equations. The equations were linearized by considering only perturbations around the steady state operating point and putting them into the standard state-space form:

$$\dot{\mathbf{x}}(t) = \mathbf{A}\,\mathbf{x}\,(t) + \mathbf{B}\mathbf{u}(t) \qquad (1)$$

$$\mathbf{y}(t) = \mathbf{C}\,\mathbf{x}\,(t)$$

where:

**u**, **y** and **x** are the input, output and state vectors, respectively.

**A**, **B** and **C** are constant-coefficient matrices.

By neglecting factors such as the heat capacity of heat transfer surfaces, the state-space evaporator model can be reduced to lower order without significant loss in accuracy. The discussion and results that follow are based on a model with a fifth-order state vector and a sixth-order input vector made up of three control variables and three disturbance variables (see Nomenclature table).

**Simplicity vs complexity**

Earlier investigators using the pilot plant evaporator did not adopt the approach outlined above, but instead derived their own models directly. These models ranged from first-order to fifth-order, and incorporated such physical assumptions as zero heat capacity in the heat transfer surfaces. Models obtained by reduction of the tenth-order model were in good agreement with those derived directly from physical assumptions (Ref. 5,6).

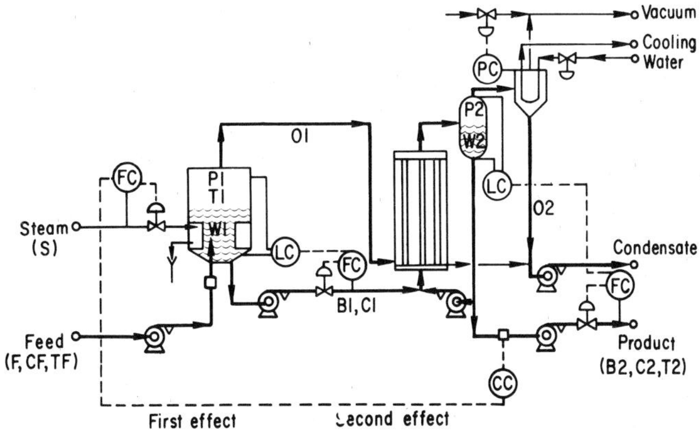


*Figure 1. Schematic diagram of the double-effect, pilot plant evaporator at the University of Alberta. The control system shown is a single variable, multiloop configuration used as the base case. Over 50 process measurements and all final control elements are interfaced to an IBM 1800 digital computer. The evaporator is normally run under DDC or special multivariable control algorithms. Letter symbols in this and other figures are explained in the Nomenclature table.*

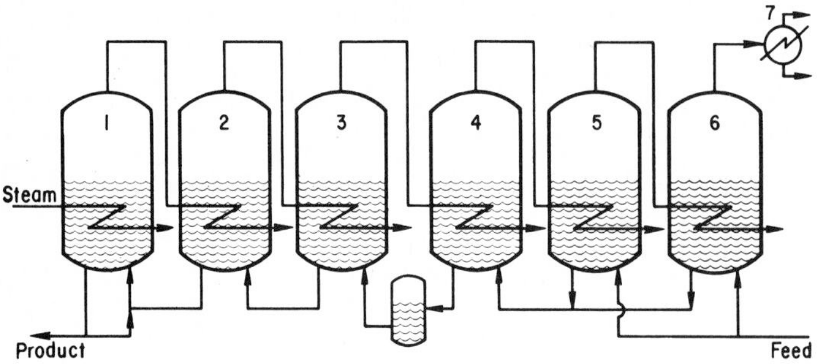


*Figure 2. A schematic diagram of a typical six-effect industrial evaporator used to derive generalized material and energy balance equations.*

Our experience with modeling has been that it is better to err on the side of complexity than to introduce assumptions prematurely or unnecessarily. Derivation of a rigorous model gives the engineer better insight and perspective and sometimes shows that the sum of certain factors is significant even though it would appear reasonable to neglect each when they are considered individually.

Most control engineers have access to a digital computer to assist with the design of control systems and frequently the final installation involves an on-line, real-time computer. Therefore, although complexity is not desirable for its own sake, it is no longer reasonable to reject techniques simply because they cannot be calculated by hand or implemented with conventional instruments.

### Process and parameter estimation

Techniques for process and parameter estimation are frequently required to modify the model so it will better approximate the actual performance of the process. This is a very broad and complex area that has recently been reviewed by Nieman *et al* (Ref. 12).

Quasi-linearization plus linear programming techniques were used by Nieman (Ref. 7,8) to adjust selected parameters of the theoretical nonlinear fifth-order state-space model so it would be in better agreement with the open-loop response data from the evaporator. Figure 3 compares the theoretical model response, the fitted model response and the actual process response for an equivalent disturbance. The change in product concentration here is much larger than occurs in most control studies and therefore tends to overemphasize the difference between the models. Although there is a significant error in the response of the theoretical model, it was adequate for most of the design methods discussed below.

### Multivariable feedback control

The control objective of the pilot plant evaporator was defined as maintaining the output concentration, $C2$, at a constant value in spite of disturbances in feed conditions. This can be accomplished using conventional, cascaded, single-variable control loops as illustrated in Figure 1. That is, $C2$ is controlled by manipulating the inlet steam $S$; and the holdup $W2$ is controlled by manipulating the outlet flow, $B2$. Similarly $W1$ is controlled by $B1$. The pairing of

### Nomenclature

| **Process variables** | | **Steady state values** |
|---|---|---|
| State vector, **x**: | W1 first effect holdup | 14.6 kg |
| | C1 first effect concentration | 4.85% |
| | H1 first effect enthalpy | 335 kj/kg |
| | W2 second effect holdup | 18.8 kg |
| | C2 second effect concentration | 9.64% |
| Control vector, **u**: | S steam flowrate | 0.86 kg/min |
| | B1 first effect bottoms | 1.5 kg/min |
| | B2 second effect bottoms | 0.77 kg/min |
| Load vector, **d**: | F feed flowrate | 2.27 kg/min |
| | CF feed concentration | 3.2 % |
| | HF feed enthalpy | 3.2 kj/kg |
| Output vector, **y**: | equal to $(W1, W2, C2)^T$ | |
| Other variables: | 01 overhead vapor from first effect | 0.77 kg/min |
| | 02 overhead vapor from second effect | 0.73 kg/min |
| | P1 pressure in first effect | 68.9 kN/m² |
| | P2 pressure in second effect | 0.38 m Hg |
| | TF temperature of feed | 88 C |
| | T1 temperature in first effect | 107 C |
| | T2 temperature in second effect | 71 C |
| Subscripts: | FB feedback | |
| | FF feedforward | |
| | I integral | |
| | m model | |
| | SP setpoint | |
| | ss steady state | |
| Superscripts: | T matrix/vector transpose | |
| | i counter for time intervals | |
| Abbreviations: | CC Concentration controller | |
| | LC Level controller | |
| | FC Flow controller | |
| | DDC Direct digital control | |
| | P+I Proportional-plus-integral | |

variables in this manner may be obvious, but it can also be derived from a sensitivity analysis (Ref. 5,13). Experimental studies showed that reasonable control could be obtained with this approach (Ref. 2,4); that feed flow was the most serious disturbance; that there were strong interactions between variables; that simple feedforward compensation would give significant improvement (Ref. 4,14); and that the process is nonlinear in nature.

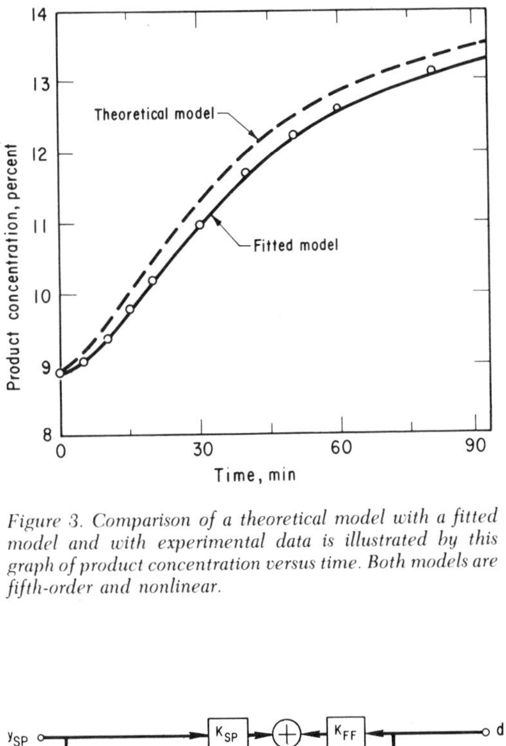


*Figure 3. Comparison of a theoretical model with a fitted model and with experimental data is illustrated by this graph of product concentration versus time. Both models are fifth-order and nonlinear.*

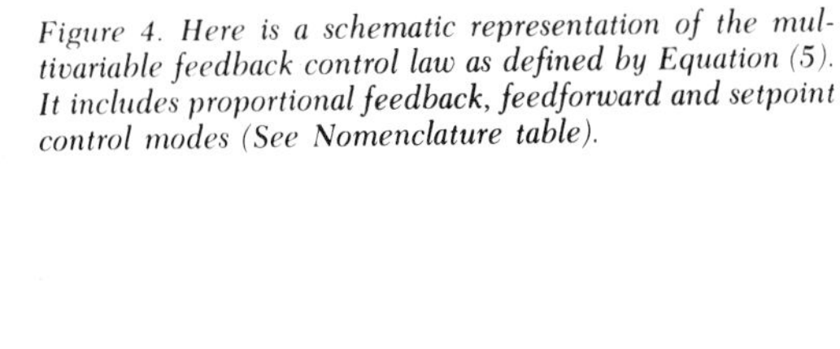

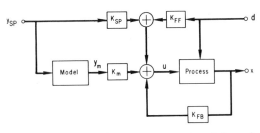

*Figure 4. Here is a schematic representation of the multivariable feedback control law as defined by Equation (5). It includes proportional feedback, feedforward and setpoint control modes (See Nomenclature table).*

Substitution of direct digital control algorithms for conventional industrial controllers showed that equivalent performance could be obtained with DDC, with due consideration of such additional factors as sampling time, filter constants and limits on input and output signals.

It is significant, relative to later discussion, that the fifth-order linear evaporator model is *not* suitable for determining values of the controller constants in the single-variable control system shown in Figure 1. The multiloop control configuration shown there was chosen by the user, based on experience and intuition. In multivariable control the situation is significantly different.

The approach taken to develop a multivariable control system was the discrete version of the "linear-quadratic-Gaussian" control problem originally developed by Kalman (Ref. 15), and recently the subject of an extended review (Ref. 16). The first step is to derive a suitable linear, discrete, state-space model of the process in the following form:

$$\mathbf{x}\,[(n+1)T] = \phi(T)\mathbf{x}\,(nT) + \Delta(T)\mathbf{u}(nT) \qquad (2)$$

$$\mathbf{y}(nT) = \mathbf{C}\,\mathbf{x}\,(nT) \qquad n = 0,1,2\ldots$$

The fundamental matrix $\phi(T)$ and the coefficient matrix $\Delta(T)$ can be derived directly from the coefficient matrices in Equation (1) and the two models are equivalent at the sampling points if $\mathbf{u}$ is constant over the sampling interval.

The control objective is then defined as a summed quadratic function of the state and control variables which can be written in simplified form as:

$$J = \sum_{i=1}^{N} [(\mathbf{e}_i{}^T \mathbf{Q}(\mathbf{e}_i) + (\mathbf{u}_{i-1})^T \mathbf{R}(\mathbf{u}_{i-1})] \qquad (3)$$

where:

i identifies the time interval.

**e** is the process error (actual minus desired value).

N is a large integer (N>>O)

**Q** and **R** are weighting matrices of constants specified by the user.

The optimization problem of finding the control output **u** that will minimize the performance index $J$ can be solved using the techniques of dynamic programming, and results in a feedback control law of the form:

$$\mathbf{u}(nT) = \mathbf{K}_{FB}\mathbf{x}\,(nT) \qquad (4)$$

where $\mathbf{K}_{FB}$ is a matrix of proportional control constants. For a given model this is essentially a synthesis procedure in which the control engineer specifies the matrices **Q** and **R** in Equation (3) and the optimization procedure generates both the form of the control law and the control constants (Ref. 5,6).

It is, however, extremely difficult to modify the problem to accommodate constraints on the state and control variables. Note that if $\mathbf{K}_{FB}$ is diagonal, then

Equation (4) represents a multiloop proportional control system—that is, a set of independent, single-variable control loops.

The multivariable procedure has been extended by the authors and coworkers to include integral feedback control, feedforward control (Ref. 17) and provision for model following during setpoint changes (Ref. 6). The complete control law, illustrated in Figure 4, is given by:

$$\mathbf{u}(nT) = \mathbf{K}_{FB}\mathbf{x}(nT)+\mathbf{K}_m\mathbf{y}_m(nT)+ \mathbf{K}_{SP}\mathbf{y}_{SP}(nT) + \mathbf{K}_{FF}\,\mathbf{d}(nT) \quad (5)$$

Note that this one method generates the multivariable equivalent of conventional single-variable control techniques. However, more importantly, when applied to the evaporator the multivariable control law proved to be practical and robust, and gave significantly better control than the multiloop approach.

Figure 5 shows a direct comparison of the effect of a 20 percent increase in feed flow rate on the product concentration $C2$. It is obvious that multivariable control is significantly better than conventional methods. The economic advantage of the tighter control would depend on the particular application.

It is interesting that the multivariable feedback controller was designed using the same fifth-order model that was *not* suitable for the design of single-variable systems. Even though some of the elements of $\mathbf{K}_{FB}$ are an order of magnitude larger than the proportional control constants used experimentally with the multiloop system, the theoretically designed control law gave excellent control when applied experimentally (Ref. 6).

## Optimal servo control

Although the multivariable feedback control laws discussed in the previous section could handle setpoint changes, they were designed primarily to regulate a process about a set of constant operating conditions. A different approach is advantageous when it is desired to make a grade change on a production unit or to change the process from one set of operating conditions to another in an "optimum" manner.

Nieman and Fisher have shown (Ref. 9,10) that for a discrete, linear, time-invariant system such as defined by Equation (2), it is possible to formulate the optimization problem so it can be solved by a standard linear programming package. The problem must also have a linear performance criteria such as minimum time to implement the change or minimum sum-of-the-absolute error between actual and desired values. This technique also readily accommodates constraints on the state and control variables, and on their rates of change. Implementation consists of outputting the precalculated values of $[\mathbf{u}(nT),\ n = 1,2,3,\ldots N]$ at the appropriate control interval. On-line calculations are not required and in simple cases the implementation can be done manually.

When applied in an open-loop configuration, this technique is sensitive to modeling errors and unanticipated disturbances. Real-time calculation or adjustment of the bang-bang switching times can significantly reduce this problem. However, for practical industrial applications it is recommended that the optimal open-loop control policy $\mathbf{u}^\circ$, for a specified change, should be added to the existing feedback control scheme. This is shown in Figure 6 by the dotted box. The calculated optimal trajectory $\mathbf{y}^\circ$ is introduced as the setpoints of the feedback controllers.

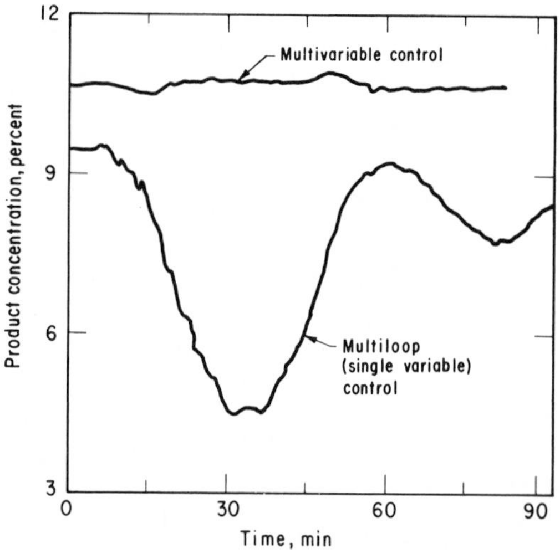


*Figure 5. Comparison of experimental response data from the evaporator, under multivariable control and under multiloop proportional-plus-integral feedback control.*

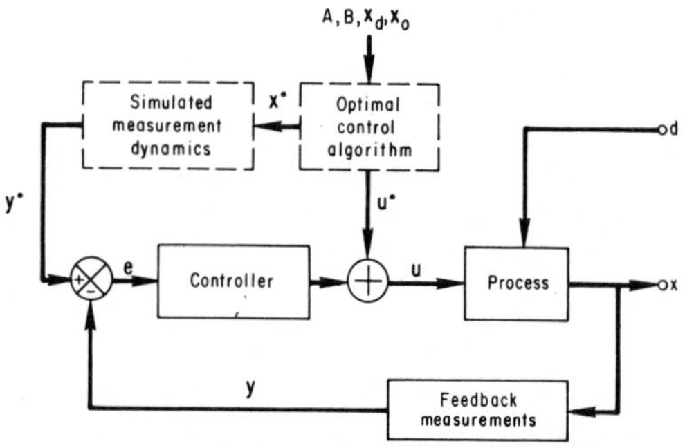


*Figure 6. Schematic diagram illustrating how an optimal open-loop servo control policy can be added to an existing multiloop or multivariable feedback control system.*

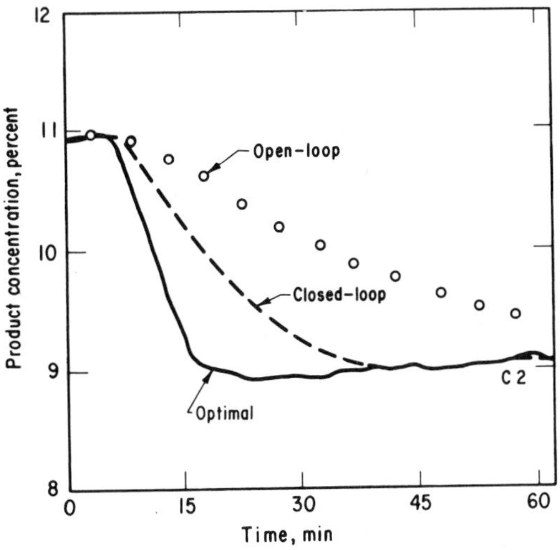


*Figure 7. Comparison of open-loop, closed-loop and optimal response data for experimental runs of the evaporator. The open-loop response is to a step change in inlet steam flow; closed-loop response to a step change in the DDC setpoint; and optimal response is based on bang-bang switching of the steam flow.*

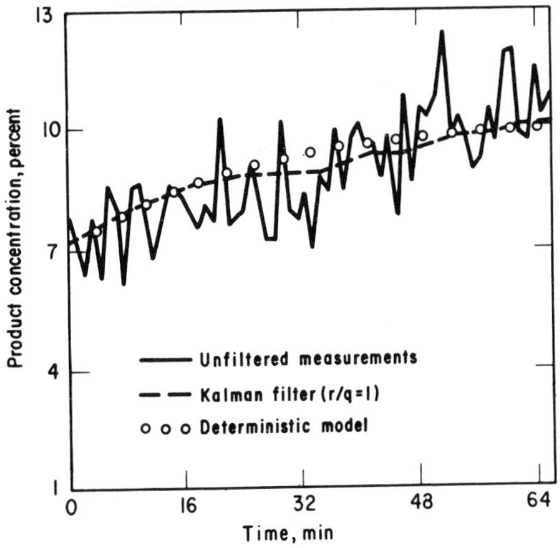


*Figure 8. Simulated data illustrates the performance of a discrete Kalman filter designed for the evaporator. The solid line is the filter input, the dashed line is the filter output, and the circles depict noise-free response of the model.*

If the calculated and actual process responses are identical, the transient proceeds as in the optimal open-loop case. However, if modeling errors or unexpected disturbances are present, then the feedback system, although suboptimal, tends to correct for these errors (Ref. 10).

The degree of improvement that can be achieved in the response of the evaporator product concentration is illustrated in Figure 7. Here, the optimal response time of the evaporator is about half of that obtained using conventional DDC control. Techniques that can produce this degree of improvement in pilot plant applications justify careful consideration by industry.

**Optimal filtering and estimation**

Two problems that commonly arise in industrial applications are noise and unmeasured variables. Optimal multivariable feedback controllers, in particular, have relatively high gains which make them sensitive to measurement noise and which require that estimates of *all* the state variables of the process be available.

If a state-space model of the process is available and the characteristics of the noise are known, it is possible to design an optimal filter that will minimize the error between the actual and measured values of the process variables. A Kalman filter (Ref. 18) was implemented on the evaporator and at each sampling interval an estimate of the state vector $\hat{\mathbf{x}}$ was calculated from:

$$\hat{\mathbf{x}}[(n+1)T] = \bar{\mathbf{x}}(nT) + \mathbf{K}[\mathbf{y}(nT) - \mathbf{C}\bar{\mathbf{x}}(nT)] \qquad (6)$$

**K** is a matrix of constants

$\bar{\mathbf{x}}$ is the value of the state variables calculated from the process model.

In the ideal case the matrix **K** is calculated off-line from the process model and measured noise characteristics. But in applications to the pilot plant evaporator, the assumed noise characteristics were treated as design parameters and adjusted to modify the filter response for different process conditions.

Figure 8 shows a comparison of measured, filtered and model output values of the product concentration from a simulated evaporator run, starting with an initial error of 30 percent in C2 and returning to setpoint. In the ideal case, the filtered values would coincide with the calculated values.

This filter has been evaluated experimentally in a series of open-loop and closed-loop runs using different process conditions, disturbances and filters (Ref. 18). The conclusions reached are as follows:

- The filter is robust and practical and gives true values of the process state variables.
- If tightly tuned to reject measurement noise, the filter becomes sensitive to unmeasured disturbances. If the disturbances are measured, its performance is excellent.
- The performance of the filter is relatively insensitive to errors in the model parameters. (For example,

changes of ±25 percent in the holdups in both evaporator effects produce negligible changes in the filter outputs).

• The filter can be extended or augmented to provide estimates of unmeasured state variables and/or constant biases in the noise signals. (It can also be extended to provide continuously updated estimates of process parameters that change with time.)

The Kalman optimal multivariable filter, as defined by Equation (6), was found to be much better in removing noise and introduced less attenuation and phase shift than single-variable exponential filters (Ref. 19).

**Other control approaches**

Adaptive control techniques are particularly attractive for industrial application because the controller parameters are adjusted on-line to compensate for uncertainties in the process model and for changing process conditions. A multivariable, model-reference adaptive control system has recently been successfully applied to the pilot plant evaporator (Ref. 20).

A generalized computer program developed over the last few years makes use of network diagram techniques (as in CPM and PERT methods) to define interactions and sequencing in discrete control activities (Ref. 21,22). This program has been successfully used to automate startup of the evaporator.

The control projects summarized in this article demonstrate that modern control techniques such as optimal, multivariable feedback control, servo control and filtering produce significantly better results than conventional methods when applied to a pilot plant evaporator.

Computer applications must grow beyond simple emulation of conventional control instruments and/or automation of existing procedures, and reach out to encompass new approaches and new techniques.

**References**

1. Stackhouse, G. E., "A New Refractometer for Continuous Black Liquor Analysis," ISA/CPPA International Symposium on Pulp and Paper Process Control, Vancouver, 1969, pp. 187-194.

2. Fehr, M., "Computer Control of an Evaporator," M.Sc. thesis, University of Alberta, Dept. of Chemical & Petroleum Engineering, 1969.

3. Wilson, A. H., "A Feedforward Control System for a Double Effect Evaporator," M.Sc. thesis, University of Alberta, Dept. of Chem. & Pet. Eng., 1966.

4. Jacobson, B. A. and Fisher, D. G., "Computer Control of a Pilot Plant Evaporator," *The Chemical Engineer*, (in press), England.

5. Newell, R. B., "Multivariable Computer Control of an Evaporator," Ph.D. thesis, University of Alberta, Dept. of Chem. & Pet. Eng., 1971.

6. Newell, R. B. and Fisher, D. G., "Experimental Evaluation of Optimal Multivariable Regulatory Controllers with Model-Following Capabilities," *Automatica*, Vol. 8, No. 3, 1972, pp. 247-262.

7. Nieman, R. E., "Application of Quasilinearization and Linear Programming to Control and Estimation Problems," Ph. D. thesis, University of Alberta, Dept. of Chem. & Pet. Eng., 1971.

8. Nieman, R. E. and Fisher, D. G., "Parameter Estimation Using Quasilinearization and Linear Programming," *Canadian Journal of Chemical Engineering*, Vol. 50, No. 6, 1972, pp. 802-806.

9. Nieman, R. E. and Fisher, D. G., "Experimental Evaluation of Time-Optimal, Open-Loop Control," *Transactions of the Institution of Chemical Engineers*, Vol. 51, No. 2, 1973, pp. 132-140.

10. Nieman, R. E. and Fisher, D. G., "Experimental Evaluation of Optimal Multivariable Servo Control in Conjunction with Conventional Regulatory Control," *Chemical Engineering Communications* (in press).

11. Newell, R. B. and Fisher, D. G., "Model Development, Reduction and Experimental Evaluation for a Double Effect Evaporator," *Industrial Engineering Chemical Process Design & Development*, Vol. 11, No. 2, 1972, pp. 213-221.

12. Nieman, R. E., Fisher, D. G. and Seborg, D. E., "A Review of Process Identification and Parameter Estimation Techniques," *International Journal of Control*, Vol. 13, No. 2, 1971, pp. 209-264.

13. Ritter, R. A. and Andre, H., "Evaporator Control System Design," *Canadian Journal of Chemical Engineering*, Vol. 48, No. 6, 1970, p. 696.

14. Nisenfeld, A. E. and Hoyle, D. L., "Dynamic Feedforward Control of Multi-effect Evaporators," *Instrumentation Technology*, Vol. 17, No. 2, 1970, pp. 49-54.

15. Kalman, R. E., "A New Approach to Linear Filtering and Prediction Theory," *Transactions ASME, Journal of Basic Engineering*, Vol. 83, 1960, p. 35; Vol. 83, 1961, p. 95.

16. *IEEE Transactions on Automatic Control*, Vol. AC-16, No. 6, 1971.

17. Newell, R. B., Fisher, D. G. and Seborg, D. E., "Computer Control Using Optimal, Multivariable Feedforward Feedback Algorithms," *AIChE Journal*, Vol. 18, No. 5, 1972, pp. 976-984.

18. Hamilton, J. C., Seborg, D. E. and Fisher, D. G., "An Experimental Evaluation of Kalman Filtering," *AIChE Journal* (in press).

19. Hamilton, J. C., "An Experimental Investigation of State Estimation in Multivariable Control Systems," M.Sc. thesis, University of Alberta, Dept. of Chem. & Pet. Eng., 1972.

20. Oliver, W. K., Seborg, D. E. and Fisher, D. G., "Model Reference Adaptive Control Based on Liapunov's Direct Method," *Chemical Engineering Communications* (in press).

21. Hayward, S. W., "Alberta Discrete Activity Monitor System (ADAMS)," M.Sc. thesis, University of Alberta, Dept. of Chem. & Pet. Eng., 1972.

22. Hayward, S. W. and Disher, D. G., "Alberta Discrete Activity Monitor Program (ADAMS)," 22nd Canadian Chemical Engineering Conference, Toronto, September 1972.

---

D. GRANT FISHER is professor and chairman, and DALE E. SEBORG is an associate professor in the Department of Chemical Engineering at the University of Alberta in Edmonton, Canada. Article is based on a paper presented at the 2nd International Symposium on Pulp & Paper Process Control, Montreal, Canada, 1973.

# Section 2: MODELLING, SIMULATION AND DESIGN

CONTENTS:



COMMENTS:

The first step in the application of most modern control techniques is to develop a suitable mathematical model of the process. In some cases this can be done by deriving the appropriate equations from first principles and calculating the parameters from physical property data and/or design correlations. In other cases the parameters must be determined by using actual plant data. When it is impractical to derive the equations analytically, then "process identification techniques" can be applied in experimental tests on the actual process.

Analog, digital or hybrid simulation of all or part of the system of interest is frequently helpful in order to give the control engineer a "feel" for system performance and to compare alternative design strategies. The procedure followed for each major control application is different but normally involves a sometimes iterative sequence of modelling, design, simulation and evaluation stages.

The first paper describes the development, reduction and evaluation of dynamics models of the pilot plant evaporator. Several different models are compared and it is concluded that the decision about which model is "best" depends on the particular way it is to be used in the final application.

Simulation on a digital computer is one way of comparing alternative models and paper 2.2 describes the GEneral, Multipurpose, Simulation and COntrol PackagE (GEMSCOPE) developed and used at the University of Alberta. Most modern control techniques are too complicated for "paper and pencil" calculation so that an interactive computer aided system such as GEMSCOPE is essential. The examples used to demonstrate different features of the program provide a general overview and summary of several projects which are discussed in sections 4 and 5.

Paper 2.3 discusses the reduction of discrete linear, time-invariant state space models which is one of the options in GEMSCOPE. Papers in section 4 include an evaluation of control systems designed using reduced-order models and also include an application of the same techniques to eliminate selected state variables from multivariable feedback control laws.

Hybrid simulation is an alternative to digital simulation and has been used to evaluate model-reference adaptive control as described in paper 2.4. (The simulated and experimental results from the evaporator application are included in section 4.)

Extensive reviews of process identification and parameter estimation techniques and their application to chemical and physical processes are presented in references [1] - [3]. Reference [4] describes the use of quasilinearization plus linear programming techniques for off-line estimation of model parameters including those of the evaporator.

REFERENCES:

[1] R.E. Nieman, D.G. Fisher and D.E. Seborg, "A Review of Process Identification and Parameter Estimation Techniques", International Journal of Control, 13, 209-264 (1971).

[2] K.J. Åström and P. Eykhoff, "System Identification - A Survey", Automatica, 7, 123-162 (1971).

[3] I. Gustavsson, "Survey of Applications of Identification in Chemical and Physical Processes", Automatica, 11, 3-24 (1975).

[4] R.E. Nieman and D.G. Fisher, "Parameter Estimation Using Linear Programming and Quasilinearization", Canadian Journal of Chemical Engineering, 50, 802-806 (1972).

# Model Development, Reduction, and Experimental Evaluation for an Evaporator

**Robert B. Newell**[1] **and D. Grant Fisher**[2]
*Department of Chemical and Petroleum Engineering, University of Alberta, Edmonton 7, ALT, Canada*

A generalized approach to the modeling of multieffect evaporators is presented which separates the development of dynamic equations from the specification of evaporator configuration. The result is a modular approach which is effective and convenient to use. A tenth-order nonlinear, dynamic model of a double-effect pilot plant evaporator was derived using this approach and then simplified and linearized to produce a fifth-order state-space model which gave generally good comparisons with experimental open-loop responses. Lower-order linear models were also developed but gave satisfactory results only in specific applications. The performance of models in the design and experimental implementation of conventional, inferential, feedforward, and multivariable optimal regulatory, and state-driving control systems is also examined.

This work is part of a continuing project which has been under way at the University of Alberta since 1964. The overall objective has been to develop modeling and control system design techniques that are of potential interest to industry and to evaluate them by application to the computer-controlled pilot-plant units. Initially the interest in developing mathematical models was for open-loop simulations and as an aid for the design of conventional control systems. However, more recently the emphasis has been on state-space models for use in the design of optimal multivariable controllers, state estimators, etc. The effect of model simplifications, such as linearization and the reduction of model order, on the design and performance of control systems has been of particular interest.

The first part of this paper contains a brief review of pertinent literature and presents a generalized approach to modeling industrial evaporators. The second part of the paper deals specifically with some of the models that have been developed for a double-effect evaporator and their performance in several different experimental applications.

Mathematical models of evaporator systems reported in the literature have used both empirical and theoretical approaches. Johnson (1960) presented a variety of empirical models of differing complexity and fitted parameters with experimental data from his falling-film evaporator. Nisenfeld and Hoyle (1970) considered simple empirical models for feedforward control and used two first-order lags and a time delay to dynamically compensate a six-effect evaporation process.

Theoretical derivations have also been presented in the literature. Anderson et al. (1961) derived a six-equation model for a single-effect, and simplified it to three differential equations. This was done by essentially neglecting vapor space and heat transfer dynamics. A frequency response comparison between the model and the equipment proved inconclusive. Manczak (1967) presented an analytical procedure to determine the dynamic properties of single and multiple effects. His relations were extensively linearized which resulted in a small range of applicability about the operating point. Zavorka et al. [1967] developed a general model for a single-effect of a commercial sugar evaporator. This model was extended to a triple-effect system after simplification and included nonlinear relations for the heat transfer coefficients in terms of solution concentrations and liquid levels. However, their analysis omitted a general heat balance on the solution and assumed that vaporization was proportional to the heat transferred to the liquid. Andre and Ritter (1968) presented a direct derivation of a five-equation model for an earlier configuration of the same double-effect evaporator used for experimental work in this study. Holzberger (1970) derived a model for a two-tank flash evaporator and obtained transient responses by simulation.

Model reduction is frequently required to produce simpler models for control applications, and many different approaches have been suggested. Procedures for reducing system order while maintaining the significant dynamic modes have been presented by Marshall (1966), Davison (1966), Chen and Shieh (1968), and Wilson et al. (1972). This approach can be intuitively approximated with state-space models by setting the derivatives of the first-order equations with small time constants equal to zero. Fitting methods depend on choosing a reduced-model form and fitting it to data generated from the more complex model. This is a similar problem to defining the form of a mathematical model and determining the parameters by fitting experimental data and enters the field of parameter identification which has been surveyed by Nieman et al. (1971) and Åström and Eykhoff (1971).

## Development of a General Evaporator Model

There are only a few basic designs for a single-effect evaporator and these can, in general, be relatively easily modeled using overall material and energy balances and basic thermodynamic relationships. However, industrial installations are more complicated because they normally involve a number of evaporator units interconnected in series and/or in parallel. Examination of a number of typical installations, such as that shown in Figure 1, suggests that the configuration and parameter values may differ widely from one installation to another,

[1] Present address, MFD Division, Shell Internationale Petroleum, Martschappy N.V., Den Haag, Netherlands
[2] To whom correspondence should be addressed.

Ind. Eng. Chem. Process Des. Develop., Vol. 11, No. 2, pp. 213-221 (1972).

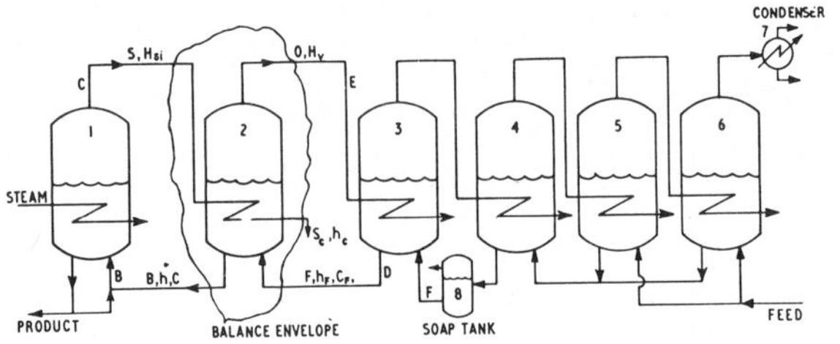


**Figure 1. Schematic diagram of a typical six-effect evaporator showing the envelope used to write the material and energy balance equations for the *i*th effect (Nisenfeld and Hoyle 1970)**

but the form of the dynamic equations for each component of the evaporator system usually remains the same.

Therefore the approach adopted was to completely separate the dynamic modeling of pieces of evaporator equipment from the specification of how these pieces were interconnected. The result is a series of dynamic building blocks which can be used linked together, following a cookbook procedure, to give a complete dynamic model. This approach is ideal for evaluating different evaporator configurations, but should be put in more efficient form for applications involving a large number of simulations.

**Dynamic Models of Evaporator Components.** For dynamic modeling purposes, it is convenient to break up each effect of an evaporator system, such as shown in Figure 1, into the following subsystems: steam chest, heat transfer surface, solution holdup, vapor space, and associated heat capacity of vessel, piping, etc. In addition, there are auxiliary units such as the condenser(s), soap tanks, etc., that must be modeled. The following subsections consider the development of dynamic models for the first three subsystems listed above. The derivation of equations to describe the other units is similar and has been presented in detail by Newell (1971).

*Steam Chest.* If the density and temperature gradients are assumed negligible, the following lumped-parameter balances can be written for the steam chest of the *i*th effect:

$$V^i_s \frac{d\rho^i_s}{dt} = S^i - S^i_c \tag{1}$$

$$V^i_s \frac{d}{dt}(\rho^i_s H^i_s) = S^i H^i_{si} - S^i_c h^i_c - Q^i_s - L^i_s \tag{2}$$

where

$Q^i_s = U^i_v A^i (T^i_c - T^i_w) = S^i_c(\lambda^i_s + \delta^i_s)$

$T^i_c = T^i_s - T^j_{bp}$

$T^j_{bp}$ = boiling point rise or superheat in connected unit

$\delta^i_s$ = correction for superheated steam and/or subcooled condensate

The assumption of constant volume, $V_s{}^i$, implies a constant condensate level in the steam chest. When the steam chest is connected to the vapor space of the previous unit by a low resistance line, then $V^i_s$ could include the vapor space volume —i.e., $V^i_s + V^j$—and hence reduce the total number of equations required.

The equations for the vapor space in each effect and for the vapor side of the condenser are very similar in form. An equation of state can be used to relate steam density, temperature, and pressure when required, for example, when it is desired to control the pressure in the vapor space.

*Heat Transfer Surface.* Assuming negligible temperature gradients in the vapor space, in the tube walls, and in the solution, a lumped-parameter energy balance for the heat transfer surface between the heating medium and the solution may be written as:

$$W^i_w C^i_{pw} \frac{dT^i_w}{dt} = U^i_v A^i (T^i_c - T^i_w) - U^i_L A^i (T^i_w - T^i) \tag{3}$$

The heat transfer coefficients may be assumed constant or expressed as general functions of temperature, concentration, and/or condensing or circulation rates. Use of an effective tube area would allow for heat transfer through downcomers and tube sheets.

A similar equation applies to the heat transfer surface in the condenser and the same approach could be used to account for the effects of heat capacity in the vessel walls and in the associated piping.

*Solution Holdup.* The material and energy balances on the solution in each effect of the evaporator assume perfect mixing. While this would generally be close to the case, large evaporator units with viscous solutions can develop significant concentration and temperature gradients within the solution. This would be particularly true where the circulation rate is low compared to the feed rate and where the vessel contains dead zones. The applicable mass, solute, and energy balances are

$$\frac{dW^i}{dt} = F^i - B^i - O^i \tag{4}$$

$$\frac{d}{dt}(W^i C^i) = F^i C^i_F - B^i C^i \tag{5}$$

$$\frac{d}{dt}(W^i h^i) = F^i h^i_F - B^i h^i - O^i H^i_v + Q^i - L^i + \phi^i \tag{6}$$

where $Q^i = U^i_L A^i (T^i_w - T^i)$ and $\phi^i$ = heat of solution effects.

This derivation has not considered boiling dynamics and assumes the vapor and solution to be in phase equilibrium at all times. In practice, the effect of pressure changes can be exceedingly complex. An increase in pressure will increase the boiling point which could stop the boiling and hence decrease the heat transfer coefficient, particularly for natural circulation units. This would cause the pressure and condensing temperature in the steam chest to rise above the normal steady-state values and hence propagate the disturbance through the heating vapor. However, except in the event of direct pressure control on an effect, the pressure changes are dependent on heat transfer dynamics. Therefore, if the sensible heat in the solution is small compared to the heat load, the heating dynamics may be fast enough to prevent a total stop in boiling. However, this depends on the equipment in question. The present model will give the correct response to pressure changes provided that the attainment of phase equilibrium is more rapid than the pressure changes.

**Multieffect Model Building.** Additional information is necessary to construct a complete process model. First, for each evaporator effect, condenser, or other unit of equipment, the following is required:

(1) The general model dynamic relations—e.g., equations 1-6—for an evaporator effect which define $\rho_s$, $H_s$, $T_w$, $W$, $C$, $h$, and possibly $\rho$ and $H$

(2) Empirical property relationships, or equations of state, which relate the physical properties of the solution, the solvent vapor, and of the liquid holdups to the system state—e.g., $h$ as a function of $T$ and $C$

(3) Physical parameters of the equipment—e.g., $V$, $V_s$, $W_w$, $A$, $C_{pw}$

(4) Operating variables, such as heat losses and heat transfer coefficients, which may be functions of the state of the unit.

Secondly, at least one algebraic configuration relationship must be written for each process stream connecting the units. At this stage the independent variables of the process will become apparent. The complete model and configuration relations for the six-effect evaporation process described by Nisenfeld and Hoyle (1970) have been developed by Newell (1971). A schematic flow sheet of the process appears in Figure 1 and an example of the configuration statements for the streams entering and leaving the second effect are given below:

$$\text{Stream C: } S^2 = O^1,\ H^2_{st} = H^1_v$$

$$\text{Stream E: } S^3 = O^2,\ H^3_{st} = H^2_v$$

$$\text{Stream D: } F^2 = B^3,\ C^2_F = C^3$$

$$h^2_F = (B^3h^3 - L_D)/F^2$$

$$\text{Stream B: } F^1 = B^2 + \alpha B^1$$

$$C^1_F = (B^2C^2 + \alpha B^1C^1)/F^1$$

$$h^1_F = (B^2h^2 + \alpha B^1h^1 - L_B)/F^1$$

where $\alpha$ is the fraction of the bottoms flow, $B^1$, which is recycled and $L_B$ and $L_D$ are the stream heat losses. Transport delays due to long pipes and/or low flow rates can be included in the configuration relations.

The profusion of differential and algebraic equations that result from the modeling procedure described above can be handled directly by such sophisticated digital simulation programs as the Continuous Systems Modeling Program (CSMP) on the IBM 360 series computers. No simplifying substitutions would be necessary although they would improve computational efficiency.

To arrive at a single set of differential equations in the form of a standard state-space model, the algebraic relations must be substituted into the differential relations. These equations can then be linearized analytically. If the relationships are to be linearized numerically by differences about the operating conditions then these substitutions are not necessary.

This cookbook approach demonstrates the power and flexibility of developing general unit models. However, the order of resulting models is generally high and some form of model reduction becomes necessary. This is demonstrated in the following sections where a tenth-order nonlinear model of an existing pilot plant is developed using this general approach and then simplified.

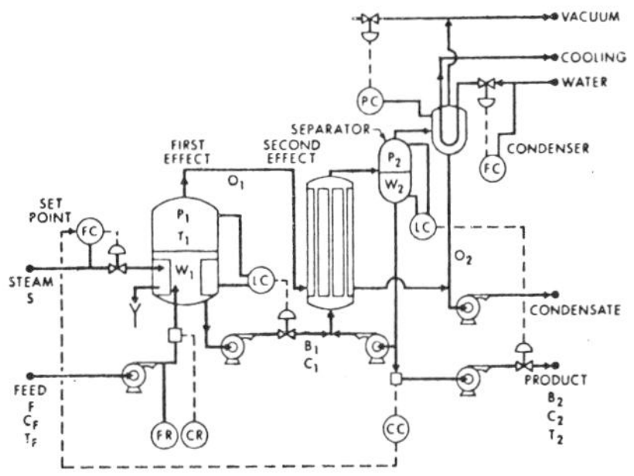


Figure 2. Schematic diagram of the double-effect evaporator pilot plant at the University of Alberta which was used for the experimental model evaluation and control studies reported in this paper

Evaporator Variables and Steady States

| | Feed | 1st Effect | 2nd Effect |
|---|---|---|---|
| $B$ = bottoms flow rate, lb/min | | 3.3 | 1.7 |
| $C$ = soln concn, % wt/wt | 3.2 | 4.85 | 9.64 |
| $F$ = feed flow rate, lb/min | 5.0 | | |
| $h$ = liquid enthalpy, Btu/lb | 162 | 194 | |
| $S$ = steam flow rate, lb/min | 1.9 | | |
| $W$ = soln holdup, lb | | 30 | 35 |
| $O$ = overhead vapor flow, lb/min | | 1.7 | 1.6 |
| $T$ = temp, °F | 190 | 225 | 160 |
| $P$ = pressure, psia | | $<25$ | 7.5 |

#### Modeling a Pilot Plant Evaporator

The equipment on which this study is based is a pilot-plant scale double-effect evaporator in the Department of Chemical and Petroleum Engineering at the University of Alberta. Figure 2 shows the major features of the process equipment including the level controllers used in this study. Details of the equipment are available in theses by Andre (1966), Newell (1971) and others.

**Evaporator.** The evaporator has a complex feed system to permit operation of the equipment in a cyclic fashion or the introduction of load changes and disturbances in the feed conditions. Controlled flows of concentrated triethylene glycol solution and of water are temperature-controlled by use of steam heaters, and then mixed in the proportions necessary to produce a feed of the desired flow rate and concentration. However, this equipment is not included in the model because it has a relatively fast dynamic response and does not interact directly with the state variables of the main evaporator.

The first effect is a short-tube vertical calendria-type unit with natural circulation. The 9-in. diam unit has an operating holdup of 2 to 4 gal and its 32 18-in. long, 3/4-in. o.d. stainless steel tubes give about 10 ft$^2$ of heat transfer area. The second effect is a long-tube vertical effect set up for either natural or forced circulation. It has 5 ft$^2$ of heat transfer surface made up of three 6-ft long, 1-in. o.d. tubes. The capacity of its circulating system is about 3 gal.

The equipment is fully instrumented and can be controlled either by Foxboro electronic controllers or under Direct

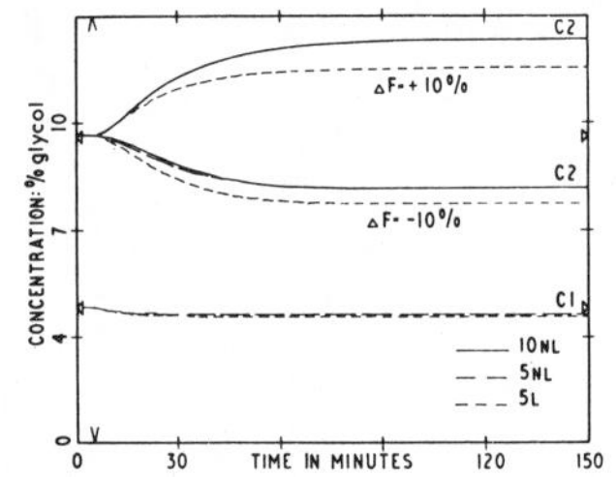


Figure 3. Simulated comparison of the open-loop concentration responses calculated from the tenth-order nonlinear (10NL), fifth-order nonlinear (5NL) and fifth-order linearized (5L) models for a 10% step in feed flow rate. The triangles on the ordinate axis indicate the initial steady state and on the time axis indicate the time at which the step disturbance was introduced

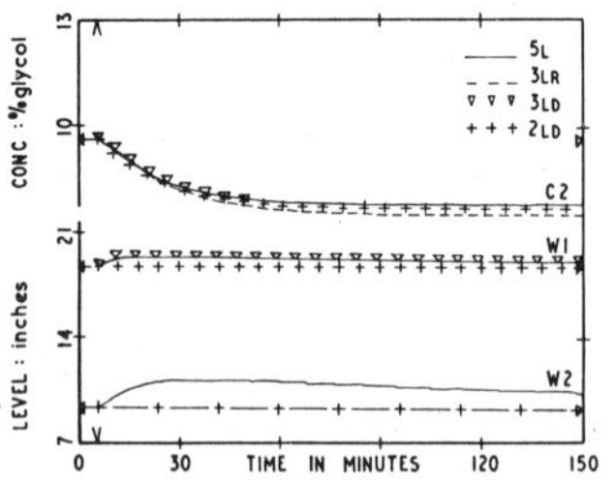


Figure 4. Simulated comparison of the responses from the fifth-order linearized (5L), third-order reduced (3LR), third-order derived (3LD), and second-order derived (2LD), models for a 10% increase in feed flow rate. Where not shown, the response of the 3LD model was the same as the 2LD model

Digital Control (DDC) from an IBM 1800 Data Acquisition and Control Computer. The sophisticated and complete interfacing between plant and computer enables implementation of complex, multivariable control schemes using the on-line computing capabilities of the digital machine.

**Tenth-Order Nonlinear Model (10NL).** The cookbook procedure for producing general model relations was used for the double-effect evaporator in Figure 2. The derivation was made assuming negligible heat-of-solution effects, zero bp rise, saturated steam in all vapor spaces, identical conditions throughout the steam chest and connected vapor space, no subcooling of the steam condensate streams, constant heat transfer coefficients, constant heat losses, zero concentration of solute in the overhead vapor streams, and negligible heat capacity in the evaporator vessels and piping. Transport delays between the effects were also neglected. The algebraic configuration relationships were substituted into the differential equations and 10 nonlinear differential equations resulted (Model 10NL). This model, as detailed by Newell (1971), involves a state vector $\mathbf{x}$, and an input vector, $\mathbf{u}$, defined as:

$$\mathbf{x}^T = (H^1{}_s,\ T^1{}_w,\ W^1,\ H^1,\ C^1,\ T^2{}_w,\ W^2,\ H^2,\ C^2,\ T_{wc})$$
$$\mathbf{u}^T = (F^1,\ C^1{}_F,\ h^1{}_F,\ S^1,\ H^1{}_{si},\ B^1,\ B^2,\ F_{cw},\ T^3{}_{ci}) \qquad (7)$$

However, for use in modern design and control techniques, it is advantageous to have a linear state-space model of lower dimension. Model 10NL was therefore linearized numerically,

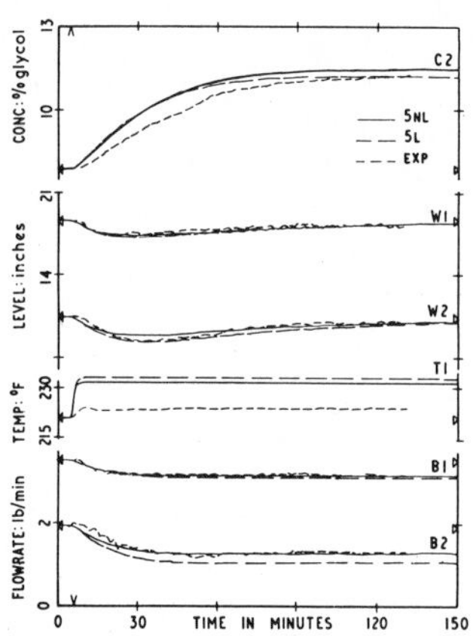


Figure 5. Comparison of experimental open-loop response data with the fifth-order models for a 20% increase in inlet steam flow

using central differences, to give a standard linear, time-invariant, tenth-order, state-space model of the form:

$$\dot{\mathbf{x}} = \mathbf{A}\mathbf{x} + \mathbf{B}\mathbf{u}$$
$$\mathbf{v} = \mathbf{C}\mathbf{x} \qquad (8)$$

**Fifth-Order Models (5NL and 5L).** Lower-order models can be produced by starting with the tenth-order state-space model and applying model reduction techniques such as those developed by Marshall (1966) or Davison (1966). A more intuitive approach, which has been shown to give essentially the same fifth-order evaporator model, is to make physically justifiable approximations which reduce some of the faster dynamic relationships to their steady-state form.

Consider the following points relative to the equipment and its operation. First, because of the relatively small heat capacity of the three sets of tube walls and the small volume of the first-effect steam chest, the dynamics of the walls and

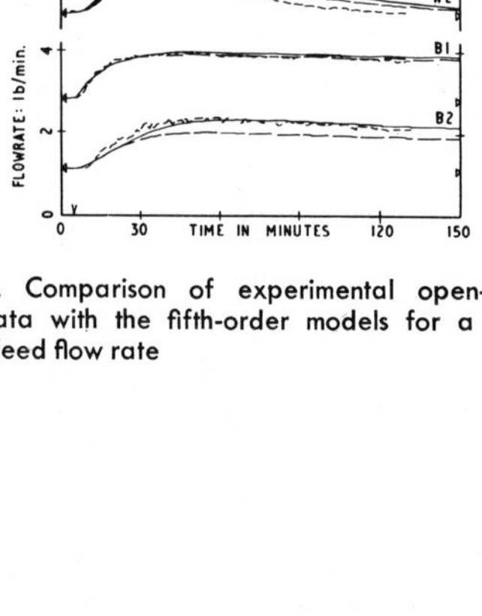

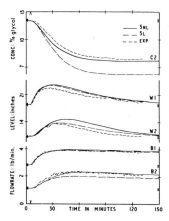

Figure 6. Comparison of experimental open-loop response data with the fifth-order models for a 20% increase in feed flow rate

steam temperatures are fast compared with those of the solution. As a result, the steady-state relationships can be used in these areas and represent a fair approximation. Second, since the evaporator is operated with very tight control on the pressure in the second effect, the second-effect temperature can be considered constant. As a result of these approximations, Model 10NL can be reduced to the five nonlinear first-order differential equations (Model 5NL) given by Equations 9–13.

As indicated by Figures 3–7, Model 5NL is in close agreement with Model 10NL and gives a reasonable representation of the open-loop response of the evaporator. Andre and Ritter (1968) obtained essentially the same set of equations by a direct derivation, but the physical construction and operating conditions of the evaporator have been changed significantly since their work.

Model 5NL has been linearized analytically, and by numerical techniques, to give a model which will describe small deviations in process conditions relative to the steady-state conditions. The model variables were also normalized by dividing by the steady-state values so that the variables are in the normalized, perturbation form $(x - x_{ss})/x_{ss}$. When numeric values are specified for the parameters, the result is Model 5L, as defined by Equation 19. Full details are available (Newell, 1971).

A FORTRAN program was written which will generate the coefficient matrices in Equation 19 for any consistent set of steady-state operating conditions. This program is normally run off-line, but the necessary data could be read directly from the process and the appropriate linearized model generated on-line. This allows a type of adaptive modeling which can be used over the full range of steady states at which the process can operate.

*First Effect.*

$$\frac{dW^1}{dt} = F^1 - B^1 - O^1 \tag{9}$$

$$W^1\frac{dC^1}{dt} = F^1(C^1_F - C^1) + O^1C^1 \tag{10}$$

$$W^1\frac{dh^1}{dt} = F^1(h^1_F - h^1) - O^1(H^1_v - h^1) + Q^1 - L^1 \tag{11}$$

*Second Effect.*

$$\frac{dW^2}{dt} = B^1 - B^2 - O^2 \tag{12}$$

$$W^2\frac{dC^2}{dt} = B^1(C^1 - C^2) + O^2C^2 \tag{13}$$

where

$$S^1\lambda^1_s = U^1A^1(T^1_s - T^1) \tag{14}$$

$$O^1 = (Q^2 + L^2)/(H^1_v - h^2_c) \tag{15}$$

$$Q^1 = U^1A^1(T^1_s - T^1) \tag{16}$$

$$Q^2 = U^2A^2(T^1 - T^2) \tag{17}$$

$$O^2\left(H^2_v - h^2 + \frac{\partial h^2}{\partial C^2}C^2\right) = Q^2 - L^2 + B^1(h^1 - h^2) + \frac{\partial h^2}{\partial C^2}B^1(C^2 - C^1) \tag{18}$$

*Model 5L.*

$$\begin{Bmatrix} \dot W^1 \\ \dot C^1 \\ \dot h^1 \\ \dot W^2 \\ \dot C^2 \end{Bmatrix} = \begin{bmatrix} 0 & -0.00156 & -0.1711 & 0 & 0 \\ 0 & -0.1419 & 0.1711 & 0 & 0 \\ 0 & -0.00875 & -1.102 & 0 & 0 \\ 0 & -0.00128 & -0.1489 & 0 & 0.00013 \\ 0 & 0.0605 & 0.1489 & 0 & -0.0591 \end{bmatrix} \begin{Bmatrix} W^1 \\ C^1 \\ h^1 \\ W^2 \\ C^2 \end{Bmatrix} + \begin{bmatrix} 0 & -0.143 & 0 \\ 0 & 0 & 0 \\ 0.392 & 0 & 0 \\ 0 & 0.108 & -0.0592 \\ 0 & -0.0486 & 0 \end{bmatrix} \begin{Bmatrix} S^1 \\ B^1 \\ B^2 \end{Bmatrix} + \begin{bmatrix} 0.2174 & 0 & 0 \\ -0.074 & 0.1434 & 0 \\ -0.036 & 0 & 0.1814 \\ 0 & 0 & 0 \\ 0 & 0 & 0 \end{bmatrix} \begin{Bmatrix} F^1 \\ C_F^1 \\ h_F^1 \end{Bmatrix} \quad \begin{Bmatrix} W^1 \\ W^2 \\ C^2 \end{Bmatrix} = \begin{bmatrix} 1 & 0 & 0 & 0 & 0 \\ 0 & 0 & 0 & 1 & 0 \\ 0 & 0 & 0 & 0 & 1 \end{bmatrix} \begin{Bmatrix} W^1 \\ C^1 \\ h^1 \\ W^2 \\ C^2 \end{Bmatrix} \tag{19}$$

**Second- and Third-Order Models (3LR, 3LD, and 2LD).** Although a fifth-order model can be readily handled by a digital computer, one of lower order can be implemented more economically and will very often be just as effective for control purposes. Consideration was therefore given to even lower-order models.

Since the five-equation model has eigenvalues of the same order of magnitude and it is required to preserve the physical significance of the variables, further reduction of order can only be achieved by making assumptions, such as the follow-

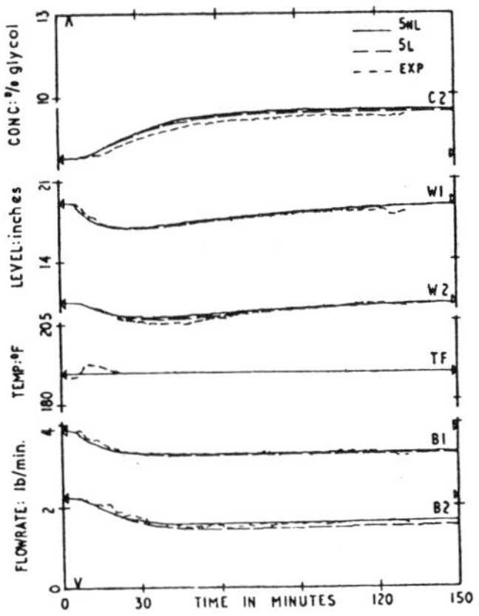


Figure 7. Comparison of experimental open-loop response data with the fifth-order models for a 10% decrease in feed flow rate

ing, regarding the equipment's operating conditions: constant holdup in the second effect (due to physical constraints it must be within $\pm 10\%$ of the steady-state value, so the assumption is reasonable in this case); constant first-effect temperature; insignificant enthalpy dependence on concentration; and no heat losses. As might be expected, these assumptions are less valid than those made previously.

Assuming that the change in solution enthalpy between effects is small compared to the heat load, the following relation follows from Equations 12, 15, and 18:

$$k^2 = \frac{O^2}{O^1} = \frac{\bar{H}^1_v - \bar{h}^2_c}{\bar{H}^2_v - \bar{h}^2} \tag{20}$$

Similarly a constant $k^1$ can be defined using Equations 11, 14, and 16, and assuming the feed temperature is close to that of the first effect.

$$k^1 = \frac{O^1}{S^1} = \frac{\bar{\lambda}^1_s}{\bar{H}^1_v - \bar{h}^1} \tag{21}$$

Analytical linearization of the remaining differential Equations 9, 10, and 13, results in the following Model 3LR

$$\frac{dW^1}{dt} = F^1 - B^1 - k^1S^1 \tag{22}$$

$$\bar{W}^1\frac{dC^1}{dt} = (\bar{C}^1_F - \bar{C}^1)F^1 + \bar{F}^1C^1_F + (k^1\bar{S}^1 - \bar{F}^1)C^1 + k^1\bar{C}^1S^1 \tag{23}$$

$$\bar{W}^2\frac{dC^2}{dt} = (\bar{C}^1 - \bar{C}^2)B^1 + \bar{B}^1C^1 + (k^1k^2\bar{S}^1 - \bar{B}^1)C^2 + k^1k^2\bar{C}^2S^1 \tag{24}$$

An alternative approach to finding a simple model is to make basically the same assumptions but develop the model relationships directly from mass balances. This direct approach gives the same basic Equations 22–24 but with the following expressions for the $k^i$ constants:

$$k^1 = \frac{O^1}{S^1} = \frac{\bar{F}^1 - \bar{B}^1}{\bar{S}^1} \tag{25}$$

$$k^2 = \frac{O^2}{O^1} = \frac{\bar{B}^1 - \bar{B}^2}{\bar{F}^1 - \bar{B}^1} \tag{26}$$

Since Equations 25 and 26 are defined directly in terms of average values of the independent variables it would be expected that they are more accurate than Equations 20 and 21, and this is confirmed in the next section. A standard state-space third-order model can also be derived with the state, control, and disturbance vectors all of dimension three and $\mathbf{y} = \mathbf{x}$. This special structure is useful in some design techniques.

The third-order models can be reduced to second-order by assuming that the first-effect level remains constant. If $\dot{W}^1 = 0$, then Equation 22 can be used to eliminate $B^1$ from Equation 24 thus giving Model 2L. The process can be operated to fit Model 2L by using very tight liquid level control on both effects. However, under these conditions, a step change in feed flow, $F^1$, results in an immediate step change in product flow, $B^2$, and as expected, this results in larger disturbances in product concentration, $C^2$, than occurs when the liquid levels are allowed to vary.

Transfer function models can be developed for each pair of input/output variables but the one of greatest interest is the one between inlet steam and product concentration

$$G(s) = \frac{C^2(s)}{S^1(s)} = \frac{10.2\ (6.7\ s + 1)}{(9.1\ s + 1)\ (21.1\ s + 1)} \approx \frac{10.2}{(\tau s + 1)} \tag{27}$$

which indicates approximately first-order behavior.

### Model Evaluation

In engineering applications, the value of a model is not usually judged by its mathematical rigor or the elegance of the derivation but rather by how well it fulfills a specific need in comparison with other alternatives. Typical applications vary from open-loop simulations to the design of multivariable controllers based on modern control theory. In the following sections, simulated and experimental data from several projects at the University of Alberta are summarized with the objective of illustrating the performance of the evaporator models in different applications.

In experimental investigations it was necessary to keep the liquid holdups, $W1$ and $W2$, in the evaporator within certain physical limits. Therefore, even in the open-loop runs, the holdups were controlled by proportional-plus-integral feedback controllers. It should also be noted that, except where specifically noted, the model parameters which determine the dynamic response of the evaporator were calculated by a theoretical analysis and were not empirically fitted to experimental data. The use of different parameter values would certainly influence model performance and can result in improved performance in some applications (Nieman, 1971).

**Simulated Open-Loop Responses.** The evaporator performance was simulated on an IBM 360/67 computer using CSMP for the nonlinear models and discrete state-space equations for the linear models. With a time interval of 1 sec, the simulations required less than half a minute of computer time to generate a 3-hr transient.

The comparison of concentration responses presented in Figure 3 shows that the tenth-order (10NL) and fifth-order (5NL) model responses are in excellent agreement. The slightly faster response of the fifth-order model is attributed to the effect of the small time constants that were eliminated in the model reduction step. The nonlinearity of the evaporator is illustrated by the difference in the change produced in product concentration, $C2$, by a 10% step increase in feed flow vs. a 10% reduction. The linearized model (5L) has steady-state gains which are approximately the average of the up and down responses of the nonlinear models, but displays significant errors at the final steady state. The responses of the liquid holdups (levels) and the flow changes were essentially the same for all three models and hence are not plotted.

The comparison of the responses of the linearized models in Figure 4 shows that except for small differences in steady-state gain they all produce approximately equivalent transients in $C2$. Model 5L is the only one that includes the dynamics of the second-effect holdup, $W2$, and Model 2LD does not include the dynamics of $W1$ or $W2$. All models accurately represented first-effect concentration, $C1$ (not plotted). The expected effect of model order can be seen in the shape of the product concentration transients immediately following the disturbance.

It can be seen from the above comparisons that the differences in final steady-state values predicted by different models is an important measure of model performance. One would expect this to be true in general, particularly for over-

**Table I. Final Steady-State Values of Concentration and Errors-of-Closure in Material and Energy Balances After Step Changes in $F$, $C_F$, or $S$**

Closures: 1. First-effect mass balance
2. First-effect solute balance
3. Second-effect solute balance

| | Disturbance | | | | | | | | | | | | | | |
|---|---|---|---|---|---|---|---|---|---|---|---|---|---|---|---|
| | 10% step in $F^1$ | | | | | 10% step in $C_F^1$ | | | | | 10% step in $S^1$ | | | | |
| Model | $\Delta C^1$ | $\Delta C^2$ | 1 | 2 | 3 | $\Delta C^1$ | $\Delta C^2$ | 1 | 2 | 3 | $\Delta C^1$ | $\Delta C^2$ | 1 | 2 | 3 |
| 5NL+ | −0.23 | −1.42 | 0.0 | 0.0 | 0.0 | +0.49 | +0.98 | 0.0 | 0.0 | 0.0 | +0.20 | +1.75 | 0.0 | 0.0 | 0.0 |
| 5NL− | +0.32 | +2.61 | 0.0 | 0.0 | 0.0 | −0.49 | −0.96 | 0.0 | 0.0 | 0.0 | −0.19 | −1.31 | 0.0 | 0.0 | 0.0 |
| 5L+ | −0.27 | −1.85 | 0.0 | 0.8 | 3.9 | +0.49 | +0.97 | 0.1 | 0.2 | 0.1 | +0.19 | +1.51 | 0.0 | 0.2 | 1.5 |
| 3LR+ | −0.26 | −2.13 | 0.0 | 0.1 | 0.1 | +0.41 | +1.10 | 0.0 | 0.0 | 0.0 | +0.29 | +2.32 | 0.0 | 0.0 | 0.0 |
| 3LD+ | −0.15 | −1.94 | 0.0 | 0.1 | 0.2 | +0.49 | +0.96 | ... | ... | ... | +0.25 | +1.94 | 0.0 | 0.0 | 0.2 |
| 2LD+ | −0.15 | −1.94 | ... | 0.0 | 0.1 | +0.48 | +0.96 | ... | ... | ... | +0.25 | +1.94 | ... | 0.0 | 0.2 |

damped responses. Table I tabulates the open-loop concentration changes and the errors of closure in the material balances using the final steady-state values predicted by the various models after step changes in the input variables. The concentration changes are in units of wt % glycol and can be compared relative to initial steady-state values of 4.85% and 9.64% for $C1$ and $C2$, respectively. The errors of closure are given as a percentage of the average balance term. The linearized Model 5L predicted steady-state values that were close to the average of the changes predicted by the nonlinear model for both up and down steps in the input variables. (Experimental results are close to those predicted by Model 5 NL.) The second- and third-order derived models, 2LD and 3LD, gave reasonable steady-state predictions of $C1$ and $C2$ for step changes in feed flow or feed composition. However, the assumption made in these models that evaporation is proportional to steam flow, is not strictly true and hence there is a significant error in the prediction of the effect of changes in the steam flow, $S$. In general, the models obtained by reduction of higher-order models—e.g., 5L—showed larger errors of closure in the material and energy balances than directly derived models.

**Experimental Open-Loop Responses.** The bottoms flow rates, $B1$ and $B2$, were used to control the liquid levels in the first and second effects respectively so that only four independent variables were left for use as forcing functions. These were steam and feed flow rate, feed concentration, and feed enthalpy. Experimental runs were carried out with positive and negative steps of 10 and 20% of steady state in all four variables.

The results showed reasonable agreement between the experimental and the linearized five-equation model, 5 LR, for step changes of up to 10% in feed flow rate and 20% in the other three variables. These results are exemplified by Figures 5–7 showing the evaporator's response to 20% increases in steam and feed flow and a 10% step decrease in feed flow rate, respectively. It was apparent in all the results that the models responded faster than the equipment. This was most probably due to the neglect of several small time constants in the initial model reduction and the neglect of the small time delays between the effects and equivalent time delays due to incomplete mixing in the effects.

Other specific shortcomings of the model were evident in some of the experimental runs. The data for $T1$ in Figure 5 shows that the models predicted a much larger change in first-effect temperature than was observed in practice with changes in steam flow. The temperatures (pressures) in the evaporator are influenced by the factors such as the presence of noncondensibles which affect heat transfer in each effect but are not present in the model. A comparison of the controlled responses of $W1$ and $W2$ in Figures 6 and 7 shows that 20% disturbances in feed flow produce relatively large changes in liquid level and factors such as the difficulty of measuring the level of a boiling liquid and translating this to an equivalent holdup—i.e., estimating effective density—leads to significant differences between simulated and observed responses. The plot of feed temperature, $TF$, in Figure 7 shows an experimental deviation due to interactions between feed flow and feed temperature in the preheaters (they were not modeled). The model errors discussed in this paragraph are worst case rather than typical results.

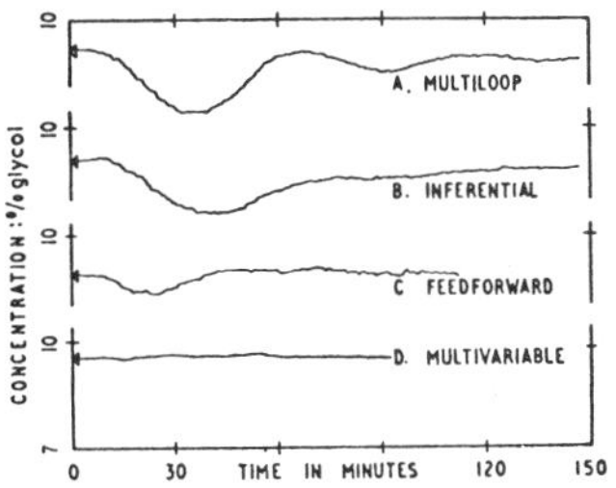


Figure 8. Comparison of experimental responses of the evaporator product concentration to 20% increases in feed flow rate when controlled by multiloop conventional control (as per Figure 2), inferential control, feedforward-plus-feedback control and optimal multivariable feedback control

**Multiloop Conventional Control.** The fifth-order models were used by Andre and Ritter (1968) and Newell (1971) in developing the conventional multiloop control system shown in Figure 2. Model 5L was also programmed on an EAI 590 hybrid computer and an attempt was made to determine suitable P + I controller constants by simulating the closed-loop system. However, it was found that the controller gains that gave the best simulated performance were about an order of magnitude too high to be used experimentally on the evaporator. This was attributed to time delays, dynamics, and nonlinearities of the process that were neglected in the model. The multiloop scheme shown in Figure 2 was implemented and tuned experimentally by

Jacobson (1970) and Figure 8*A* gives a typical response to a 20% step increase in feed flow.

**Inferential Control.** Inferential feedback control schemes use estimated or calculated values rather than directly measured values of the controlled variable and, in general, one would expect the performance to be poorer or equal to that of a conventional controller. Figure 8*B* illustrates an evaporator response with inferential control of *C*2 and can be compared directly with Figure 8*A*. The improved response in Figure 8*B* is due to the fact that the model response leads that of the process and gives the inferential controller some anticipatory characteristics. If the model is improved by adding a time delay so that the model agrees more closely with the open-loop process response, then the performance of the inferential controller deteriorates. Is a poor model better than an accurate one?

Current investigations are using Model 5L as the basis for the design of Kalman filters and Luenberger type observers and it appears adequate for these purposes.

**Feedforward Control.** If the disturbance to a process can be measured, then it is theoretically possible, in many cases, to model the process and design a feedforward controller that will eliminate all effects of the disturbances on the controlled variables. Figure 8*C* shows how the addition of simple proportional feedforward control to the feedback scheme used in Figure 8*A* results in significant improvements. Feedforward control of the evaporator has been studied by Fehr (1969) Jacobson (1970) and Newell et al. (1972). In applications like this, even a model that consists of only a knowledge of the steady-state gains can significantly improve control.

**Optimal Multivariable Control.** Several techniques exist for designing multivariable feedback controllers but they all require an adequate model of the process. Newell and Fisher (1972) discuss the design and application of discrete optimal controllers based on a linear time-invariant model of the process and a quadratic performance index. Figure 8*D* represents a typical response when such a controller is used on the evaporator. It is significant that Model 5L, which is theoretical and was *not* suitable for the design of conventional controllers, resulted in a controller design that gave the best experimental control yet achieved on the evaporator. The excellent performance of the linear model is probably due to the fact that the optimal controller maintains the process in a very narrow band around the normal operating conditions—i.e., the point of linearization.

**State-Driving Control.** Several of the linear models have also been used as a basis for calculating the control policy, $\mathbf{u}(kT)$, that will drive the evaporator from one state, $\mathbf{x}(0)$, to another set of desired values, $\mathbf{x}_d$, while minimizing some criterion such as time required (Nieman and Fisher, 1970).

In general, it has been found that model accuracy is much more critical than in regulatory control because the model must predict the process response over a relatively wide range of operating conditions (i.e., at least from $\mathbf{x}(0)$ to $\mathbf{x}_d$); an accurate final steady-state gain is necessary to determine the control action that will maintain the undisturbed process at $\mathbf{x}_d$; and because the bang-bang manipulation of the control variables is a more severe test of the model than slowly changing inputs.

## General Discussion

The evaporator models include a number of features which make them interesting from a control study viewpoint. The liquid levels introduce two integrating states and consequently two zero eigenvalues into the model. While this is a common feature in most chemical processes, the zero eigenvalues are considered in much of the theory as special cases which are not fully treated and which many methods cannot handle. The process is also such that it cannot be decoupled by state variable feedback so that control design techniques must handle the interactions or ignore them and suffer the consequences. Other special features include the physical observability of the complete state vector, and the nonlinearities, particularly to feed flow rate and steam rate. A significant amount of experimental and simulated data are also available as outlined in the previous subsections. For these reasons, the evaporator models developed here are recommended as test problems for other control studies.

## Conclusions

This paper illustrates the use of a generalized modeling approach for a piece of process equipment and some of the inherent dangers and implications of simplifying the resulting model. Experimental and simulated results from the design and implementation of several different control schemes on a pilot-plant evaporator showed that the suitability of a model depends on its end use and that different models may be required for different applications.

## Nomenclature

$A$ = heat transfer area, $ft^2$
$C_p$ = heat capacity, Btu/lb/°F
$H$ = vapor enthalpy, Btu/lb
$L$ = heat loss, Btu/min
$Q$ = heat flow rate, Btu/min
$t$ = time, min
$U$ = heat transfer coefficient, Btu/min/°F/$ft^2$
$V$ = volume, $ft^3$
$W$ = mass, lb

GREEK LETTERS

$\lambda$ = heat of vaporization, Btu/lb
$\rho$ = density, lb/$ft^3$

SUBSCRIPTS

$c$ = condenser, condensate, condensing
$cw$ = cooling water
$F$ = feed
$i$ = inlet value
$L$ = liquid or solution
$m$ = mean value
$s$ = steam
$ss$ = steady state
$v$ = vapor
$w$ = wall of tubes

SUPERSCRIPTS

$\overline{\ }$ = mean value
$i$ = effect or unit number
$T$ = matrix transpose

## Literature Cited


Anderson, J. A., Glasson, L. W. A., Lees, F. P., *Trans. Soc. Instrum. Technol.*, **13** (1), 21 (1961).
Andre, H., PhD thesis, Department of Chemical and Petroleum Engineering, University of Alberta (1966).
Andre, H., Ritter, R. A., *Can. J. Chem. Eng.*, **46**, 259 (1968).
Åström, K. J., Eykhoff, R., IFAC Symposium on System Identification in Automatic Control, Prague, Czechoslovakia, Paper 0.1 (1970); *Automatica*, **7** (2), 123 (1971).
Chen, C. F., Sheih, L. S., *Joint Automat. Contr. Conf., 9th, Prepr.* 454 (1968).
Davison, E. J., *IEEE Trans. Automat. Contr.*, Vol. AC-11 (1), 93 (1966).

Fehr, M., MSc thesis, Department of Chemical and Petroleum Engineering, University of Alberta (1969).
Holzberger, T. W., PhD thesis, Department of Chemical Engineering, Montana State University (1970).
Jacobson, B. A., MSc thesis, Department of Chemical and Petroleum Engineering, University of Alberta (1970).
Johnson, D. E., *ISA J.*, **7** (7), 46 (1960).
Manczak, K., Preprints, IFAC Symposium on Identification in Automatic Control Systems, Prague, Czechoslovakia Section 2.2 (1967).
Marshall, S. A., *Control*, **10** (102), 642 (1966).
Newell, R. B., PhD thesis, Department of Chemical and Petroleum Engineering, University of Alberta, (1971); (available from University Microfilms, Ann Arbor, MI).
Newell, R. B., Fisher, D. G., *Automatica* (May 1972).
Newell, R. B., Fisher, D. G., Seborg, D. E., *AIChE J.*, in press, 1972.
Nieman, R. E., Fisher, D. G., Seborg, D. E., *Int. J. Contr.*, **13**, (2), 209 (1971).
Nieman, R. E., Fisher, D. G., *Nat. Conf. Automat. Contr., Prepr.*, University of Waterloo, ONT, Canada (August 1970).
Nisenfeld, A. E., Hoyle, D. L., *Instrum. Technol.*, **17** (2), 49 (1970).
Wilson, R. G., Fisher, D. G., Seborg, D. E., *Int. J. Contr.*, in press, 1972.
Zavorka, J., Sutek, L., Aguado, A., Delgado E., Preprints, IFAC Symposium on Identification in Automatic Control Systems, Prague, Czechoslovakia, Section 2.3 (1967).

RECEIVED for review December 1, 1970
ACCEPTED December 30, 1971

# Description and Applications of a Computer Program for Control Systems Design*†

## Description et Applications d'un Programme D'ordinateur pour la Conception de Systèmes de Contrôle

## Beschreibung und Anwendung eines Rechnerprogrammes für den Entwurf von Regelungen

## Описание и применение программы вычислительной мащины для разработки систем управления

D. GRANT FISHER, ROBERT G. WILSON and WALTER AGOSTINIS‡

*The development of general computer programs for control system design and simulation is useful and practical as illustrated by application to a pilot-plant evaporator.*

**Summary**—A General, Multipurpose Simulation and COntrol PackagE (GEMSCOPE) is described which assists with the design, analysis and digital simulation of linear, time-invariant dynamic systems. The process model may be defined in block diagram form with transfer functions for each component, or in standard state-space form. Options are included for generating the optimal regulatory control matrices, determining the state feedback matrix for non-interaction, reducing the order of the model, and calculating the open or closed-loop system response. GEMSCOPE has proven useful for student education and for the design of practical control systems. Experimental and simulated data obtained by applying GEMSCOPE techniques to computer controlled pilot plants at the University of Alberta demonstrate the validity of the methods and the significant improvements possible through the use of modern control techniques.

### INTRODUCTION

THE DEVELOPMENT and use of computers has reached the point where a significant number of control engineers have access to large off-line computers for use in design and analysis, and in many cases to real-time digital computers for implementation of the resulting control systems and/or optimization procedures. The Department of Chemical and Petroleum Engineering at the University of Alberta therefore started a number of computer oriented projects with both educational and research objectives [1]. These included the simulation of dynamic systems, the evaluation and development of design techniques, and the evaluation of the resulting control systems by implementation on computer-controlled pilot-plant processes.

The results of these projects include three digital simulation programs, CSAP [2], CSDAP [4] and GEMSCOPE. The latter program, which is designed to run on a large data processing computer, either as a batch job or from a time sharing terminal, is the subject of this paper.§ Many other digital simulation and design programs exist. One of the first was developed by KALMAN and ENGLAR [26] and others are described in the textbooks by FRANKS [3] and MELSA [25]. Several companies and organizations also have their own programs for internal use. This paper is therefore an example of a particular program and some of its applications rather than a unique or definitive development.

Digital computer programs, however, are seldom an end in themselves and are often difficult to evaluate on an absolute basis. Their value must

* This paper was received in February 1971, for presentation at the IFAC Symposium on Digital Simulation of Continuous Processes, Györ, Hungary (1971).

† Received 13 December 1971; revised 17 May 1972. The original version of this paper was presented at the IFAC Symposium on Digital Simulation of Continuous Processes (DISCOP) which was held in Györ, Hungary during September 1971. It was recommended for publication in revised form by Associate Editor H. Kwakernaak.

‡ Dr. D. G. Fisher is a professor, and Mr. R. G. Wilson a Ph.D. candidate, in the Department of Chemical and Petroleum Engineering at the University of Alberta, Edmonton 7, Alberta, Canada. Mr. W. Agositinis is a former graduate student.

§ At the University of Alberta GEMSCOPE runs on an IBM 360 model 67 computer, operating under the University of Michigan, time-sharing, terminal-oriented system, MTS. Copies of the program, FORTRAN listings plus limited documentation, are available for a nominal charge to cover the cost of reproduction, handling, postage, etc.

be judged by how well they meet the objective of assisting the user in his particular application and how well the conclusions and/or control systems which result apply to real, physical systems. It was this interest in the physical application of research results and the desire to expose students to the practical as well as theoretical aspects of process dynamics and control that led to the establishment, in 1967, of the department's Data Acquisition, Control and Simulation (DACS) Centre. This Centre includes an EAI 590 hybrid system plus an IBM 1800 data acquisition and control computer. The IBM 1800 operates under a multiprogramming executive system (MPX) and current applications include a direct digital control (DDC) monitor system, lab automation applications [6], multivariable control of pilot plant processes [7–10] and background execution of programs such as digital simulations [2, 4]. Experience gained through these applications has influenced the development, and is the motivation for using simulation programs such as GEMSCOPE.

## GEMSCOPE PROGRAMS AND APPLICATIONS

GEMSCOPE (GEneral Multipurpose, Simulation and COntrol PackagE) is a series of separate programs which share common disk storage files and can be executed either singly or sequentially to perform operations such as model definition, control calculations or simulation of time domain responses. From the point of view of simulation the important programs are:

(1) Specification of the programs to be executed
(2) Definition of the state-space model
(3) Calculation of the state difference equation
(4) Calculation of the time domain response.

These programs are shown in the centre column of Fig. 1. However, the usefulness of GEMSCOPE is increased significantly by the functions represented by the blocks in the right most column of Fig. 1 which include:

(5) Generation of an equivalent state-space form for models entered as transfer functions
(6) Reduction of the order of the state-space model
(7) Generation of state feedback required for non-interaction
(8) Calculation of the control matrices for optimal regulatory control
(9) Utilization of a discrete control vector for open-loop or optimal state-driving.

These methods are discussed in the following sections. Some of the "utility functions" for performing tasks like listing output data are shown in Fig. 1 but are not discussed further. The extremely important file handling functions, such as initiation, editing, listing, etc. are standard features of the MTS monitor program and hence are not included in GEMSCOPE.

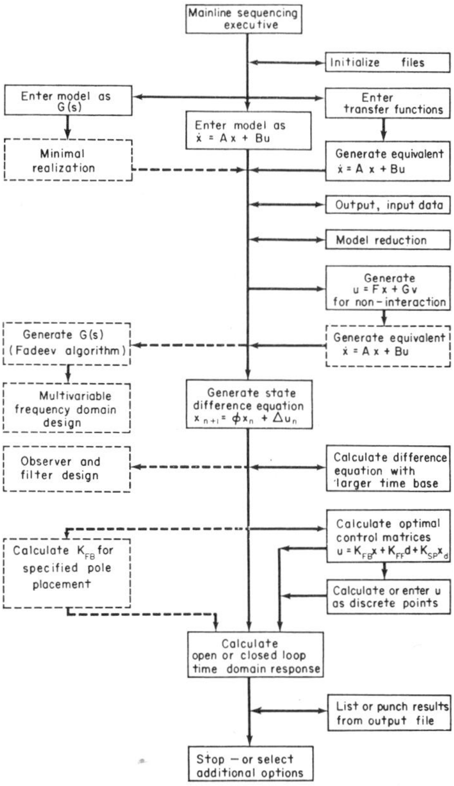


FIG. 1. Simplified flow diagram of the General, Multipurpose Simulation and COntrol PackagE (GEMSCOPE). The centre column shows the main steps for simulation, the right column some auxiliary design programs; and the left column some possible future additions.

The GEMSCOPE programs are written in FORTRAN and except for critical calculations, such as determining eigenvalues, are single precision. In their present form they consist of about 5500 source cards and occupy about 100,000 words of disk storage. Core memory requirements vary because the IBM 360/67 uses dynamic memory paging but a typical average would be 30,000 words. Central Processing Unit time requirements vary with each run but average about 40 sec to simulate one evaporator run of the type described later in this paper.

### *Job specification*

It should be emphasized that each program in GEMSCOPE is a separate mainline program and can be executed individually or in any sequence the user specifies. The mainline sequencing executive permits the user to specify which program blocks he wants executed by simply entering a series of integer code numbers, either via punched cards, if he is running GEMSCOPE as a batch program, or via a keyboard on a time sharing terminal. The executive then generates the appropriate series of job control statements and stores them in a file. Special files of job control statements are normally available for frequently used combinations of programs, or for special purposes such as student assignments, and these can be specified by a single user entry. The computer monitor system will then execute the GEMSCOPE programs in accordance with the job control statements in this file and then exit, or return to the executive. The job control languages for large time-sharing computers permit the user to specify almost all the parameters that affect the execution of his program and/or to call for standard system options or utility programs. For example he can specify core requirements, program priority, input/output device assignments, i.e. logical unit numbers, file identification, use of system or library programs etc. Thus this approach maximizes flexibility, reduces core and execution time requirements, facilities reassignment of input/output devices and files for each program block, and makes it a simple matter to add, modify or delete programs.

*State-space model specification*

The system to be simulated can be specified in standard state space form (order $\leq 30$) or, as described later, in the classical transfer function and block diagram form. For matrix input the coefficient matrices for the following equations are entered by the user or obtained from previously established input files:

$$\dot{\mathbf{x}}(t)=\boldsymbol{A}\mathbf{x}(t)+\boldsymbol{B}\mathbf{u}(t) \quad (1)$$

$$\mathbf{y}(t)=\boldsymbol{C}\mathbf{x}(t)+\boldsymbol{D}\mathbf{w}(t) \quad (2)$$

where:

$\mathbf{x}$ = state vector
$\mathbf{u}$ = input vector
$\mathbf{y}$ = output vector
$\mathbf{w}$ = input plus delayed vector
$\boldsymbol{A}, \boldsymbol{B}, \boldsymbol{C}, \boldsymbol{D}$ are constant coefficient matrices of appropriate dimension.

*State difference equation*

A program generates the discrete form of the analytical solution to equation (1) in the form:

$$\mathbf{x}((n+1)T)=\boldsymbol{\phi}(T)\mathbf{x}(nT)+\boldsymbol{\Delta}(T)\mathbf{u}(nT) \quad (3)$$

where:

$\boldsymbol{\phi}$ = fundamental or transition matrix
$\boldsymbol{\Delta}$ = coefficient matrix
$T$ = discrete time interval
$n$ = counter for time intervals.

Equation (3) agrees exactly with equation (1) at each sampling point if the input vector, $\mathbf{u}$, is constant over the sampling interval, which is true when $\mathbf{u}$ originates from a digital control computer, otherwise an approximation is involved. This discrete form is particularly convenient for digital calculation and $\boldsymbol{\phi}$ and $\boldsymbol{\Delta}$ are available for use by other programs.

The fundamental matrix, $\boldsymbol{\phi}$, can be evaluated using either a power series expansion or the Cayley-Hamilton technique as described by AGOSTINIS [2] or in textbooks such as OGATA [13] and SAUCEDO and SCHIRING [14]. Implementation of the latter technique includes algorithms for determining real, complex and/or repeated eigenvalues of non-symmetric $\boldsymbol{A}$ matrices. The $\Delta$ matrix is obtained by numerical integration of $\boldsymbol{\phi}\,\boldsymbol{B}$ over the interval $T$.

The specification of the time interval, $T$, has a pronounced effect on the numerical stability of some of the algorithms, and it is usually the dominant factor affecting simulation time and determines the resolution or precision of the time domain response. Therefore to provide additional flexibility a program has been included to modify the previously calculated coefficient matrices in equation (3) so that they apply to an interval $mT$ (where $m$ is an integer) instead of $T$ and

$$\boldsymbol{\phi}(mT)=\boldsymbol{\phi}^{m}(T) \quad (4)$$

$$\Delta(mT)=(\phi^{m-1}(T)+\cdots+\boldsymbol{\phi}(T)+\boldsymbol{I})\Delta(T). \quad (5)$$

In general, since use of equations (4) and (5) is considerably faster than recalculating $\boldsymbol{\phi}$ and $\Delta$, it is most expedient to specify a small interval $T$ for the first evaluation of $\boldsymbol{\phi}$ and $\Delta$. Relatively large multiples $mT$, can be used for exploratory studies and then $m$ can be reduced until satisfactory results are obtained.

*Time domain response*

The time domain response is evaluated directly using equations (2) and (3). The initial state vector, if not supplied by the user, is initialized to zero. The elements of the input vector, $\mathbf{u}$, can be defined by selecting standard options as step, ramp, sinusoid or pulse functions or defined by a discrete time series of point values. The control variables can also be specified as originating from a control law as discussed in the section on optimal multivariable control.

*Transfer function model*

Most of the current research work in areas such as optimal control use multivariable state space models. However a great deal of practical design and analysis of control systems is still done using classical techniques based on transfer functions and block diagrams such as those shown in Fig. 2a. Therefore, to serve as a convenient link between the two approaches, a program was written which would accept a problem defined in terms of transfer functions and a coded equivalent of the user's block diagram, and produce an equivalent state space representation in the form of equations (1) and (2). There are no arbitrary restrictions on the form of the block diagram except the dimensions specified for vectors and matrices in the digital program. Time delays, isolated differentiation terms and isolated integral terms can be handled.

Pure time delays, such as the one included in Fig. 2a, are handled by defining additional elements in the input vector, **u**, equal to the original input(s) to that subsystem but delayed by the appropriate time period. Thus if a time delay, $\tau_d$, is present, then in the state space representation defined by equations (1) and (2) the input vector **u** will contain elements of the type $u_i(t-\tau_d)$ and the coefficient matrices $B$ and $D$ are modified as necessary. The printout in Fig. 2b indicates the existence of one time delay but the user does nothing except

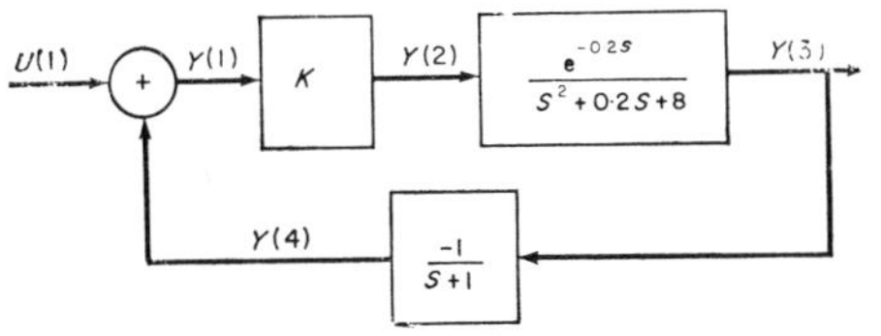


FIG. 2a. A simple example used to illustrate the ability of GEMSCOPE to accept problems defined in terms of a block diagram and transfer functions. Note the inclusion of a pure time delay element and non-dynamic blocks.

```
        * THIS IS A THIRD ORDER EXAMPLE TO TEST THE          *
        * CAYLEY-HAMILTON TECHNIQUE AGAINST THE POWER *
        * SERIES METHOD. THE EXAMPLE HAS BEEN CHOSEN      *
        * BECAUSE OF COMPLEX EIGENVALUES                    *
* THIRTEEN INPUT DATA CARDS ARE REQUIRED FOR THIS EXAMPLE *
* THE TWO INPUT CARDS DEFINING THE PROCESS TRANSFER FUNCTION ARE *
3 0 1·0 0·2 −2 8·0 0·2 1·0  **
103 2 1·0                   **
```

BLOCK DIAGRAM INPUT

| | |
|---|---|
| INPUT MATRIX FLAG ........ | 0 |
| JOB NUMBER ........ | 1 |
| PRIORITY NUMBER FOR PRINTING OUTPUT ........ | 3 |
| LOGICAL UNIT NUMBER FOR INPUT DEVICE ........ | 3 |
| LOGICAL UNIT NUMBER FOR OUTPUT DEVICE ........ | 4 |

COEFFICIENT MATRICES OF THE STATE EQUATION AND RELATED DATA

| | |
|---|---|
| NUMBER OF STATE VARIABLES ........ | 3 |
| NUMBER OF OUTPUT VARIABLES ........ | 4 |
| NUMBER OF EXTERNAL FORCING FUNCTIONS ........ | 1 |
| NUMBER OF TIME DELAYS ........ | 1 |
| NUMBER OF EXTERNAL FORCING FUNCTIONS IN U(T) ........ | 0 |
| DELAYED EXTERNAL FORCING FUNCTIONS IN....U(T) ........ | 1 |
| DELAYED STATE VARIABLES IN ........U(T) ........ | 1 |

MATRIX $A$ MATRIX $B$ MATRIX $C$ MATRIX $D$

$$\begin{bmatrix} 0{\cdot}0 & 1{\cdot}0 & 0{\cdot}0 \\ -8{\cdot}0 & -0{\cdot}2 & 0{\cdot}0 \\ 1{\cdot}0 & 0{\cdot}0 & -1{\cdot}0 \end{bmatrix} \begin{bmatrix} 0{\cdot}0 & 0{\cdot}0 \\ 2{\cdot}0 & -2{\cdot}0 \\ 0{\cdot}0 & 0{\cdot}0 \end{bmatrix} \begin{bmatrix} 0{\cdot}0 & 0{\cdot}0 & -1{\cdot}0 \\ 0{\cdot}0 & 0{\cdot}0 & -2{\cdot}0 \\ 1{\cdot}0 & 0{\cdot}0 & 0{\cdot}0 \\ 0{\cdot}0 & 0{\cdot}0 & 1{\cdot}0 \end{bmatrix} \begin{bmatrix} 1{\cdot}0 \\ 2{\cdot}0 \\ 0{\cdot}0 \\ 0{\cdot}0 \end{bmatrix}$$

COEFFICIENT MATRICES OF THE STATE DIFFERENCE EQUATION

| MATRIX $FM(T)$ | | | MATRIX $H(T)$ | |
|---|---|---|---|---|
| 0·9605E 00 | 0·9769E−01 | 0·0000E 00 | 0·9866E−02 | −0·9866E−02 |
| −0·7815E 00 | 0·9409E 00 | 0·0000E 00 | 0·1953E 00 | −0·1953E 00 |
| 0·9387E− 01 | 0·4772E−02 | 0·9048E 00 | 0·3237E−03 | −0·3237E−03 |

FIG. 2b. Selected samples of input data and printed output produced by GEMSCOPE for the system defined in Fig. 2a. The top section is an image of the first part of the input data deck; the centre is part of a summary of key parameters which define the problem and solution method; the bottom section shows the coefficient matrices for equations (1), (2) and (3).

specify the appropriate transfer function, in this example he adds the input parameters 0, 1·0, 0·2.

Figure 2b presents samples of the input data required and the output produced. A free format input routine is used which will accept "comments" if enclosed with asterisks (*) and mixed mode numeric fields terminated by a double asterisk (**). All input data and control information is printed out in annotated form so the user will have full documentation. (This is particularly important for student laboratories.) Similarly the user can, by selecting one of three levels of output documentation, get a summary of the important intermediate results such as the "coefficient matrices of the state equation and related data" shown in Fig. 2b. All results from GEMSCOPE are placed in disk files which can be punched onto cards or listed on the line printer. This example was actually run using the IBM 1800 version of GEMSCOPE (i.e. CSAP [2]), which permits results to be displayed on a graphic terminal or plotted on a digital plotter as shown in Fig. 2c. This example was run twice and the results plotted on the same graph to show the equivalence of two different GEMSCOPE options for calculating the fundamental matrix, $\boldsymbol{\phi}$, in equation (3). The same approach when used to compare different control strategies makes a very effective design and/or teaching tool.

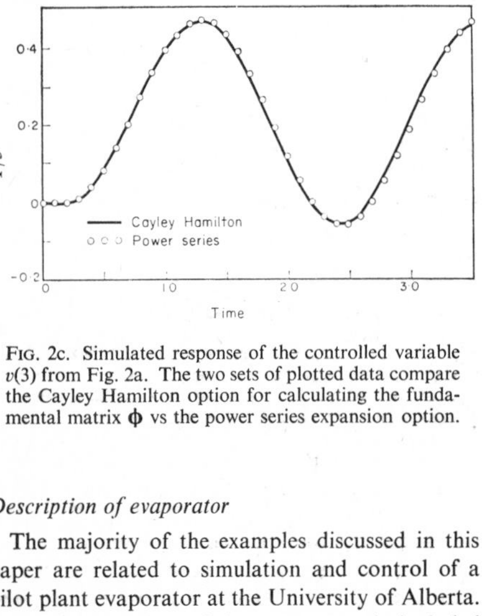


FIG. 2c. Simulated response of the controlled variable $v(3)$ from Fig. 2a. The two sets of plotted data compare the Cayley Hamilton option for calculating the fundamental matrix $\boldsymbol{\phi}$ vs the power series expansion option.

*Description of evaporator*

The majority of the examples discussed in this paper are related to simulation and control of a pilot plant evaporator at the University of Alberta. A brief description of the evaporator is included so the reader will be able to evaluate the simulated and experimental results presented with these examples.

The evaporator is a double-effect unit constructed largely of glass so the process response can be observed. It is fully instrumented with conventional industrial instruments but is also interfaced to the IBM 1800 computer for supervisory, DDC, or special control applications. The evaporator operates in a forward-feed configuration with 135 kg/hr of aqueous glycol solution and 55 kg/hr of fresh steam fed to the first effect. The second effect operates under vacuum and utilizes the overhead vapour from the first effect to further concentrate the liquid product from the first effect. Several mathematical models have been developed for the evaporator and are discussed elsewhere [12, 8]. The model used for the multivariable control studies was a fifth order, linear state-space model with output variables ($W1$, $W2$, $C2$) and input variables (B1, B2, $S$) where

$W1$ = liquid holdup in the first effect
$W2$ = liquid holdup in the second effect
$C2$ = product concentration from the second effect
$B1$ = bottoms flowrate from the first effect
$B2$ = bottoms flowrate (product) from the second effect
$S$ = steam flowrate into the first effect.

The control policy was to produce a product of constant composition, $C2$, in spite of variations in the feed flowrate, composition and/or temperature. For conventional single variable controllers the output variables were controlled by manipulating the corresponding input variable.

*Model reduction*

Model reduction is frequently desirable to reduce the computational requirements when a large series of simulations is to be done, or to eliminate states which are not of interest and/or are not physically measurable. Also when the model, or control strategies based on it, are to be implemented on computer controlled processes, then model reduction will reduce the amount of real-time calculation. Therefore a program has been included based on a method proposed by MARSHALL [15], which will reduce the order of the system defined by equation (1) by eliminating the least significant eigenvalues. It has been shown [5] that model reduction followed by conversion to the discrete form of equation (3) is equivalent to discretization followed by model reduction so the order of these steps is unimportant.

Model reduction was particularly important in the series of simulation and experimental studies related to control of the double effect evaporator at the University of Alberta. Several evaporator models have been derived but the most accurate one [12] consists of ten non-linear, ordinary, differential equations. These were linearized numerically and arranged into the state-space form represented by equations (1) and (2). The simulated open loop response of the outlet concentration from the evaporator to a 20 per cent increase in feed flow rate is shown by the solid curve in Fig. 3. It should be noted that the actual process response is slightly slower, but it is, in general, in good agreement with the simulated response. The order of the model was then reduced to five by eliminating state variables such as the temperature of the tubes which were of no interest and were expected to be of relatively minor importance. The response of the fifth order model plotted in Fig. 3 confirms that it is almost as good as the tenth order linear model. It was then decided to

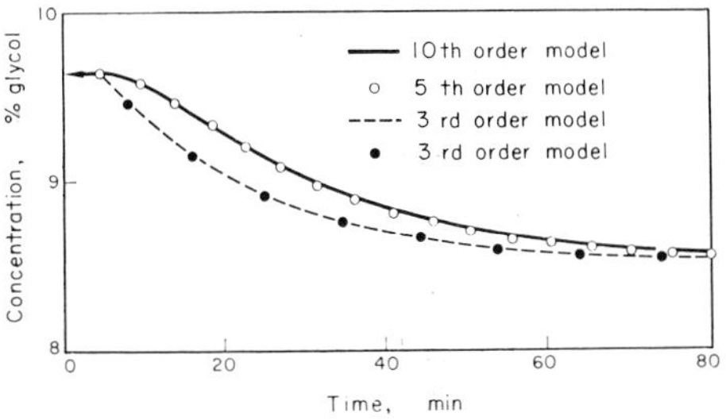


FIG. 3. Comparison of simulated, open-loop responses of the evaporator product concentration to a 20 per cent increase in feed flowrate, to illustrate the effect of model reduction.

further reduce the model by eliminating the first effect concentration and enthalpy from the state vector. This then left a third order system where the only state variables, the two liquid holdups and the product concentration, were the variables that physically had to be controlled. However the responses plotted in Fig. 3 show that the responses of the third order models are considerably different than the original tenth order model. As a check on the algorithm one third order model was generated by reduction of the tenth order model and a second by reduction of the fifth order model, but no significant difference was observed. The fifth order model was selected for use in further control studies. It should be noted that, for this example, the results using MARSHALL's method [15] of model reduction were in good agreement with results based on DAVISON's method [16], and with a fifth order linear model derived using physical assumptions based on "experience and engineering judgement".

*Non-interaction*

For the control of multivariable processes it is frequently not the dynamic relationship between a given pair of input and output variables that causes difficulty, but rather the interaction between these variables and other state variables. One design approach is to develop an augmented plant model such that there is no interaction between the new reference variables and the output variables.

The decoupling program included in GEMSCOPE is a modified version of one developed by GILBERT and PIVNICHNY [17]. The program first tests to see whether the model meets the necessary and sufficient conditions for non-interaction by state feedback as described by FALB and WOLOVICH [18]. If the model, given in the form of equations (1) and (2) can be made non-interacting the program synthesizes a control law in the following form:

$$\mathbf{u} = \boldsymbol{F}\mathbf{x} + \boldsymbol{G}\mathbf{v}. \tag{6}$$

The reference input vector $\mathbf{v}$ is related to the original process outputs by the following equations written in terms of Laplace transforms

$$\mathbf{y}(s) = \boldsymbol{H}(s)\mathbf{v}(s). \tag{7}$$

The transfer matrix $H(s)$ is diagonal and the elements are specified functions of parameters $\boldsymbol{\delta}$ and $\boldsymbol{\lambda}$. These parameters are chosen by the user so that the system defined by equation (7) has the desired dynamic response and this fixes the elements of $\boldsymbol{F}$ and $\boldsymbol{G}$. The basic theory behind the algorithm has been described by GILBERT [19]. The user then specifies the augmented model in the state space form of equations (1) and (2) and re-enters GEMSCOPE for further simulation and/or control.

The evaporator model did not meet the conditions for non-interaction and therefore, for demonstration purposes, a fifth order distillation column model developed by RAFAL and STEVENS [20] was used. The open-loop response of the linearized model to a disturbance in reflux is shown by the dashed lines in Fig. 4. In the top plot of condenser composition the dashed and solid lines are identical. The design procedure for non-interacting systems as described above was then used. Five parameter values in $\boldsymbol{H}(s)$ were specified so that the augmented system had approximately the same dynamic response as the original system, although this was just an arbitrary criterion. The response of the decoupled system to a change in $\mathbf{v}$, i.e. a new specification for overhead composition, is shown by the solid curves in Fig. 4. It can be seen that the desired change in overhead composition took place with almost the same dynamics as the original system but without any effect on the reboiler composition.

Several single variable closed-loop feedback control schemes were then added to this decoupled model and control responses simulated, but it was found, since the dynamics of the system represented by $H(s)$ were essentially first order, that arbitrarily good control could be obtained by simply increasing the controller gain. However in general, non-interacting control was significantly better than the same control applied to the original system.

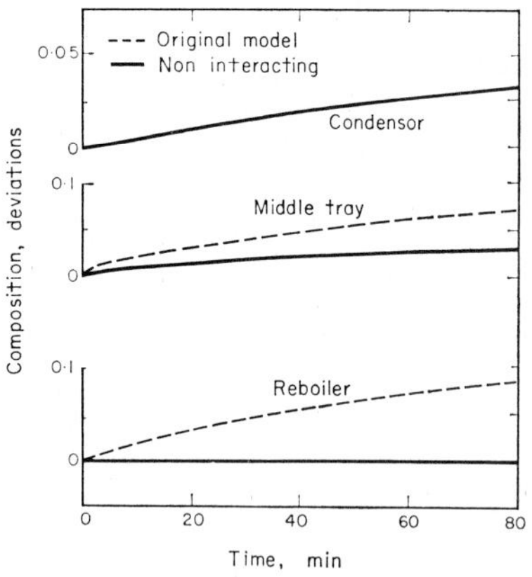


FIG. 4. Simulated composition responses from a distillation column model [20] to an increase in reflux flow. The dashed lines are from the open-loop model and the solid lines from the model plus state feedback designed to eliminate interactions between condenser and reboiler compositions.

As mentioned above the evaporator model does not meet the conditions necessary to achieve strict non-interaction via statefeedback. However, it can be made "almost non-interacting" using the computer-assisted-design (CAD) suite of programs developed at the University of Manchester Institute of Science and Technology. This approach is based on the multivariable, inverse-Nyquist-array technique developed by ROSENBROCK [21] which provides a mathematical measure of interaction in terms of "diagonal dominance"; a means of determining closed-loop stability; and a design procedure whereby pre- and post-compensator matrices can be developed which reduce interaction to an acceptable level. A single 3 × 3 precompensator matrix of constants and three conventional single variable feedback controllers designed using this approach gave excellent simulated control of the evaporator [23].

*Optimal multivariable controllers*

Optimal multivariable controllers with proportional-plus-integral state feedback, feedforward compensation for constant disturbances, $\mathbf{d}$, and provision for driving the process from its current position to setpoint values, $y_{SP}$, can be implemented using a control law of the form [8, 9]

$$\mathbf{u} = \boldsymbol{K}_{FB}\mathbf{x} + \boldsymbol{K}_{FF}\mathbf{d} + \boldsymbol{K}_{SP}\mathbf{y}_{SP}. \tag{8}$$

The integral action is generated by augmenting the state vector, $\mathbf{x}$, with variables, $\mathbf{z}$, defined by $\dot{\mathbf{z}} = \mathbf{y}$. Therefore the integral control parameters are included in $\boldsymbol{K}_{FB}$. By proper definition of augmented coefficient matrices in equation (1) it is also possible to generate control systems such that for a step change in desired output values, $\mathbf{y}_{SP}$, the process response, $\mathbf{y}$, will follow the model response, $\mathbf{y}_m$, generated by

$$\dot{\mathbf{y}}_m = \boldsymbol{G}\mathbf{y}_m + \boldsymbol{H}\mathbf{y}_{SP} \tag{9}$$

where $\boldsymbol{G}$ and $\boldsymbol{H}$ are specified by the user [8]. This approach gives the designer a direct method for specifying the dynamics of the process response and produces a control system that is less sensitive to external disturbances than "indirect model-following schemes". The resulting control law has the same form as equation (8) except that an additional term, $\boldsymbol{K}_m\mathbf{y}_m$, is added.

The computer program included in GEMSCOPE generates the coefficient matrices for equation (8) using recursive relationships developed by applying the concept of discrete dynamic programming to a model described by equation (3) and a quadratic performance criterion of the following form:

$$J = \beta^N(\mathbf{x}_N - \mathbf{x}_d)^T \boldsymbol{S}(\mathbf{x}_N - \mathbf{x}_d) + \sum_{k=1}^{N} \beta^k[(\mathbf{x}_k - \mathbf{x}_d)^T \boldsymbol{Q}(\mathbf{x}_k - \mathbf{x}_d) + \mathbf{u}_{k-1}^T \boldsymbol{R}\mathbf{u}_{k-1}] \tag{10}$$

where, $\boldsymbol{Q}$, $\boldsymbol{R}$ and $\boldsymbol{S}$ are weighting factors specified by the user. This method of generating control matrices has proven to be relatively straightforward, practical, easy to implement and has given outstanding results in both simulated and experimental control studies on the pilot plant evaporator.

Figures 5 and 6 show the simulated response of the evaporator model to a 20 per cent change in feed flow under different control schemes. The use of conventional proportional feedback control, using gains that gave reasonable control on the

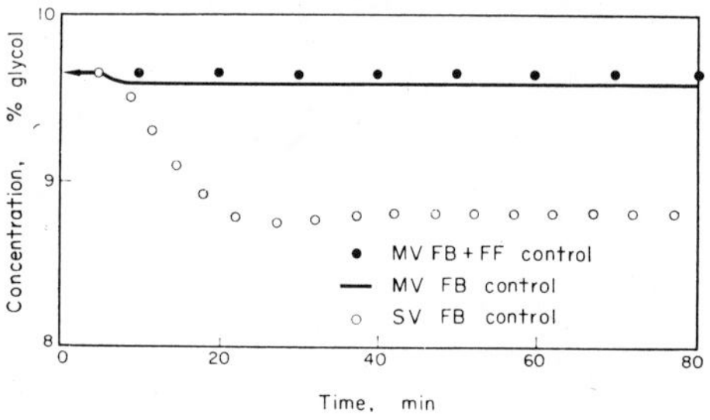


FIG. 5. Simulated comparison of single vs multivariable feedback control systems for evaporator product composition. The single variable controller used a gain that performed well experimentally. Feed-forward control eliminated the small offset left by multivariable proportional control.

actual evaporator, resulted in a significant offset as shown by "*SV FB* Control" data in Fig. 5. Use of optimal, multivariable, proportional, feedback control (*MV FB*) resulted in almost negligible offset and the addition of feedforward control (*MV FB+FF*) eliminated even this small offset. The data in Fig. 6 show similar results except that integral action has been added to both the single variable and multivariable controllers to eliminate the offset. In both cases multivariable control is clearly superior to conventional single variable loops.

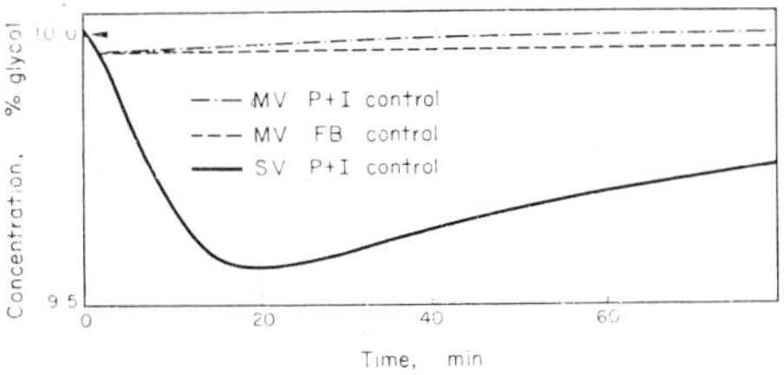


FIG. 6. Simulated comparison of single vs multivariable controllers, similar to Fig. 5, but showing the effect of integral action.

However it is not unusual to obtain good results when optimal control policies are applied, using simulation techniques, to the same process model that was used in the derivation of the optimal policy. The real test comes when the control policies are applied to real processes and the effects of non-linearities, modelling errors, noise, etc. can be evaluated. NEWELL and FISHER [8] have compared the experimental response of the evaporator under optimal multivariable proportional control, which was designed using GEMSCOPE, and conventional proportional-plus-integral direct digital control tuned experimentally. Multivariable control maintained the outlet concentration essentially constant in spite of two 20 per cent feed changes in the interval considered. A single feed disturbance under conventional single variable DDC control generally resulted in a transient that lasted for 2 or 3 hr. It is significant that the fifth order model which proved to be an excellent basis for optimal control calculations and simulations was *NOT* suitable for determining the controller constants for conventional, single-variable controllers by closed loop simulations.

For a given model and criterion the only decision left to the designer of optimal control systems is the choice of the weighting parameters in the criterion, as indicated in equation (10). The dashed curves in Fig. 7 shows the response of the outlet concentration of the evaporator, $C2$, and the liquid holdup in the first effect, $W1$, with equal weighting on all states. There are significant offsets in both state variables. Increasing the weighting in $\boldsymbol{Q}$, ($\boldsymbol{R}=\boldsymbol{S}=\boldsymbol{O}$) on $C2$ results in virtual elimination of the offset in $C2$ but increases the offset in $W1$, and the other three state variables. Since the relationship between weighting factors and the system response is not easily defined in a quantitive manner, simulation is an excellent design tool. However the criteria for optimality must be tempered by practical factors. For example high sampling frequencies and high controller gains produce the best simulated responses. However for experimental runs on the evaporator the design parameters in GEMSCOPE were defined to

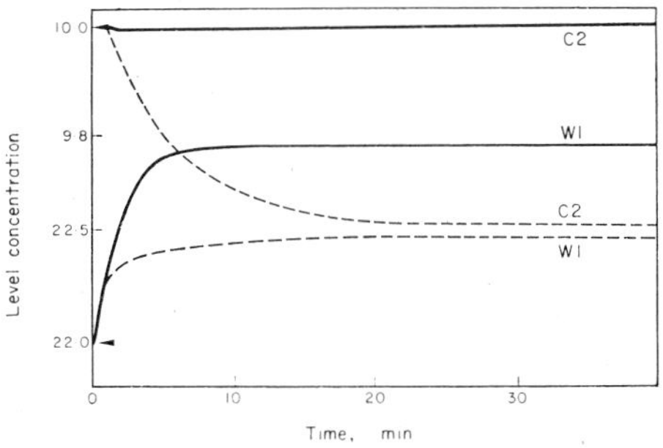


FIG. 7. Simulated responses from evaporator model $5L$ showing that the concentration response, $C2$, improves, but the liquid level response, $W1$, gets worse as the relative weighting on $C2$ in equation (10) increases.

generate longer controller intervals and lower gains. The slight loss in "simulated performance" was more than made up for by reduced real-time computing requirements and by decreased sensitivity to process noise [7, 8].

*Utilization of a discrete control vector*

Provision has been made within GEMSCOPE to carry out time domain simulation using control variables, $u_i$, specified by a discrete time series of points stored in a computer file. The values for $u_i$ can be supplied by the user or calculated by another program. Digital computer programs have been developed at the University of Alberta which use quasilinearization and/or linear programming techniques to generate the optimal control policy to drive a process from one state to another in such a manner as to minimize a criterion such as minimum time, or minimum sum of the absolute errors [10]. The solid curve in Fig. 8 shows the simulated response of the evaporator product concentration to time optimal (bang-bang) manipulation of the feed steam. The response is excellent. However, use of a similar policy on the actual evaporator results in significantly poorer performance as typified by the dashed curve. An extended series of simulated and experimental trials has been done utilizing different models,

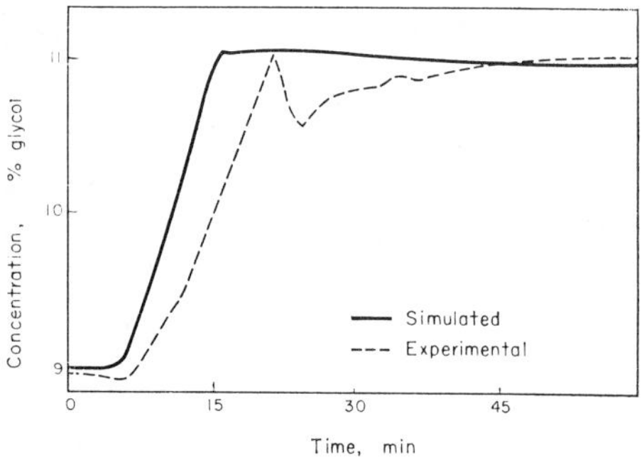


FIG. 8. A comparison of simulated, vs experimental open-loop responses of evaporator concentration to bang-bang manipulation of the steam flow, calculated to give a time optimal response.

different criteria and with and without constraints on the state and control variables [10]. The best experimental results approach that of the solid curve in Fig. 8 and are significantly better than with conventional control techniques. In general the bang-bang manipulation of the control variables and the rapid rates of change produced in the state variables severely tax a linear process model and small modelling errors can cause a significant deviation between simulated and experimental results. However the relative performance of different control policies can usually be determined by simulation.

## DISCUSSION

As suggested by the above examples, GEMSCOPE, has been used successfully at the University of Alberta for student education and the solution of practical control problems. Other design and simulation programs have been developed and used elsewhere (e.g. [3, 17, 25, 26]). However there is still a need for further development of control and numerical techniques that will extend the applications to larger systems, non-linear systems, stochastic systems, etc.

Even for the class of linear systems GEMSCOPE is far from complete and could be extended to incorporate a number of additional techniques. Some obvious additions, indicated on the left side of Fig. 1 include provision for definition of the problem in terms of a single matrix $\boldsymbol{G}(s)$; generation of minimal order state-space realization for a given $\boldsymbol{G}(s)$; generation of $\boldsymbol{G}(s)$ from a given state-space representation; and other control system design techniques such as closed-loop pole placement. In addition there are what amount to alternative design approaches based on complex or frequency domain methods, and/or formal algebraic manipulation of polynomial matrices[21,22].

As the scope of applications and the number of available techniques increase the design and simulation programs themselves grow and can overwhelm the user and/or become an end in themselves instead of just a tool. Perhaps future development requires a set of standards developed by a group of users, as is being done for direct digital control programs [24], and/or a computer manufacturer or service bureau that would take over the maintenance and operation of "general, multipurpose simulation and control packages".

Similarly there is a continuing need for comparative studies and experimental evaluation of simulation and design techniques so the user can select the best method for his application and can have confidence in the results.

## CONCLUSIONS

(1) Digital simulation is an important technique for the design and analysis of dynamic systems and programs such as GEMSCOPE are a useful and effective tool for design engineers and students.
(2) The success of most design and simulation techniques depends both on the accuracy of the model and the end use to which the results are applied.
(3) Many design techniques, such as those for optimal control of linear systems, are now well enough defined that they can be incorporated into computer algorithms and made available to designers via time-shared computer terminals.

*Acknowledgements*—The contributions of other students and staff in the department, in particular those of R. B. Newell, R. E. Nieman, H. R. Ramstead and the DACS Centre staff, are gratefully acknowledged. This work was partially supported by funds from the National Research Council of Canada in the form of research grants and scholarships.

## REFERENCES

[1] D. G. FISHER: Real time computing in the university. *Chem. Eng. Education* **5,** 24 (1971).
[2] W. AGOSTINIS and D. G. FISHER: Control systems analysis program, IBM Type IV Program Number 1800-43.2.001 (1969). Also M.Sc. thesis (1969).
[3] R. G. E. FRANKS: *Mathematical Modelling in Chemical Engineering*. Wiley, New York (1966).
[4] R. K. WOOD and R. A. FARWELL: Computerized design and analysis of control loops. *Instrum. Technol.* **18,** 42-47 (1971).
[5] R. G. WILSON, D. G. FISHER and D. E. SEBORG: Model reduction foı discrete time dynamic systems. *Int. J. Control* (in press).
[6] T. CHUANG, G. MISKO, I. G. DALLA LANA and D. G. FISHER: On-line operation of a PE 621 infrared spectrophotometer—IBM 1800 computer system. *Computers in Analytical Chemistry*, Vol. 4. Plenum (1969).
[7] R. B. NEWELL: Multivariable computer control of an evaporator. Ph.D. thesis, Department of Chemical and Petroleum Engineering, University of Alberta, Edmonton. Available from University Microfilms, Ann Arbor, Mich. (1971).
[8] R. B. NEWELL and D. G. FISHER: Experimental evaluation of optimal multivariable regulatory controllers with model following capabilities. *Automatica* **8,** 247-262 (1972).

[9] R. B. Newell, D. G. Fisher and D. E. Seborg: Computer control using optimal multivariable feedforward—feedback algorithms. *AIChE J.* **18,** 881 (1972).
[10] R. E. Nieman: Parameter and state estimation techniques applied to process control. Ph.D. thesis, Department of Chemical Petroleum Engineering, University of Alberta, Edmonton (1971).
[11] H. Andre and R. A. Ritter: Dynamic response of a double effect evaporator. *Can. J. Chem. Eng.* **46,** 259 (1968).
[12] R. B. Newell and D. G. Fisher: Model development, reduction and experimental evaluation for a double effect evaporator. *Ind. Eng. Chem Process Design & Dev.* **11,** 213 (1972).
[13] K. Ogata: *State Space Analysis of Control Systems.* Prentice-Hall, N.J. (1967).
[14] R. Saucedo and E. E. Schiring: *Introduction to Continuous and Digital Control Systems.* MacMillan, New York (1968).
[15] S. A. Marshall: An approximate method for reducing the order of a linear system. *Control* **10,** 642 (1966).
[16] E. J. Davison: A method for simplifying linear dynamic systems. *IEEE Trans. Aut. Control* **AC-11,** 93 (1966).
[17] E. G. Gilbert and J. R. Pivnichny: A computer program for the synthesis of decoupled multivariable feedback systems. *IEEE Trans. Aut. Control* **AC-14,** 652 (1969).
[18] P. L. Falb and W. A. Wolovich: On the decoupling of multivariable systems. Joint Automatic Control Conference, 8th, Preprints, 791 (1967).
[19] E. G. Gilbert: The decoupling of multivariable systems by state feedback. *SIAM J. Control* **7,** 50 (1969).
[20] M. D. Rafal and W. F. Stevens: Discrete dynamic optimization applied to on-line optimal control. *AIChE J.* **14,** 85 (1968).
[21] H. H. Rosenbrock: *State-Space and Multivariable Theory.* Thomas Nelson, London (1970).
[22] A. G. J. Macfarlane: Linear multivariable feedback theory: A survey. Invited paper at IFAC Symposium, Düsseldorf (1971). *Automatica* **8,** 455-492 (1972).
[23] R. Butland: An application and assessment of the inverse Nyquist array design technique. M.Sc. Thesis, UMIST, Manchester, England (in preparation).
[24] T. J. Williams: Computers and process control-fundamentals. *Ind. Eng. Chem.* **62,** 12, 94 (1970).
[25] J. L. Melsa: *Computer Programs for Computational Assistance in the Study of Linear Control Theory.* McGraw-Hill, New York (1970).
[26] R. E. Kalman and T. S. Englar: A user's manual for the automatic synthesis program. NASA report CR-475 (1966).

**Résumé**—On décrit un système général de simulation et de contrôle (GEMSCOPE) qui aide la conception, l'analyse et la simulation digitale de systèmes dynamiques indépendants du temps. Le modèle du procédé peut être défini schématiquement avec des fonctions de transfert pour chaque composant ou sous forme standard d'espace d'état. Des options sont comprises pour générer les matrices de contrôle régulatrices optimales, déterminant la matrice de rétroaction d'état pour la non interaction, réduisant l'ordre du modèle et calculant la réponse du système à boucle fermée ou ouverte. GEMSCOPE s'est avéré utile pour l'éducation d'étudiants et pour la conception de systèmes de contrôle pratiques. Des résultats expérimentaux et simulés obtenus en appliquant des techniques GEMSCOPE à des installations contrôlées par ordinateur à l'Université d'Alberta démontrent la validité des méthodes et les améliorations possibles grâce à l'emploi de techniques de contrôle modernes.

**Zusammenfassung**—Eine allgemeine Vielzweck-Gerätebaugruppe zur Simulation und Steuerung (GEMSCOPE) wird beschrieben, die den Entwurf, die Analyse und die digitale Simulation linearer, zeitinvarianter dynamischer Systeme erleichtert. Das Prozeßmodell kann durch ein Blockdiagramm mit Ubertragungsfunktionen für jedes Element oder in der Standardform für den Zustandsraum angegeben werden. Die Wahl für die Gewinnung der optimalen Steuermatrix, die die Rückführmatrix bei fehlender Wechselwirkung bestimmt, die Ordnung des Modells reduziert und das die Berechnung des Verhaltens des offenen oder geschlossenen Systems erlaubt ist eingeschlossen. GEMSCOPE erwies sich für die Ausbildung von Studenten und für den Entwurf von praktischen Steuersystemen als nützlich. Die bei der Anwendung der GEMSCOPE-Technik aus rechnergesteuerte Pilotanlagen an der Universität von Alberta erhaltenen experimentellen und simulierten Daten zeigen die Gültigkeit der Methoden und die bedeutsamen und möglichen Verbesserungen auf, die durch die Benutzung der modernen Steuertechnik ermöglicht werden.

**Резюме**—Описана общая многоцельная система моделирования и управления GEMSCOPE, которая помогает при разработке, анализе и циффровом моделировании линейных динамическихсистем, инвариантных по времени. Модель процесса может обыть определена в форме блоковой схемы с переходными функциями для каждого компонента или в стандартной форме простарнства состояний. Система включает могущость образования матриц оптимального регулирующего управления, определения матрици обратной свлзи состояния для невзаймодействия, снижения порядка модели и исчисления реакции систем с разомкнутым или замкнутым контуром. GEMSCOPE оказался полезным для обучения студентов и разрабтки практических систем управления. Опытные данные и данные полученные моделированием с помощью метода GEMSCOPE на опытных установках университета в Алберты, управленных вычислительной машиной, показали обоснованность методов и значительные улучшения получаемые испольэзованием современных методов управления.

# Model reduction for discrete-time dynamic systems†

R. G. WILSON, D. G. FISHER and D. E. SEBORG
Department of Chemical and Petroleum Engineering,
University of Alberta, Edmonton, Alberta

[Received 28 September 1971 ; revision received 30 November 1971]

The model reduction techniques of Marshall and Davison for linear continuous-time models are extended for use in reducing the order of linear discrete-time models. Two approaches for the reduction of high-order continuous-time models to low-order discrete-time models are presented and evaluated. The resulting reduced order discrete-time models obtained by the two approaches are shown to be equivalent. A simple numerical example is solved to demonstrate the techniques presented.

## 1. Introduction

Linear state-space models have been extensively used to mathematically describe a wide variety of physical phenomena, such as aircraft, chemical plants and refineries. In many situations, dynamic models consisting of a large number of differential or difference equations can be derived from theoretical considerations. Often, such models are so large as to be inconvenient or impractical for many purposes, including simulation and control system design and implementation. Consequently, a need exists for systematic procedures for deriving reduced-order dynamic models from high-order models.

Model reduction techniques for continuous-time state-space models have been the subject of many recent investigations with modal approaches (Marshall 1966, Davison 1966, Chidambara and Davison 1967 a, b, c, Graham 1968) receiving the most attention. By contrast, only a few investigations have considered the model reduction of discrete-time systems. Anderson (1967 a) presented a method for deriving a discrete-time reduced-order system based on a least squares fit of the response of the reduced-order model to that of the original high-order model. He has also presented revisions to his method to reduce computational difficulties (Anderson 1967 b), to account for pre-specified elements of the reduced-order system (Anderson 1967 c) and to produce better agreement at the final steady state (Anderson 1968). In this investigation the modal methods of Marshall (1966) and Davison (1966) are extended to discrete-time models.

The discrete-time model reduction techniques will be derived in § 2. In § 3 the problem of reducing a high-order continuous-time model to a low-order discrete-time model is considered. Finally, a numerical example illustrating the model reduction techniques is presented in § 4.

† Communicated by Professor D. E. Seborg.

International Journal of Control, Vol. 16, No. 3, pp. 549-558 (1972).

## 2. Model reduction for discrete-time systems

### 2.1. *Problem formulation*

Consider the stable, time-invariant, discrete-time model :

$$\mathbf{x}(j+1)=\boldsymbol{\phi}\mathbf{x}(j)+\boldsymbol{\Delta}\mathbf{u}(j) \tag{1}$$

which can be partitioned as

$$\begin{bmatrix}\mathbf{x}_1(j+1)\\ \mathbf{x}_2(j+1)\end{bmatrix}=\begin{bmatrix}\boldsymbol{\phi}_1 & \boldsymbol{\phi}_2\\ \boldsymbol{\phi}_3 & \boldsymbol{\phi}_4\end{bmatrix}\begin{bmatrix}\mathbf{x}_1(j)\\ \mathbf{x}_2(j)\end{bmatrix}+\begin{bmatrix}\boldsymbol{\Delta}_1\\ \boldsymbol{\Delta}_2\end{bmatrix}\mathbf{u}(j) \tag{2}$$

where $\mathbf{x}$ is the $n$-dimensional state vector partitioned into $\mathbf{x}_1$, the $l$ states to be retained in the reduced-order model and $\mathbf{x}_2$, the $n-l$ states to be eliminated. $\mathbf{u}$ is the $m$-dimensional system input and $\boldsymbol{\phi}$ and $\boldsymbol{\Delta}$ are $n\times n$ and $n\times m$ matrices, respectively.

The objective is to determine a reduced-order model of specified order, $l$, which is of the form

$$\mathbf{x}_1(j+1)=\boldsymbol{\phi}_{\mathrm{R}}\mathbf{x}_1(j)+\boldsymbol{\Delta}_{\mathrm{R}}\mathbf{u}(j), \tag{3}$$

where $\boldsymbol{\phi}_{\mathrm{R}}$ and $\boldsymbol{\Delta}_{\mathrm{R}}$ are constant matrices of appropriate dimensions and must be determined.

Since the model reduction techniques to be considered are based on a modal analysis, the required preliminary material will be presented.

An $n\times n$ matrix, $\mathbf{M}$, exists which transforms eqn. (1) into the Jordan canonical form (Gantmacher 1959). Define

$$\mathbf{x}(j)=\mathbf{M}\mathbf{z}(j) \tag{4}$$

or in partitioned form

$$\begin{bmatrix}\mathbf{x}_1(j)\\ \mathbf{x}_2(j)\end{bmatrix}=\begin{bmatrix}\mathbf{M}_1 & \mathbf{M}_2\\ \mathbf{M}_3 & \mathbf{M}_4\end{bmatrix}\begin{bmatrix}\mathbf{z}_1(j)\\ \mathbf{z}_2(j)\end{bmatrix}, \tag{5}$$

where $\mathbf{z}$ is the canonical state vector, $\mathbf{z}_1$ is an $l$-dimensional vector, and $\mathbf{z}_2$ is an $(n-l)$-dimensional vector. It is assumed that matrix $\mathbf{M}$ is arranged so that its columns are ordered, from left to right, in order of decreasing significance of the corresponding eigenvalues of $\boldsymbol{\phi}$. (This can always be achieved by appropriate column interchange.) Since the system in eqn. (1) is assumed to be stable, all the eigenvalues of $\boldsymbol{\phi}$ lie in the unit circle (Kuo 1963). Thus, the columns of $\mathbf{M}$ on the left correspond to the eigenvalues of $\boldsymbol{\phi}$ which are nearest to one (in absolute value) and those on the right correspond to eigenvalues of $\boldsymbol{\phi}$ nearest to zero. If matrix $\boldsymbol{\phi}$ has $n$ distinct eigenvalues, the columns of $\mathbf{M}$ are then the eigenvectors of $\boldsymbol{\phi}$ and $\mathbf{M}$ is referred to as the Modal Matrix. When eqn. (5) is substituted into eqn. (2), the Jordan canonical form results :

$$\begin{bmatrix}\mathbf{z}_1(j+1)\\ \mathbf{z}_2(j+1)\end{bmatrix}=\begin{bmatrix}\boldsymbol{\alpha}_1 & \mathbf{0}\\ \mathbf{0} & \boldsymbol{\alpha}_2\end{bmatrix}\begin{bmatrix}\mathbf{z}_1(j)\\ \mathbf{z}_2(j)\end{bmatrix}+\begin{bmatrix}\boldsymbol{\delta}_1\\ \boldsymbol{\delta}_2\end{bmatrix}\mathbf{u}(j), \tag{6}$$

where $\boldsymbol{\alpha}$, a block diagonal matrix, and $\boldsymbol{\delta}$, a matrix, are defined as

$$\boldsymbol{\alpha} = \begin{bmatrix} \alpha_1 & \mathbf{0} \\ \mathbf{0} & \alpha_2 \end{bmatrix} = \mathbf{M}^{-1}\boldsymbol{\phi}\mathbf{M}, \tag{7}$$

$$\boldsymbol{\delta} = \begin{bmatrix} \boldsymbol{\delta}_1 \\ \boldsymbol{\delta}_2 \end{bmatrix} = \mathbf{V}\boldsymbol{\Delta} \tag{8}$$

and $\mathbf{V}$ is defined as

$$\mathbf{V} = \begin{bmatrix} \mathbf{V}_1 & \mathbf{V}_2 \\ \mathbf{V}_3 & \mathbf{V}_4 \end{bmatrix} = \mathbf{M}^{-1}. \tag{9}$$

Because of the way the columns of $\mathbf{M}$ have been arranged, the Jordan canonical matrix, $\boldsymbol{\alpha}$, has the eigenvalues of $\boldsymbol{\phi}$ nearest to one in the $l \times l$ matrix, $\boldsymbol{\alpha}_1$, and those nearest to zero in the $n-l \times n-l$ matrix, $\boldsymbol{\alpha}_2$. Hence, it follows from eqn. (6), that $\mathbf{z}_1$ represents the slow modes of the system and $\mathbf{z}_2$ represents the fast modes. That is, after an input disturbance, $\mathbf{z}_2$ comes to its final value faster than does $\mathbf{z}_1$. It is also apparent from eqn. (6) that the modes represented by $\mathbf{z}_1$ and $\mathbf{z}_2$ are non-interacting. For systems with $n$ linearly independent eigenvectors, $\boldsymbol{\alpha}$ is a diagonal matrix and $\mathbf{z}_1$ and $\mathbf{z}_2$ will always be non-interacting. However, for systems with less than $n$ linearly independent eigenvectors, the partitioning of the original system in eqn. (2) must be such that it does not split a Jordan block in $\boldsymbol{\alpha}$, so that $\mathbf{z}_1$ and $\mathbf{z}_2$ will remain non-interacting.

### 2.2. *Marshall's analysis*

In Marshall's analysis for continuous systems (Marshall 1966) the reduced model is derived by neglecting the dynamics of the fast modes of the original high-order model. This approach retains the significant system modes and offers the advantages of simplicity and zero steady-state error for constant inputs. Graham (1968) has shown that the model reduction method developed independently by Chidambara (Chidambara and Davison 1967 b) is equivalent to Marshall's method (Marshall 1966). In this section, Marshall's analysis will be extended to the reduction of discrete-time models.

Equation (2) implies the relation

$$\mathbf{x}_1(j+1) = \boldsymbol{\phi}_1\mathbf{x}_1(j) + \boldsymbol{\phi}_2\mathbf{x}_2(j) + \boldsymbol{\Delta}_1\mathbf{u}(j). \tag{10}$$

An expression for $\mathbf{x}_2(j)$ can be obtained by combining eqns. (4) and (9) to give

$$\mathbf{x}_2(j) = \mathbf{V}_4^{-1}(\mathbf{z}_2(j) - \mathbf{V}_3\mathbf{x}_1(j)). \tag{11}$$

As discussed in § 2.1, $\mathbf{z}_2(j)$ responds much faster than $\mathbf{z}_1(j)$ because the eigenvalues in $\boldsymbol{\alpha}_2$ are closer to zero than those of $\boldsymbol{\alpha}_1$. As an approximation, assume that $\mathbf{z}_2(j)$ immediately attains the new steady-state value (i.e. instantaneous response). Thus, from eqn. (6),

$$\mathbf{z}_2(j+1) = \mathbf{z}_2(j) = (\mathbf{I} - \boldsymbol{\alpha}_2)^{-1}\boldsymbol{\delta}_2\mathbf{u}(j). \tag{12}$$

Now, combining eqns. (10), (11) and (12) gives the reduced-order system of eqn. (3) with

$$\boldsymbol{\phi}_R = \boldsymbol{\phi}_1 - \boldsymbol{\phi}_2 \mathbf{V}_4^{-1} \mathbf{V}_3, \tag{13}$$

$$\boldsymbol{\Delta}_R = \boldsymbol{\Delta}_1 + \boldsymbol{\phi}_2 \mathbf{V}_4^{-1} (\mathbf{I} - \boldsymbol{\alpha}_2)^{-1} \boldsymbol{\delta}_2. \tag{14}$$

Thus, Marshall's analysis for continuous-time models can be extended to provide a straightforward method for reducing the order of discrete-time models. For constant inputs, the reduced and original systems reach the same steady-state value of $\mathbf{x}_1$. For other types of sustained inputs (e.g. a ramp or a sinusoid) there is a continuing error between the responses of the reduced and original systems (Graham 1968). However, this error is often small.

### 2.3. *Davison's analysis*

Davison's analysis for continuous systems (Davison 1966) is based on the assumption that the contribution of the insignificant eigenvalues of the original system to the system response is small and may be neglected. In Davison's analysis a reduced system results in which the dominant modes are excited in the same proportion as in the original system.

The solution to eqn. (1) can be written as

$$\mathbf{x}(j+1) = \boldsymbol{\phi}^j \mathbf{x}(0) + \sum_{k=0}^{j-1} \boldsymbol{\phi}^k \boldsymbol{\Delta} \mathbf{u}(k), \tag{15}$$

where $\mathbf{x}(0)$ is the initial condition for vector $\mathbf{x}$. Following Davison (1966), consider the case with $\mathbf{x}(0) = \mathbf{0}$. Then

$$\mathbf{x}(j+1) = \sum_{k=0}^{j-1} \boldsymbol{\phi}^k \boldsymbol{\Delta} \mathbf{u}(k). \tag{16}$$

Using the theorem in the Appendix and eqn. (7), eqn. (16) can be written as

$$\mathbf{x}(j+1) = \mathbf{M} \left( \sum_{k=0}^{j-1} \boldsymbol{\alpha}^k \mathbf{M}^{-1} \boldsymbol{\Delta} \mathbf{u}(k) \right). \tag{17}$$

Equation (17) becomes, after partitioning,

$$\begin{bmatrix} \mathbf{x}_1(j+1) \\ \mathbf{x}_2(j+1) \end{bmatrix} = \begin{bmatrix} \mathbf{M}_1 & \mathbf{M}_2 \\ \mathbf{M}_3 & \mathbf{M}_4 \end{bmatrix} \left\{ \sum_{k=0}^{j-1} \left\{ \begin{pmatrix} \boldsymbol{\alpha}_1^k & \mathbf{0} \\ \mathbf{0} & \boldsymbol{\alpha}_2^k \end{pmatrix} \begin{pmatrix} \mathbf{V}_1 & \mathbf{V}_2 \\ \mathbf{V}_3 & \mathbf{V}_4 \end{pmatrix} \boldsymbol{\Delta} \mathbf{u}(k) \right\} \right\}. \tag{18}$$

As an approximation, assume that the eigenvalues in $\boldsymbol{\alpha}_2$ have no effect on the system response, and therefore can be neglected (i.e. $\boldsymbol{\alpha}_2 = \mathbf{0}$). Eqn. (18) becomes,

$$\begin{bmatrix} \mathbf{x}_1(j+1) \\ \mathbf{x}_2(j+1) \end{bmatrix} = \begin{bmatrix} \mathbf{M}_1 \\ \mathbf{M}_3 \end{bmatrix} \left\{ \sum_{k=0}^{j-1} \{ \boldsymbol{\alpha}_1^k (\mathbf{V}_1, \mathbf{V}_2) \boldsymbol{\Delta} \mathbf{u}(k) \} \right\}. \tag{19}$$

Define an $l \times 1$ vector $\boldsymbol{\xi}$ as

$$\boldsymbol{\xi} = \sum_{k=0}^{j-1} \{ (\boldsymbol{\alpha}_1^k)(\mathbf{V}_1, \mathbf{V}_2) \boldsymbol{\Delta} \mathbf{u}(k) \}$$

and eqn. (19) becomes

$$\begin{bmatrix} \mathbf{x}_1(j+1) \\ \\ \mathbf{x}_2(j+1) \end{bmatrix} = \begin{bmatrix} \mathbf{M}_1 \\ \\ \mathbf{M}_3 \end{bmatrix} \boldsymbol{\xi}. \tag{20}$$

Rearranging eqn. (20) gives

$$\left. \begin{aligned} \boldsymbol{\xi} &= \mathbf{M}_1^{-1}\mathbf{x}_1(j+1) \\ \text{and} \qquad \mathbf{x}_2(j+1) &= \mathbf{M}_3\mathbf{M}_1^{-1}\mathbf{x}_1(j+1). \end{aligned} \right\} \tag{21}$$

As in Davison (1966), $\boldsymbol{\phi}_R$ and $\boldsymbol{\Delta}_R$ are calculated by the following analysis. Equation (21) is substituted into the unforced form of eqn. (10) (i.e. $\mathbf{u}(j)=\mathbf{0}$) to give

$$\mathbf{x}_1(j+1) = (\boldsymbol{\phi}_1 + \boldsymbol{\phi}_2\mathbf{M}_3\mathbf{M}_1^{-1})\mathbf{x}_1(j),$$

so that

$$\boldsymbol{\phi}_R = \boldsymbol{\phi}_1 + \boldsymbol{\phi}_2\mathbf{M}_3\mathbf{M}_1^{-1}. \tag{22}$$

To calculate $\boldsymbol{\Delta}_R$, the solutions of the original model and the reduced-order model are compared with $\mathbf{x}(0)=\mathbf{0}$.

The solution of the original model in eqn. (1) for $\mathbf{x}_1$, ignoring $\boldsymbol{\alpha}_2$, is

$$\mathbf{x}_1(j+1) = \mathbf{M}_1 \left( \sum_{k=0}^{j-1} \{\boldsymbol{\alpha}_1^k(\mathbf{V}_1\boldsymbol{\Delta}_1 + \mathbf{V}_2\boldsymbol{\Delta}_2)\mathbf{u}(k)\} \right). \tag{23}$$

The solution of the reduced-order model is

$$\mathbf{x}_1(j+1) = \sum_{k=0}^{j-1} \{\boldsymbol{\phi}_R^k\boldsymbol{\Delta}_R\mathbf{u}(k)\}.$$

However, using the equations, $\boldsymbol{\phi} = \mathbf{M}\boldsymbol{\alpha}\mathbf{V}$ and $\mathbf{V}\mathbf{M} = \mathbf{I}$, it can be shown that

$$\boldsymbol{\phi}_R = \mathbf{M}_1\boldsymbol{\alpha}_1\mathbf{M}_1^{-1},$$

so this reduced-order solution becomes

$$\mathbf{x}_1(j+1) = \mathbf{M}_1 \left( \sum_{k=0}^{j-1} \{\boldsymbol{\alpha}_1^k\mathbf{M}_1^{-1}\boldsymbol{\Delta}_R\mathbf{u}(k)\} \right). \tag{24}$$

Equating eqns. (23) and (24) gives the following expression for $\boldsymbol{\Delta}_R$ :

$$\boldsymbol{\Delta}_R = \mathbf{M}_1(\mathbf{V}_1\boldsymbol{\Delta}_1 + \mathbf{V}_2\boldsymbol{\Delta}_2) = \mathbf{M}_1\boldsymbol{\delta}_1. \tag{25}$$

Thus, a simple, reduced-order, discrete-time system has been derived using Davison's modal approach. This analysis could also be extended to include the various modifications of Davison's method (Davison 1968, Chidambara and Davison 1967 b).

It can be shown, using the partitioned form of the expression $\mathbf{V}\mathbf{M} = \mathbf{I}$, that the transition matrix in eqn. (22) is equivalent to the transition matrix in eqn. (13), obtained by Marshall's analysis.

## 3. Reduction of high-order continuous models to low-order discrete models

Models of physical systems are often represented as large sets of linear differential equations, in the general form,

$$\dot{\mathbf{x}} = \mathbf{A}\mathbf{x} + \mathbf{B}\mathbf{u}, \tag{26}$$

where $\mathbf{x}$ and $\mathbf{u}$ are as defined in § 2.1 and $\mathbf{A}$ and $\mathbf{B}$ are matrices of appropriate dimensions. For both simulation studies and the design and implementation of computer control systems, it may be desirable to have available a low-order discrete-time model.

Two general approaches for deriving a low-order discrete-time model from a high-order continuous-time model are shown in the figure. The high order model can first be reduced and then the resulting low-order model discretized (Approach I), or alternatively, the high-order model can be discretized and this discrete-time model can then be reduced (Approach II). The continuous-time, reduced-order methods of Marshall (1966) and Davison (1966) will be needed for Approach I and will now be summarized.

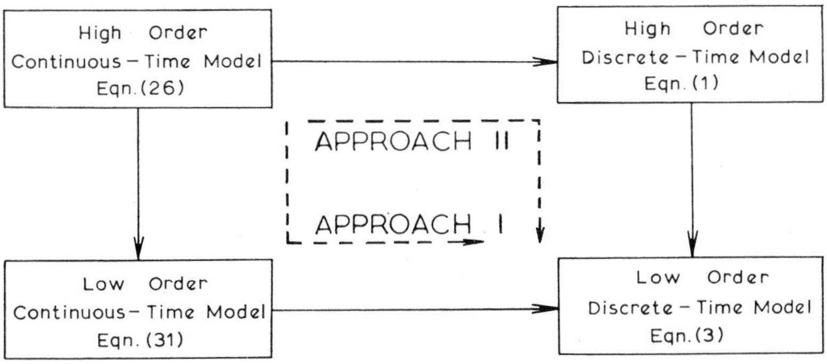


Two approaches for reducing a high-order continuous model to a low-order discrete model.

The transformation matrix $\mathbf{M}$ exists (see Appendix) which transforms eqn. (26) into the Jordan canonical form as

$$\mathbf{x} = \mathbf{M}\mathbf{z}, \tag{27}$$

$$\dot{\mathbf{z}} = \mathbf{J}\mathbf{z} + \mathbf{P}\mathbf{u} = \begin{pmatrix} \mathbf{J}_1 & \mathbf{0} \\ \mathbf{0} & \mathbf{J}_2 \end{pmatrix} \begin{pmatrix} \mathbf{z}_1 \\ \mathbf{z}_2 \end{pmatrix} + \begin{pmatrix} \mathbf{P}_1 \\ \mathbf{P}_2 \end{pmatrix} \mathbf{u}, \tag{28}$$

where

$$\mathbf{J} = \mathbf{M}^{-1}\mathbf{A}\mathbf{M}, \tag{29}$$

$$\mathbf{P} = \mathbf{M}^{-1}\mathbf{B} = \mathbf{V}\mathbf{B}. \tag{30}$$

To calculate the continuous-time reduced-order model, eqn. (26) must be arranged as discussed in § 2.1 and partitioned in a manner similar to eqn. (2). The columns of matrix $\mathbf{M}$ are arranged as discussed in § 2.1, except that the columns on the left now correspond to the eigenvalues of $\mathbf{A}$ with real parts nearest to zero, while those on the right correspond to the eigenvalues of $\mathbf{A}$ with real parts nearest to $-\infty$.

The reduced-order model is of the form

$$\dot{\mathbf{x}}_1 = \mathbf{A}_R \mathbf{x}_1 + \mathbf{B}_R \mathbf{u}, \tag{31}$$

where Marshall's method (1966) gives

$$\mathbf{A}_R = \mathbf{M}_1 \mathbf{J}_1 \mathbf{M}_1^{-1}, \tag{32}$$

$$\mathbf{B}_R = \mathbf{B}_1 - \mathbf{A}_2 \mathbf{V}_4^{-1} \mathbf{J}_2^{-1} \mathbf{P}_2, \tag{33}$$

and Davison's method (1966) yields

$$\mathbf{A}_R = \mathbf{A}_1 + \mathbf{A}_2 \mathbf{M}_3 \mathbf{M}_1^{-1}, \tag{34}$$

$$\mathbf{B}_R = \mathbf{M}_1 \mathbf{P}_1. \tag{35}$$

Approaches I and II differ in the order in which the individual steps of model reduction and discretization are carried out. Since both steps consist of linear operations, intuitively one expects that both paths will result in an identical reduced-order discrete-time model. This equivalence can be shown analytically by making use of the definitions of $\boldsymbol{\phi}$ and $\boldsymbol{\Delta}$ in terms of $\mathbf{A}$ and $\mathbf{B}$ as

$$\left.\begin{aligned} \boldsymbol{\phi} &= \exp(\mathbf{A}T), \\ \boldsymbol{\Delta} &= \int_0^T \exp(\mathbf{A}(T-\tau))\mathbf{B}\, d\tau, \end{aligned}\right\} \tag{36}$$

and $\boldsymbol{\phi}_R$ and $\boldsymbol{\Delta}_R$ in terms of $\mathbf{A}_R$ and $\mathbf{B}_R$ as

$$\left.\begin{aligned} \boldsymbol{\phi}_R &= \exp(\mathbf{A}_R T), \\ \boldsymbol{\Delta}_R &= \int_0^T \exp(\mathbf{A}_R(T-\tau))\mathbf{B}_R\, d\tau. \end{aligned}\right\} \tag{37}$$

Since the required proof is straightforward but requires lengthy matrix manipulation, it will not be presented here. Alternatively the equivalence of Approaches I and II will be demonstrated by the numerical example in § 4.

Since the order of model reduction and discretization does not affect the resulting reduced-order discrete-time model, one should consider which approach is the most convenient for a particular application. If the only requirement is the derivation of a low-order discrete-time model from the original high-order continuous-time model, then Approach I should be followed, since its computational requirements are less. The continuous-time reduction and the discrete-time reduction involve the same number of matrix operations. However, the derivation of the high-order discrete-time system using eqn. (36) requires many more calculations than the derivation of the low-order discrete-time system in eqn. (37). Thus, Approach I is computationally more efficient than Approach II. However, if the availability of a high-order, discrete-time model is desirable, perhaps for the evaluation of proposed control systems, Approach II should be used since the high-order, discrete-time model must be derived in any event.

It is interesting to compare the results of the discrete-time model reduction with the results of the continuous-time reduction. It can be shown by elementary matrix algebra, using the partitioned forms of the expressions $\boldsymbol{\phi} = \mathbf{M}\boldsymbol{\alpha}\mathbf{V}$ and $\mathbf{MV} = \mathbf{I}$, that the discrete transition matrix in eqn. (13) resulting from Marshall's analysis is equivalent to the following :

$$\boldsymbol{\phi}_{\mathrm{R}} = \boldsymbol{\phi}_1 - \boldsymbol{\phi}_2 \mathbf{V}_4^{-1} \mathbf{V}_3 = \mathbf{M}_1 \boldsymbol{\alpha}_1 \mathbf{M}_1^{-1}. \tag{38}$$

Thus, by comparing eqn. (38) with (32), eqn. (14) with (33), eqn. (22) with (34) and eqn. (25) with (35), it can be seen that there is a direct, term-by-term correspondence between the discrete-time and the continuous-time results.

## 4. Example

A simple problem considered by Chidambara and Davison (1967 b) and by Graham (1968) will be used to demonstrate both the discrete-time model reduction techniques and the equivalence of the two approaches shown in the figure. Consider the second-order continuous system in eqn. (39), which is to be reduced to a first-order discrete system in $x_1$,

$$\dot{\mathbf{x}} = \begin{pmatrix} \dot{x}_1 \\ \dot{x}_2 \end{pmatrix} = \begin{pmatrix} 0 & 1 \\ -4 & -5 \end{pmatrix} \mathbf{x} + \begin{pmatrix} 0 \\ 1 \end{pmatrix} u. \tag{39}$$

The eigenvalues of eqn. (39) are $-1$ and $-4$ so that

$$\mathbf{J} = \begin{bmatrix} -1 & 0 \\ 0 & -4 \end{bmatrix}$$

and

$$\mathbf{M} = \begin{bmatrix} 1/3 & -1/3 \\ -1/3 & 4/3 \end{bmatrix}, \quad \mathbf{V} = \begin{bmatrix} 4 & 1 \\ 1 & 1 \end{bmatrix}.$$

The equivalent discrete-time system, with sampling period $T$ is

$$\mathbf{x}(j+1) = 1/3 \begin{bmatrix} 4 \exp(-T) - \exp(-4T), & \exp(-T) - \exp(-4T) \\ -4 \exp(-T) + 4 \exp(-4T), & -\exp(-T) + 4 \exp(-4T) \end{bmatrix}$$

$$\times \mathbf{x}(j) + 1/3 \begin{bmatrix} 3/4 - \exp(-T) + 1/4 \exp(-4T) \\ \exp(-T) - \exp(-4T) \end{bmatrix} u(j). \tag{40}$$

Equation (40) has eigenvalues of $\exp(-T)$ and $\exp(-4T)$.

### 4.1. *Marshall's analysis*

Marshall's analysis for Approach I, using eqns. (32) and (33), gives,

$$\mathbf{A}_{\mathrm{R}} = \mathbf{M}_1 \mathbf{J}_1 \mathbf{M}_1^{-1} = -1,$$

$$\mathbf{B}_{\mathrm{R}} = \mathbf{B}_1 - \mathbf{A}_2 \mathbf{V}_4^{-1} \mathbf{J}_2^{-1} \mathbf{P}_2 = 1/4.$$

Thus, the low order continuous-time model is

$$\dot{x}_1 = -x_1 + u/4 \tag{41}$$

with the corresponding discrete-time model as

$$x_1(j+1) = \exp(-T)x_1(j) + [1 - \exp(-T)]u(j)/4. \tag{42}$$

For Approach II, using eqns. (13), (14) and (40),

$$\boldsymbol{\phi}_R = \boldsymbol{\phi}_1 - \boldsymbol{\phi}_2 \mathbf{V}_4^{-1} \mathbf{V}_3 = \exp(-T),$$

$$\boldsymbol{\Delta}_R = \boldsymbol{\Delta}_1 + \boldsymbol{\phi}_2 \mathbf{V}_4^{-1}(\mathbf{I} - \boldsymbol{\alpha}_2)^{-1} \boldsymbol{\delta}_2 = [1 - \exp(-T)]/4,$$

and the reduced-order, discrete-time model from Approach II is

$$x_1(j+1) = \exp(-T)x_1(j) + [1 - \exp(-T)]u(j)/4, \tag{43}$$

which is identical to eqn. (42).

### 4.2. *Davison's analysis*

Davison's analysis, for Approach I, using eqns. (34) and (35), gives

$$\mathbf{A}_R = \mathbf{A}_1 + \mathbf{A}_2 \mathbf{M}_3 \mathbf{M}_1^{-1} = -1,$$

$$\mathbf{B}_R = \mathbf{M}_1 \mathbf{P}_1 = 1/3.$$

The low-order, continuous-time model is

$$\dot{x}_1 = -x_1 + u/3, \tag{44}$$

with the corresponding discrete-time model as

$$x_1(j+1) = \exp(-T)x_1(j) + [1 - \exp(-T)]u(j)/3. \tag{45}$$

For Approach II, using eqns. (22), (25) and (40),

$$\boldsymbol{\phi}_R = \boldsymbol{\phi}_1 + \boldsymbol{\phi}_2 \mathbf{M}_3 \mathbf{M}_1^{-1} = \exp(-T),$$

$$\boldsymbol{\Delta}_R = \mathbf{M}_1(\mathbf{V}_1 \boldsymbol{\Delta}_1 + \mathbf{V}_2 \boldsymbol{\Delta}_2) = [1 - \exp(-T)]/3,$$

and the reduced-order, discrete-time model from Approach II is

$$x_1(j+1) = \exp(-T)x_1(j) + [1 - \exp(-T)]u(j)/3, \tag{46}$$

which is identical to eqn. (45).

## 5. Conclusions

Existing model reduction techniques of Marshall and Davison for continuous-time systems have been extended for use with discrete-time systems. Two approaches for reducing a high-order continuous-time model to a low-order discrete-time model have been evaluated. The reduction of a continuous-time model followed by discretization is computationally more efficient, but analytically equivalent to, discretizing the high-order system and then reducing the high-order discrete model using the methods outlined in this paper.

ACKNOWLEDGMENT

The financial support of the National Research Council of Canada is gratefully acknowledged.

**Appendix**

Gantmacher (1959, p. 98) presented a theorem that is used in § 2.3 and § 3. It states :

If two matrices **A** and **B** are similar and **W** transforms **A** into **B**,

$$\mathbf{B}=\mathbf{W}^{-1}\mathbf{A}\mathbf{W},$$

then the matrices $f(\mathbf{A})$ and $f(\mathbf{B})$ are also similar and **W** transforms $f(\mathbf{A})$ into $f(\mathbf{B})$,

$$f(\mathbf{B})=\mathbf{W}^{-1}f(\mathbf{A})\mathbf{W}.$$

For the continuous-time system of eqn. (26), from eqn. (29) we have

$$\mathbf{J}=\mathbf{M}^{-1}\mathbf{A}\mathbf{M}.$$

From the above theorem, and the definition of $\boldsymbol{\phi}$ as $\boldsymbol{\phi}=\exp(\mathbf{A}T)$, it follows that

$$\exp(\mathbf{J}T) \doteq \mathbf{M}^{-1}\exp(\mathbf{A}T)\mathbf{M}=\mathbf{M}^{-1}\boldsymbol{\phi}\mathbf{M}=\boldsymbol{\alpha}.$$

Thus, the same transformation matrix, **M**, will transform both the discrete-time and the continuous-time systems into their corresponding Jordan canonical forms, and **M** can be calculated using either the discrete-time system or the continuous-time system.

REFERENCES

ANDERSON, J. H., 1967 a, *Proc. Instn elect. Engrs*, **114,** 1014 ; 1967 b, *Electron. Lett.*, **3,** 469 ; 1967 c, *Ibid.*, **3,** 504 ; 1968, *Ibid.*, **4,** 75.

CHIDAMBARA, M. R., and DAVISON, E. J., 1967 a, *I.E.E.E. Trans. Automn Control*, **12,** 119 ; 1967 b, *Ibid.*, **12,** 213 ; 1967 c, *Ibid.*, **12,** 799.

DAVISON, E. J., 1966, *I.E.E.E. Trans. Automn Control*, **11,** 93 ; 1968, *Ibid.*, **13,** 214.

GANTMACHER, F. R., 1959, *The Theory of Matrices* (New York : Chelsea Publishing Co.).

GRAHAM, E. U., 1968, Ph.D. Thesis, Carnegie Institute of Technology, Pittsburgh, Pa., U.S.A.

KUO, B. C., 1963, *Analysis and Synthesis of Sampled-Data Control Systems* (Englewood Cliffs, N.J. : Prentice-Hall, Inc.).

MARSHALL, S. A., 1966, *Control*, **10**, 642.

# Hybrid simulation of a computer-controlled evaporator

*by*

W. Kent Oliver
Dale E. Seborg
D. Grant Fisher
Department of Chemical Engineering
*University of Alberta*
Edmonton, Alberta, Canada T6G 2G6

After receiving his B.Sc in chemical engineering from the University of Calgary in 1969, WILLIAM KENT OLIVER worked in the oil- and gas-processing industry for Chevron Standard Oil Company as a gas engineer. He then returned to graduate work at the University of Alberta in 1970, receiving an MSc in chemical engineering in 1972. It was during this period that the research for the following paper was conducted. After graduation he worked for Syncrude Canada Ltd. until he moved to his present position as an instructor at Northern Alberta Institute of Technology. Mr. Oliver is married and lives in Edmonton, Alberta. His recreational interests include tennis, golf, and curling.

DALE E. SEBORG received his BSc degree from the University of Wisconsin and his PhD degree from Princeton University, both in chemical engineering. He joined the University of Alberta in 1968 and has performed research in the areas of process control, computer control, and parameter estimation. He is a member of the AIChE and CSChE and is currently president of the northern Alberta section of ISA. His current position is associate professor of chemical engineering and director of the Data Acquisition, Control and Simulation Centre in the Department of Chemical Engineering.

D. GRANT FISHER received his BE and MSc degrees from the University of Saskatchewan, worked for four years with Union Carbide Canada Ltd. where he attained the position of Group Leader--Polyolefin Design, and then enrolled at the University of Michigan and completed his PhD thesis on the dynamic response of heat exchangers. Since joining the Department of Chemical and Petroleum Engineering at the University of Alberta in 1964, he has continued research work in process control and computer applications and has taught courses in applied mathematics, engineering design, process control and computer applications. He is currently Professor and Chairman of the Department of Chemical Engineering there.

## INTRODUCTION

In the evaluation of computer control techniques, a hybrid computer simulation offers several advantages over a strictly analog or strictly digital simulation. The hybrid simulation serves as a more realistic representation of the actual control problem since the analog computer can be used to simulate the continuous plant, while the digital computer closely imitates the operation of the process-control computer. Thus, most of the problems arising from signal conversion and sampling operations that would occur in the actual application become readily apparent from the hybrid simulation. Furthermore, if similar programming languages can be used in the hybrid simulation and in the actual process application, the preparation of the simulation control program greatly reduces the amount of time required to debug the corresponding process programs.

In this investigation, a hybrid simulation was used to evaluate two multivariable computer control systems for a pilot-scale double-effect evaporator. The analog computer was used to simulate a linear five-state-variable dynamic model of the evaporator; the digital computer performed the control calculations for the two multivariable control systems: optimal feedback control and model-reference adaptive control.

This paper describes the hybrid simulation and its use in evaluating the multivariable control schemes. Simulation results are presented to show the potential of these design approaches for industrial applications, and experimental data from the actual evaporator are included as a validation of the results obtained from the simulation.

Simulation, Vol. 12, No. 9, pp. 77-84 (1974).

## EVAPORATOR AND MATHEMATICAL MODEL

A schematic diagram of the pilot-scale double-effect evaporator at the University of Alberta is shown in Figure 1. The process variables shown in the figure and their nominal steady-state values are given in the *Notation* section. The evaporator is normally operated with a feed stream of approximately 2.27 kg/min of 3% aqueous triethylene glycol. The primary control objective is to maintain the product concentration, C2, at a constant value of approximately 10% in spite of disturbances in the feed conditions. The liquid holdups, W1 and W2, must also be maintained within acceptable operating limits. The primary control (manipulated variables) are the inlet steam flow, S, the bottom flow from the first effect, B1, and the product flow, B2. It is assumed that the feed flow, feed concentration, and feed temperature are measurable but uncontrollable disturbance variables. The evaporator is heavily instrumented and can be controlled by either conventional analog controllers or an IBM 1800 digital computer.

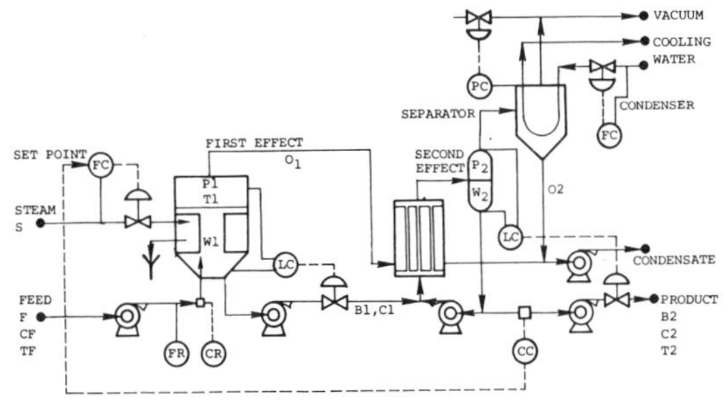


Figure 1 - Schematic diagram of the pilot-plant double-effect evaporator and conventional multiloop control system

The evaporator model used in this study is a five-state linear time-invariant model of the standard state-space form:[1]

$$\dot{\underline{x}}_p(t) = \underline{\underline{A}}\,\underline{x}_p(t) + \underline{\underline{B}}\,\underline{u}(t) + \underline{\underline{D}}\,\underline{d}(t) \qquad (1)$$

$$\underline{y}(t) = \underline{\underline{C}}\,\underline{x}_p(t)$$

where $\underline{x}_p$ represents the five state variables, $\underline{u}$ represents the three control variables, $\underline{d}$ represents the three disturbance variables, and $\underline{y}$ represents the three output variables. These vectors and the numerical values of constant matrices $\underline{\underline{A}}$, $\underline{\underline{B}}$, $\underline{\underline{C}}$, and $\underline{\underline{D}}$ are given in the appendix

The state-space model in Equation 1 was derived by linearizing material and energy balances about the normal steady-state operating conditions. The model can be considered to be theoretical rather than empirical since no attempt was made to adjust the parameters of the model to provide a better fit to experimental data. Further details concerning the derivation of the evaporator model have been reported by Newell and Fisher[2] and by Ritter and André.[3]

Figure 2 presents a comparison of simulated and experimental responses and illustrates the approximate nature of the theoretical process models.

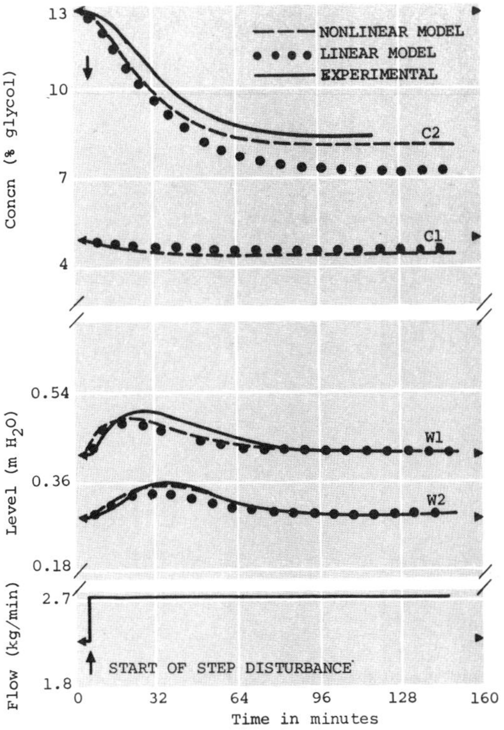


Figure 2 - Comparison of experimental open-loop response data with fifth-order linear and nonlinear models for a 20% increase in feed flow

Despite the lack of good agreement between the linearized process model in Equation 1 and the experimental data (top of Figure 2), this model was used to design the two multivariable controllers. The chief reason for initially considering the linear model rather than the nonlinear model is that the linear model greatly simplifies the design of the control system. Fortunately, the resulting control systems performed quite well on the actual evaporator, as will be described later in this paper. If the performance of these control systems had not been satisfactory, then it would have been necessary to design a suitable controller for the nonlinear model. Since the simulation employed a hybrid computer, it would have been relatively straightforward to substitute the nonlinear process model for the linear process model. However, believing that an adequate control system could be designed using the linear model, we decided to proceed on this basis.

## MULTIVARIABLE CONTROL TECHNIQUES

The hybrid simulation was used to evaluate two types of multivariable computer control systems. The first method, optimal multivariable feedback control, consisted of a discrete version of a proportional-integral controller which employed constant control matrices, $\underline{\underline{K}}_{FB}$ and $\underline{\underline{K}}_I$ (see Equation 8). Multivariable proportional-integral control was of special interest because this control configuration has the ability to eliminate offsets in the controlled variables and can be used for servo (setpoint) control as well as for regulatory control. The second method, model-reference adaptive control, provides for the continual adjustment of the feedback control gains to compensate for changing process conditions or to generate on-line satisfactory multivariable control constants.

*Optimal controller design*

The design of optimal controllers has received considerable attention in recent years.[1,4-7] The basic objective in this approach is to design a controller which is "optimal" in the sense that the resulting closed-loop system response minimizes a specified performance index such as response time or integral error squared. To design an optimal proportional-integral controller for the double-effect evaporator, it is convenient to define a new variable $\underline{z}$ as

$$\underline{z} = \int_0^t \underline{y}\, dt \tag{2}$$

where $\underline{y}$ is a $q$-dimensional subset of $\underline{x}_p$ given by

$$\underline{y} = \underline{\underline{C}}\, \underline{x}_p \tag{3}$$

Then the first equation in (1) can be combined with the differential form of Equation 2, $\dot{\underline{z}} = \underline{\underline{C}}\, \underline{x}_p$, to give

$$\begin{bmatrix} \dot{\underline{x}}_p(t) \\ \dot{\underline{z}}(t) \end{bmatrix} = \begin{bmatrix} \underline{\underline{A}} & \underline{\underline{0}} \\ \underline{\underline{C}} & \underline{\underline{0}} \end{bmatrix} \begin{bmatrix} \underline{x}_p(t) \\ \underline{z}(t) \end{bmatrix} + \begin{bmatrix} \underline{\underline{B}} \\ \underline{\underline{0}} \end{bmatrix} \underline{u}(t) + \begin{bmatrix} \underline{\underline{D}} \\ \underline{\underline{0}} \end{bmatrix} \underline{d}(t) \tag{4}$$

Since the computer control system is to be implemented using a digital computer, the optimal controller was designed using the discrete version of Equation 4:

$$\underline{x}_p'(k+1) = \underline{\underline{\Phi}}\, \underline{x}_p'(k) + \underline{\underline{\Delta}}\, \underline{u}(k) + \underline{\underline{\Theta}}\, \underline{d}(k); \quad k = 0,1,2.... \tag{5}$$

where $\underline{x}_p'^{T} \equiv [\underline{x}_p, \underline{z}]^T$, T denotes the vector transpose, and $\underline{\underline{\Phi}}$, $\underline{\underline{\Delta}}$, and $\underline{\underline{\Theta}}$ are defined in the appendix. Equation 5 can be derived from Equation 4 using well-known techniques.[1,4,5,6]

A convenient and useful performance index for the system in Equation 5 is the quadratic form

$$J = \underline{x}_p'^{T}(N)\, \underline{\underline{S}}\, \underline{x}_p'(N) + \sum_{k=1}^{N-1} [\underline{x}_p'^{T}(k)\, \underline{\underline{Q}}\, \underline{x}_p'(k) + \underline{u}^T(k-1)\, \underline{\underline{R}}\, \underline{u}(k-1)] \tag{6}$$

where $\underline{\underline{S}}$, $\underline{\underline{Q}}$, and $\underline{\underline{R}}$ are constant weighting matrices which are specified by the designer. This performance index has been widely used for several reasons.[1,4,5,6] The first is that the designer can tailor the dynamic behavior of the resulting closed-loop system by appropriate choices of these weighting matrices.[4,5,6] Second, when the quadratic performance index in Equation 6 is employed, the optimal control law for the case where $N = \infty$ is of the feedback form:

$$\underline{u}(k) = \underline{\underline{K}}\, \underline{x}_p'(k) \tag{7}$$

where $\underline{\underline{K}}$ can be calculated *a priori* using well-known dynamic optimization techniques.[5,6,7] Substituting the discrete form of Equation 2 into Equation 7 gives the desired proportional-integral control law:[4]

$$\underline{u}(k) = \underline{\underline{K}}_{FB}\, \underline{x}_p(k) + \underline{\underline{K}}_I \sum_{j=0}^{k} \underline{y}(j); \quad k = 0,1,2,3.... \tag{8}$$

The numerical values of the constant controller matrices, $\underline{\underline{K}}_{FB}$ and $\underline{\underline{K}}_I$, for a sampling interval of 64 seconds are presented in the appendix.

*Model-reference adaptive control by Liapunov's Direct Method*

Model-reference adaptive control (MRAC) is a well-known control technique in which the controller parameters are adjusted automatically to compensate for varying process conditions.[5,8-10] The primary objective in MRAC is to have the process variables follow the response of the reference model as closely as possible. The reference model is specified *a priori* and represents the desired closed-loop behavior. The adaptive control system then attempts to reduce the error between the process response and the reference model response when both are subjected to the same input signals. A block diagram of an MRAC system is shown in Figure 3.

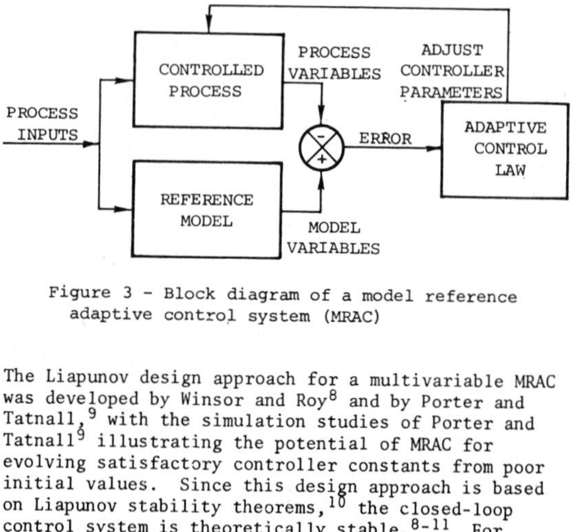


Figure 3 - Block diagram of a model reference adaptive control system (MRAC)

The Liapunov design approach for a multivariable MRAC was developed by Winsor and Roy[8] and by Porter and Tatnall,[9] with the simulation studies of Porter and Tatnall[9] illustrating the potential of MRAC for evolving satisfactory controller constants from poor initial values. Since this design approach is based on Liapunov stability theorems,[10] the closed-loop control system is theoretically stable.[8-11] For simplicity, only proportional feedback MRAC will be presented here, although adaptive feedforward, integral, and setpoint modes were also investigated as part of this project.[11,12]

In this study, a feedback control law of the form of Equation 9 is assumed for purposes of controlling the process:

$$\underline{u}(t) = \underline{\underline{K}}_{FB}(t)\, \underline{x}_p(t) \tag{9}$$

The feedback control matrix, $\underline{\underline{K}}_{FB}(t)$, has time-varying elements which are adjusted by the adaptive algorithm. Substituting Equation 9 into Equation 1 gives the closed-loop system:

$$\dot{\underline{x}}_p(t) = (\underline{\underline{A}} + \underline{\underline{B}}\, \underline{\underline{K}}_{FB}(t))\underline{x}_p(t) + \underline{\underline{D}}\, \underline{d}(t) \tag{10}$$

A reference model of the same order as the process is then specified which has the form:

$$\dot{\underline{x}}_m(t) = \underline{\underline{A}}_m \underline{x}_m(t) + \underline{\underline{D}}_m \underline{d}(t) \qquad (11)$$

As noted earlier, the reference model represents the desired closed-loop behavior of the process.

Based on Liapunov stability considerations, an adaptive control law can be derived[11] for adapting the elements of matrix $\underline{\underline{\Gamma}}$, where $\underline{\underline{\Gamma}}$ is defined as

$$\underline{\underline{\Gamma}}(k) \equiv \underline{\underline{B}}\, \underline{\underline{K}}_{FB}(k) \qquad (12)$$

The discrete form of the adaptive algorithm expressed in terms of $\gamma_{ij}$, an element of $\underline{\underline{\Gamma}}$, and $\underline{e} \equiv \underline{x}_m - \underline{x}_p$ is[11]

$$\gamma_{ij}(k) = x_{pj}(k)\, \underline{e}(k)^T \underline{P}_i \xi_{ij} \Delta t + \gamma_{ij}(k-1) \qquad (13)$$

where $\Delta t$ is the sampling interval, $\xi_{ij}$ is the adaptive loop gain (a design parameter), and $\underline{P}_i$ is the $i$th column of $\underline{\underline{P}}$, the Liapunov matrix which satisfies Equation 14.

$$\underline{\underline{A}}_m^T \underline{\underline{P}} + \underline{\underline{P}}\,\underline{\underline{A}}_m = -\underline{\underline{I}} \qquad (14)$$

In Equation 14, $\underline{\underline{I}}$ is the $n \times n$ identity matrix.

In employing this MRAC technique, the following control calculations are made at each sampling instant. First, $\underline{\underline{\Gamma}}$ is updated using Equation 13 and current values of the state variables of the process and reference model. Then the current feedback control matrix, $\underline{\underline{K}}_{FB}(k)$ is calculated using a least-squares solution to Equation 12.[11,12] Finally, the control variables are calculated from Equation 9. Further details concerning the derivation and application of this adaptive control law have been reported elsewhere.[11,12]

## HYBRID COMPUTER SIMULATION

An EAI 590 hybrid computer system (Figure 4) was employed for the evaporator simulation, with the EAI 580 analog computer providing the solution to the open-loop evaporator model in Equation 1. This process model was expressed in terms of normalized perturbation variables (all states, controls, and disturbances normalized about the actual steady-state values), which proved to be convenient for the analog simulation since ±100% deviations in all variables were permitted before amplifiers saturated. Consequently, magnitude scaling was not required. The entire hybrid simulation was time-scaled by a factor of 60 to allow 1 second of computer time to represent 1 minute of real time and to permit interaction with the user. The analog circuit of Figure 5 was used to simulate the linear model in Equation 1. (The potentiometer settings are tabulated in the appendix.)

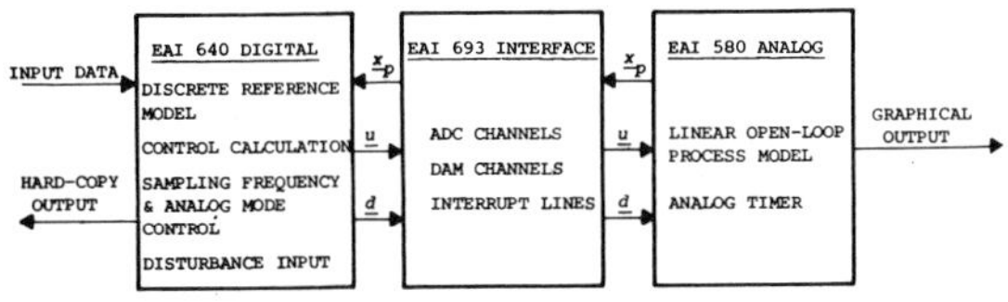


**Figure 4 - Block diagram for the hybrid computer simulation of a computer controlled evaporator**

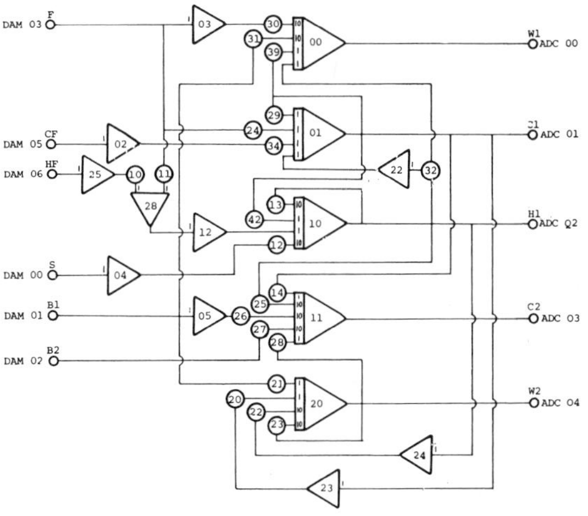


Figure 5 - Analog circuit diagram for the evaporator simulation

The EAI 640 digital computer controlled the hybrid simulation and performed the multivariable control calculations. In both control schemes, the following steps were executed by the digital computer:

(1) Initialize the simulation run by accepting input data from the operator, including control matrices, control configuration, sampling frequency, and type of disturbance.

(2) Control the operation of the analog simulation, e.g., set pots, mode control.

(3) During the run, sample $\underline{x}_p(t)$ and $\underline{d}(t)$, execute the control algorithms and complete the control loop by sending the calculated values of the manipulated variables to the analog simulation via digital-to-analog multiplying converters (DAMs).

(4) After the run is terminated, provide documentation for the run and reinitialize for an additional run.

Since rapid digital execution was not essential, a control interval of 1.1 seconds was selected to allow standard floating-point FORTRAN programming to be used. More rapid "scaled-fraction" FORTRAN[13] (all calculations limited to ±1) or assembler programs could have been used if digital execution time had been more critical in the hybrid simulation.

## RESULTS

The bulk of this investigation was performed with the hybrid computer rather than the actual evaporator due to the relatively long experimental response times of the evaporator (see Figure 2). The conclusions drawn from the simulation study were then experimentally confirmed using the pilot plant evaporator and an IBM 1800 process control computer. The IBM 1800 control programs were written in FORTRAN and executed the same steps as the digital program in the hybrid simulation. More details concerning the real-time programs may be found in a paper by Newell and Fisher[4] and the thesis of Oliver.[12]

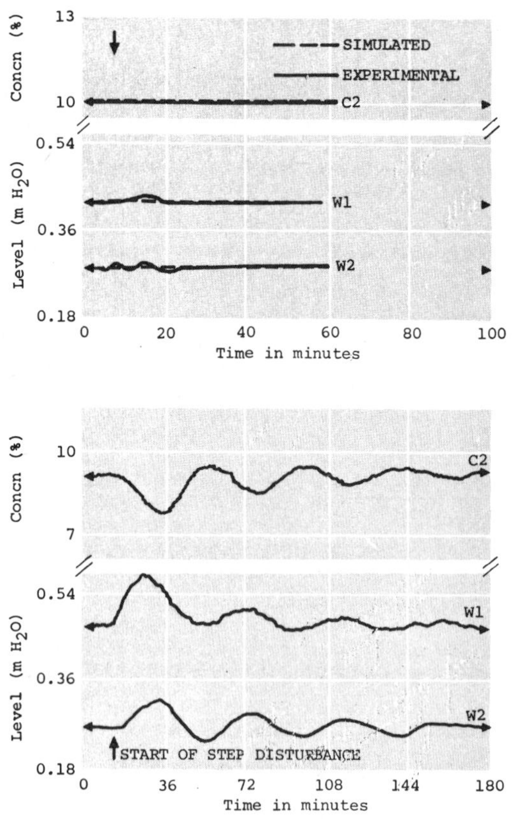


Figure 6 - Top: Comparison of experimental and simulated responses for multivariable proportional-integral control of an evaporator and a 20% increase in feed flow
Bottom: Experimental response for conventional multiloop control and a feed flow disturbance of -20%

The top half of Figure 6 compares the hybrid simulation results with the experimental results for an optimal proportional-integral control configuration. Although the experimental and simulation responses differ somewhat due to modelling errors, the general trends are the same and conclusions drawn from the simulation results concerning the excellence of the control and the elimination of offsets are supported by this and other experimental runs.[4] In Figures 6-10, the arrows along the horizontal axis denote the start of a step disturbance and the arrows along the vertical axis denote the steady-state of interest.

The experimental response in the bottom half of Figure 6 illustrates that a conventional multiloop control scheme results in significantly worse control since the response exhibits larger deviations and a longer response time. (The multiloop control system consisted of three single-variable proportional-integral controllers in the configuration shown in Figure 1). The superior performance of the multivariable controller becomes quite dramatic when it is realized that this controller was designed using a linearized, theoretical, process model and did not require on-line tuning. By contrast, the linearized model was not suitable for designing the multiloop controller since the controller gains found by simulation were much larger than those which could be experimentally implemented. Thus online tuning was required for the multiloop controller but not for the multivariable controller.

In Figure 7 the simulated responses for setpoint changes of 50% in W1 and 20% in C2 (two separate runs) demonstrate the effectiveness of the multivariable controller in implementing changes in operating conditions. These results also illustrate the interactions that are present in the multivariable process since the setpoint change in W1 introduces an undesired transient in C2.

Next, a series of simulation runs was made to determine whether MRAC could be used to evolve suitable controller constants on-line starting with very poor initial control policies. The results in Figure 8 indicate that MRAC enables the process to quickly recover from the feed flow disturbance and to closely follow the reference model response despite the very poor initial control policies of positive feedback control and open-loop control. By contrast, if MRAC were not used, these initial control policies would result in very poor control of the process since positive feedback control would result in an unstable response and the open-loop control policy would allow the liquid holdups, W1 and W2, to drift badly since these variables are not self-regulating.

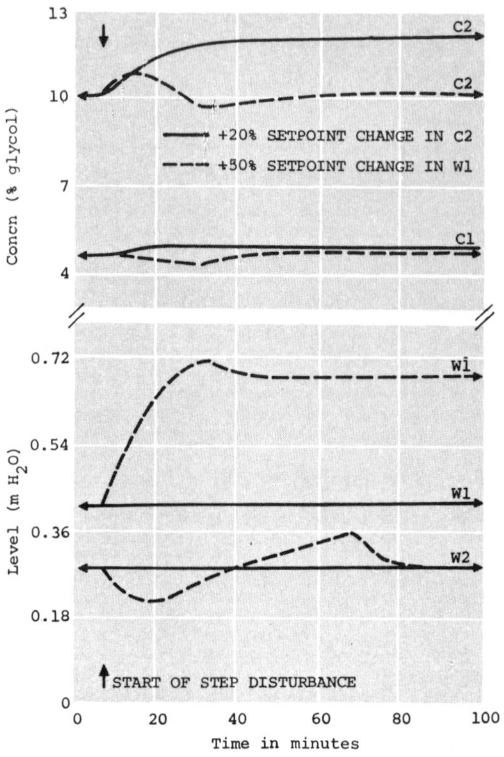


Figure 7 - Simulation results for multivariable proportional-integral control and setpoint changes in W1 and C2

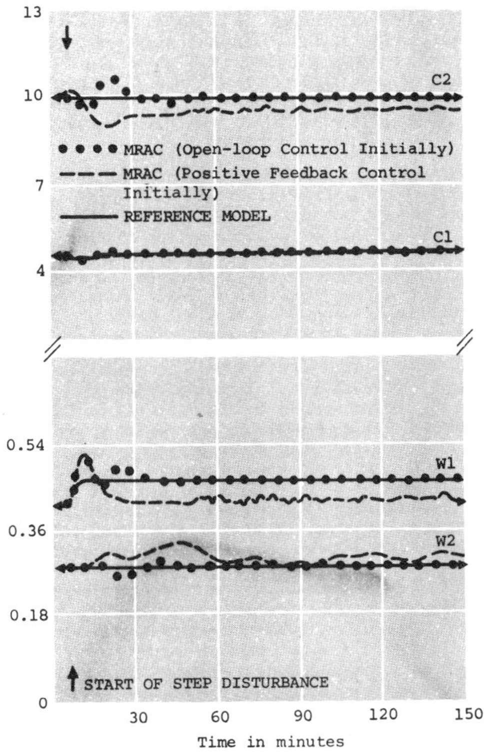


Figure 8 - Comparison of MRAC and reference-model responses to a +50% feed flow disturbance when very poor initial control policies are used ($\xi_{ij}$ = 10)

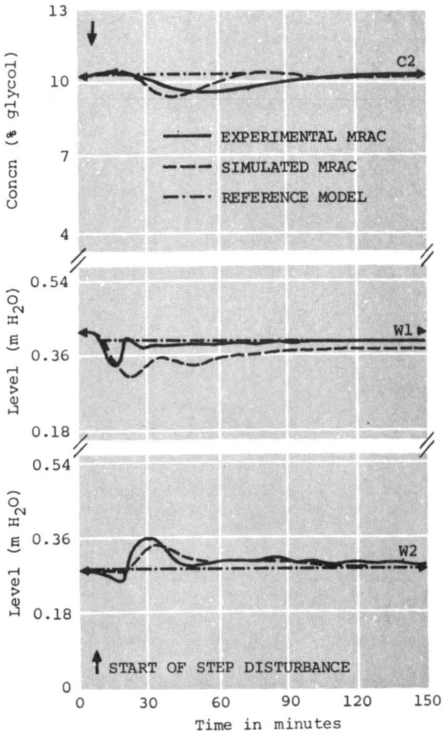


Figure 9 - Comparison of experimental and simulated model-reference adaptive control for a -20% step change in feed flow ($\xi_{ij}$ = 1, no control initially for MRAC)

Thus, these initial control policies provided a severe test of the MRAC system's ability to recover from an unfavorable starting point.

The experimental MRAC response in Figure 9 supports the simulation results and demonstrates that the model-following is quite good. Differences between the simulated and experimental results are due to modelling errors which are present since the actual evaporator is nonlinear but a linearized process model was used to design the MRAC system.

Figure 10 illustrates the gradual improvement in control that occurs when a disturbance repeatedly upsets the process. During the first step-disturbance in feed flow, MRAC enables the process to recover from the poor initial control policy of no control, but the process response deviates significantly from the reference model response. After the second step disturbance, these deviations become negligible and the model-following is very good. Experimental confirmation of these hybrid simulation results has been reported elsewhere.[11,12]

CONCLUSIONS

A hybrid computer simulation has proved to be very useful in the evaluation of two computer control strategies for a pilot-scale double-effect evaporator. Two multivariable controllers, one for optimal proportional-integral control and one for model-reference adaptive control, were designed using a linearized process model. They resulted in excellent control of the evaporator as simulated by either the linear or the nonlinear process model. Experimental response data from the actual evaporator confirmed the simulation results, although some differences did arise due to modelling errors.

This study demonstrates that a hybrid simulation offers several advantages in the evaluation of the computer control techniques used in the course of this investigation. It illustrates the problems associated with sampling and signal conversion, while allowing the time scale to be adjusted easily. Hybrid simulation also allows the control-system designer to employ a sequential approach to real process-control problems. For example, in this study the control system designs were based on a linearized process model rather than the original nonlinear model in order to simplify the design problem. If the resulting control systems had been unsatisfactory, the simulation could have been easily expanded to incorporate more involved design procedures such as tuning the previously designed controllers using the nonlinear process model. However, this was not necessary.

ACKNOWLEDGEMENT

Financial support from the National Research Council

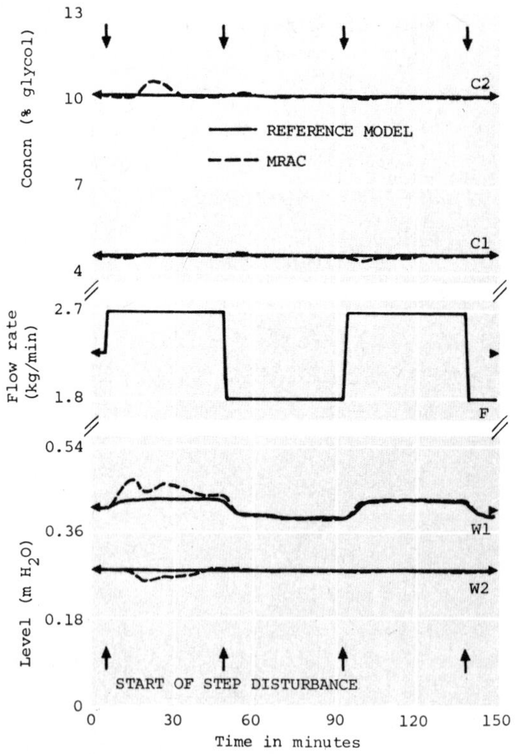


Figure 10 - Simulated MRAC for sustained feed-flow disturbances (+20% change in feed flow every 45 minutes, $\xi_{ij}$ = 10, no control initially)

of Canada and the assistance provided by the staff of the Data Acquisition, Control and Simulation (DACS) Centre in the Department of Chemical Engineering is gratefully acknowledged. Mr. R.G. Wilson provided invaluable assistance in the multivariable proportional-integral control study.

## REFERENCES


1 OGATA, K.
*State Space Analysis of Control Systems*
Prentice-Hall Englewood Cliffs, New Jersey 1967

2 NEWELL, R.B. FISHER, D.G.
*Model Development, Reduction, and Experimental Evaluation for an Evaporator*
*Industrial Engineering Chemistry Process Design and Development* vol. 11 1972 p. 213

3 RITTER, R.A. ANDRÉ, H.
*Evaporator Control System Design*
*Canadian Journal of Chemical Engineering*
vol. 48 1970 p. 696

4 NEWELL, R.B. FISHER, D.G.
*Experimental Evaluation of Optimal Multivariable Regulatory Controllers with Model Following Capabilities*
*Automatica* vol. 8 1972 p. 247

5 EVELEIGH, V.W.
*Adaptive Control Systems and Optimization Techniques"*
McGraw-Hill New York 1967

6 SAGE, A.P.
*Optimum Systems Control*
Prentice-Hall Englewood Cliffs, New Jersey 1968

7 RAY, W.H. SZEKELEY, J.
*Process Optimization*
Wiley New York 1973

8 WINSOR, C.A. ROY, R.J.
*Design of Model Reference Adaptive Control Systems by Liapunov's Second Method*
*IEEE Transactions on Automatic Control*
vol. AC-13 1968 p. 204

9 PORTER, B. TATNALL, M.L.
*Performance Characteristics of Multivariable Model Reference Adaptive Systems Synthesized by Liapunov's Direct Method*
*International Journal of Control* vol. 10 1969 p. 241

10 GRAYSON, P.G.
*The Status of Synthesis Using Liapunov's Method*
*Automatica* vol. 3 1965 p. 91

11 OLIVER, W.K. SEBORG, D.E. FISHER, D.G.
*Model Reference Adaptive Control Based on Liapunov's Direct Method*
*Chemical Engineering Communications* (in press)

12 OLIVER, W.K.
*Model Reference Adaptive Control: Hybrid Computer Simulation and Experimental Verification*
MSc thesis University of Alberta, Edmonton, Alberta, Canada 1972

13 OLIVER, W.K.
*Model Reference Adaptive Control: Hybrid Computer Simulation and Experimental Verification*
FORTRAN Run-Time Library EAI Publication #00 827 8623-3 Electronic Associates, Inc. West Long Branch, New Jersey 1970


## NOTATION

(a) PROCESS VARIABLES

| | | Normal steady-state value |
|---|---|---|
| | <u>State vector $\underline{x}$:</u> | |
| W1 | First-effect holdup | 20.8 kg |
| C1 | First-effect concentration | 4.59% glycol |
| H1 | First-effect enthalpy | 441 kJ/kg |
| W2 | Second-effect holdup | 19.0 kg |
| C2 | Second-effect concentration | 10.11% glycol |
| | <u>Control vector $\underline{u}$:</u> | |
| S | Steam flowrate | 0.91 kg/min |
| B1 | First-effect bottoms flowrate | 1.58 kg/min |
| B2 | Second-effect bottoms flowrate | 0.72 kg/min |
| | <u>Disturbance vector $\underline{d}$:</u> | |
| F | Feed flowrate | 2.26 kg/min |
| CF | Feed concentration | 3.2% glycol |
| HF | Feed enthalpy | 365 kJ/kg |

Output vector $\underline{y}$:

$\underline{y}^T$ [W1, W2, C2]

(b) ADDITIONAL PROCESS VARIABLES (cf. Figure 1)

- $O_1$ Overhead from first effect
- $O_2$ Overhead from second effect
- $P_1$ Pressure in first effect
- $P_2$ Pressure in second effect
- $T_F$ Temperature of feed
- $T_1$ Temperature of first effect
- $T_2$ Temperature of second effect

APPENDIX

(1) Evaporator model: $\dot{\underline{x}} = \underline{\underline{A}}\underline{x} + \underline{\underline{B}}\underline{u} + \underline{\underline{D}}\underline{d}$

$$\underline{\underline{A}} = \begin{bmatrix} 0 & -.0011 & -.1255 & 0 & 0 \\ 0 & -.0755 & .1255 & 0 & 0 \\ 0 & -.0060 & -.7741 & 0 & 0 \\ 0 & -.0012 & -.1448 & 0 & .0001 \\ 0 & .0393 & .1448 & 0 & -.0380 \end{bmatrix}$$

$$\underline{\underline{B}} = \begin{bmatrix} 0 & -.0766 & 0 \\ 0 & 0 & 0 \\ .2160 & 0 & 0 \\ 0 & .0795 & -.0381 \\ 0 & -.0414 & 0 \end{bmatrix} \qquad \underline{\underline{C}} = \begin{bmatrix} 1 & 0 & 0 & 0 & 0 \\ 0 & 0 & 0 & 1 & 0 \\ 0 & 0 & 0 & 0 & 1 \end{bmatrix}$$

$$\underline{\underline{D}} = \begin{bmatrix} .1098 & 0 & 0 \\ -.0333 & .0766 & 0 \\ -.0188 & 0 & .0911 \\ 0 & 0 & 0 \\ 0 & 0 & 0 \end{bmatrix}$$

The discrete matrices $\underline{\underline{\Phi}}$, $\underline{\underline{\Delta}}$, and $\underline{\underline{\Theta}}$ are evaluated numerically from

$$\underline{\underline{\Phi}} = \int_0^T e^{\underline{\underline{A}}t}\, dt$$

$$\underline{\underline{\Delta}} = \int_0^T e^{\underline{\underline{A}}(T-t)}\, \underline{\underline{B}}\, dt$$

$$\underline{\underline{\Theta}} = \int_0^T e^{\underline{\underline{A}}(T-t)}\, \underline{\underline{D}}\, dt$$

(2) Control matrices

$$\underline{\underline{K}}_{FB} = \begin{bmatrix} 8.209 & -1.239 & -3.640 & .1391 & -15.44 \\ 4.542 & .3714 & .5527 & -1.284 & 9.070 \\ 4.238 & 1.166 & -.0591 & 12.26 & 14.22 \end{bmatrix}$$

$$\underline{\underline{K}}_{I} \quad \begin{bmatrix} 1.277 & .02747 & -1.432 \\ .7931 & -.2981 & .8948 \\ .6475 & 1.936 & 1.306 \end{bmatrix}$$

(3) Potentiometer settings for Figure 5

| Potentiometer Identification | Coefficient |
|---|---|
| 10 | $d_{33}$ |
| 11 | $-d_{31}$ |
| 12 | $b_{31}/10$ |
| 13 | $-a_{33}/10$ |
| 14 | $-a_{42}$ |
| 20 | $a_{52}$ |
| 21 | $-b_{52}$ |
| 22 | $a_{53}/10$ |
| 23 | $-a_{55}$ |
| 24 | $-d_{21}$ |
| 25 | $-a_{43}/10$ |
| 26 | $b_{42}/10$ |
| 27 | $-b_{43}/10$ |
| 28 | $-a_{45}$ |
| 29 | $-a_{22}$ |
| 30 | $d_{11}/10$ |
| 31 | $-b_{12}/10$ |
| 32 | $a_{23}$ and $-a_{13}$ |
| 34 | $d_{22}$ |
| 39 | $-a_{12}$ |
| 42 | $-a_{32}$ |

# Section 3: CONVENTIONAL CONTROL

CONTENTS:



COMMENTS:

The evaporator pilot plant is a very flexible unit that can be operated in a number of different configurations. However all the results presented in the study are concerned with a double-effect, forward-feed configuration. Although the feed system is not shown in the simplified flow diagrams, the process equipment includes a rather complex, two-stream blending unit which can be used to introduce independent, arbitrary disturbances in feed flowrate, feed temperature or feed concentration.

It was arbitrarily assumed that the control objective was to control the product concentration at a constant value and to keep the entire unit within its feasible operating range. It was also assumed that the feed variables were disturbances and that the fresh steam input could be regulated. Sensitivity analyses, simulation studies and actual operating experience showed that the three main control loops should involve:

1) control of product concentration (measured by an in-line refractometer) by a cascaded loop that adjusts the setpoint of the steam input flow.
2) control of the liquid level in the first effect by direct or cascaded manipulation of the liquid bottoms flow from that unit.
3) control of the liquid level in the second effect by direct or cascaded manipulation of the product flow.

These control loops (plus additional control loops for regulation of the process vacuum, cooling water flow etc.) were implemented using conventional, industrial electronic instruments and also via Direct Digital Control (DDC). Several of the research projects from 1965 to 1968 included evaluations and comparisons of different single variable control strategies and/or analog vs DDC implementations. Since a number of similar studies have been published in the literature, this section includes only those topics that are the basis for, or compare directly with, the multivariable techniques presented in later sections.

# Computer Control of a Pilot Plant Evaporator

*This paper summarises some of the theoretical and practical points associated with operating a pilot plant evaporator, under computer control, during the last three years. Experimental results are included to illustrate conventional DDC, the effect of sampling time, inferential control, and multivariable feed forward control.*

**By D. G. Fisher† and B. A. Jacobson†‡**

### Introduction

Computer control is becoming increasingly popular in the process industries at both the direct digital control (DDC) and supervisory levels. Industrial applications already range from small mini computers dedicated to DDC of a processing unit, to hierarchies of computers which implement control, supervisory and plant information functions. However, most of the control applications have been based on classical, single-variable feedback control techniques and have not taken maximum advantage of new approaches based on modern control theory. Some of the reasons why these new methods have not been adopted include; lack of experience with experimental applications; a shortage of people trained in both the theoretical and practical aspects of the design of such systems; and the lack of a suitable process model. A number of continuing projects,[6,7,20] concerned with the development of process control techniques and their implementation on pilot plant units, are under way in the Department of Chemical and Petroleum Engineering at the University of Alberta, in Canada. These projects have both educational and research objectives and it is hoped the results will help bridge the gap between theory and practice and contribute toward greater industrial acceptance of computer control techniques.

The purpose of this paper§ is to document the results obtained during the conversion of a pilot plant evaporator from conventional analog control to an equivalent DDC system and the extension of this basic control system to include more advanced multivariable techniques such as inferential and feedforward control. Comments are also included on the experience gained operating an industrial type DDC program on a multi-programmed IBM 1800 system. The following sections contain a brief discussion of: the evaporator pilot plant; a fifth order state-space model of the evaporator; state estimation and filtering techniques; typical results obtained using proportional-plus-integral DDC algorithms; and some of the experimental results obtained using feedforward and inferential control techniques.

†*Department of Chemical Engineering, University of Alberta, Edmonton, Alberta, Canada.*

‡*Present address:* c/o *Imperial Oil Enterprises Ltd., Sarnia Refinery, Sarnia, Ontario, Canada.*

§*An earlier version of this paper was presented as Paper 52(a) at the 68 National Meeting, American Institute of Chemical Engineers, Houston, Texas.*

### Experimental Equipment

A simplified sketch of the evaporator used for this study is shown in Fig. 1. It is a double-effect, forward-feed unit which operates with a nominal feed-rate of 300 lb/h of 3% triethylene glycol and 120 lb/h of fresh steam. The liquid flow-rate from the bottom of the first effect, *B*1, is fed to the second effect which operates under vacuum and utilises the overhead vapour from the first effect as a heating medium. The final product stream, *B*2, is normally controlled at about 10% glycol. The other variables and their normal steady-state values are defined in the nomenclature section.

The overall control objective is to maintain the output variables at constant values by manipulating the input steam flow to compensate for disturbances in the feed flow, composition, or temperature. The evaporator is completely instrumented with conventional Foxboro electronic instruments but is also interfaced to an IBM 1800 computer. Over 50 process variables from the evaporator are available to the computer programmer. The evaporator is normally controlled using an industrial type DDC monitor program which can handle over 100 loops from different processes. Other control programs can obtain process data directly but normally use the DDC monitor for both input and output operations. This simplifies programming, reduces system conflicts and presents the process operator with a uniform method for displaying measurements, changing parameters, or taking over manual control for any variable. Since full-time process operators are not available to run the evaporator, the computer is used to monitor 24 h/day, unattended operation and shutdown the unit in the event of severe upsets. Further information on the computing facilities and some of the projects which make use of them is available.[6,7,20]

### Process Model

The evaporator shown in Fig. 1 is essentially two lumped parameter systems in series and hence is relatively straightforward to model by writing the appropriate material and energy balance equations. Several models have been

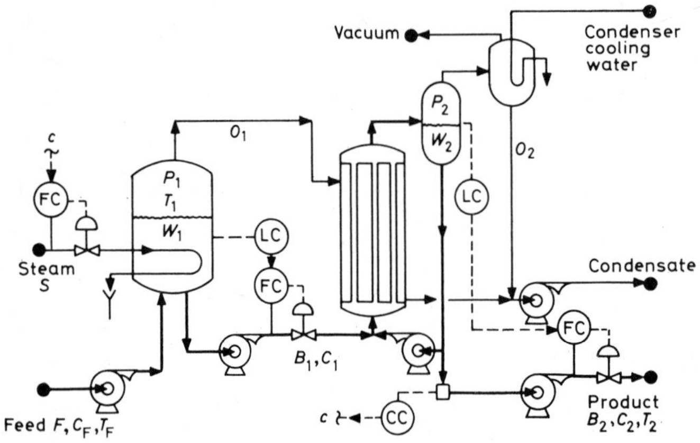


FC: flow control
PC: pressure control
LCC: level control
C: composition control

Fig. 1.—*Flow diagram of the double effect evaporator at the University of Alberta*

TABLE I.—*Five Equation Evaporator Model as Defined by Equation* (1)

$$\begin{vmatrix} W1(n) \\ C1(n) \\ h1(n) \\ W2(n) \\ C2(n) \end{vmatrix} = \begin{vmatrix} 1.0 & -0.0008 & -0.1043 & 0.0 & 0.0 \\ 0.0 & 0.8960 & 0.0978 & 0.0 & 0.0 \\ 0.0 & -0.0047 & 0.3450 & 0.0 & 0.0 \\ 0.0 & -0.0007 & -0.0936 & 1.0 & 0.0001 \\ 0.0 & 0.0440 & 0.0938 & 0.0 & 0.9544 \end{vmatrix} \begin{vmatrix} W1(n-1) \\ C1(n-1) \\ h1(n-1) \\ W2(n-1) \\ C2(n-1) \end{vmatrix} + \begin{vmatrix} -0.1107 & 0.0 & -0.0182 \\ 0.0 & 0.0 & 0.0175 \\ 0.0 & 0.0 & 0.1828 \\ 0.0890 & -0.0468 & -0.0163 \\ -0.0412 & 0.0 & 0.0163 \end{vmatrix} \begin{vmatrix} B1(n-1) \\ B2(n-1) \\ S(n-1) \end{vmatrix} + \begin{vmatrix} 0.1648 & -0.0 & -0.0085 \\ -0.0546 & 0.1049 & 0.0082 \\ -0.0169 & -0.0003 & 0.0856 \\ 0.0015 & -0.0 & -0.0076 \\ -0.0028 & 0.0025 & 0.0076 \end{vmatrix} \begin{vmatrix} F(n-1) \\ CF(n-1) \\ hF(n-1) \end{vmatrix}$$

$$\begin{vmatrix} W1(n) \\ W2(n) \\ C2(n) \end{vmatrix} = \begin{vmatrix} 1.0 & 0.0 & 0.0 & 0.0 & 0.0 \\ 0.0 & 0.0 & 0.0 & 1.0 & 0.0 \\ 0.0 & 0.0 & 0.0 & 0.0 & 1.0 \end{vmatrix} \begin{vmatrix} W1(n) \\ C1(n) \\ h1(n) \\ W2(n) \\ C2(n) \end{vmatrix}$$

derived[1,2,5,14] but the most complete one consists of ten non-linear ordinary differential equations plus associated property relations.[14] For use in control applications this model was linearised numerically, reduced to fifth-order and, assuming piecewise constant **u** and **d**, was converted to the following discrete form:

$$\mathbf{x}(n) = \varphi(T)\,\mathbf{x}(n-1) + \Delta(T)\,\mathbf{u}(n-1) + \theta(T)\,\mathbf{d}(n-1)$$
$$\mathbf{y}(n) = C\mathbf{x}(n) \qquad (1)$$

where:

**x**, **u**, **d**, **y** are the state, control, disturbance and output vectors, respectively (see nomenclature).

$\varphi$, $\Delta$, $\theta$ are constant coefficient matrices that can be derived from the continuous model (see Table I).

$T$ is the sampling or control interval.

$n$ is a counter for time intervals.

The fifth-order discrete model defined in Table I was found to adequately represent the open-loop response of the evaporator to changes of 10 to 20 percent in the disturbance variables and was convenient for use in multi-variable control studies.[8,12]

## Filtering and State Estimation

The evaporator model described in the previous section was used, in conjunction with measured values of the process variables, to predict the state and output variables for use in the inferential and feedforward control algorithms. Two problems existed; filtering of noise in the measured variables, and estimation of unmeasured state variables. A simple exponential digital noise filter had proven adequate in previous studies and was used since it was available as a standard feature of the DDC system. Optimal observer[10] and filtering[9] techniques have been reported in the literature and are currently being implemented at the University of Alberta but this work is based on the following approach:

(1). The digital computer obtained **u**, **d**, and **y** from process measurements

(2). The state vector, **xc**, was calculated from the following equation:

$$\mathbf{xc}(n) = \varphi\mathbf{xs}(n-1) + \Delta\mathbf{u}(n-1) + \theta\mathbf{d}(n-1) \qquad (2)$$

(3). The calculated state vector, **xc**, and the measured, filtered, state vector, **xm**, were "exponentially" combined using a matrix, $\alpha$, of fifteen components:

$$\mathbf{xs}(n) = \mathbf{xc}(n) + \alpha[\mathbf{xm}(n) - \mathbf{xc}(n)] \qquad (3)$$

(4). The output variables for use in the control algorithms were calculated as:

$$\mathbf{y}(n) = C\,\mathbf{xs}(n) \qquad (4)$$

It should be noted that if $\alpha = \mathbf{0}$ in equation (3) the estimated value of the current state, **xs**, is equal to the calculated value, **xc**, and if $\alpha = 1$ it is equal to the measured value, **xm**. For $0 < \alpha_{ii} < 1.0$ the estimated value of the state $\mathbf{xs}_i$ is a function of both the measured and calculated (model) values. Equation (3) can also be interpreted as a special case of a discrete Kalman filter with an infinite observation interval and hence for proper choice of $\alpha$ it is "optimal". However, in this work $\alpha$ was chosen to be diagonal and the elements $\alpha_{ii}$ were chosen by experience. (Since the concentration in the first effect, $C1$, was not measured $\alpha_{22}$ was always set to zero and for the inferential control of $C2$ discussed in a later section $\alpha_{55}$ was also set to zero.)

This "exponential combination" of measured and calculated values smoothed the influence of process noise and, since the model contained strong interactions, it partially compensated for modelling errors. It also proved useful as a means of "instrument back-up". Whenever one of the measured state variables was determined to be unreliable, the corresponding column of $\alpha$ was set to zero and hence, further control was done using calculated (or "inferred") values until the problem was corrected.

However, the method is sensitive to modelling errors. The control system will attempt to drive the output variables ($\mathbf{y} = C\mathbf{xs}$) to the desired value but if the calculated state vector **xc** is in error, the actual final state of the evaporator will not be at the desired value. This effect was observed when the same approach was used in conjunction with multi-variable setpoint changes.[12]

## Basic DDC Control

The evaporator pilot plant was operated under conventional analog control from 1964 until the IBM 1800 control computer was operational in the spring of 1968 and under analog and/or DDC since then. Several significant changes in equipment and operating conditions have been made, but the basic configuration has remained the same. The most satisfactory control scheme based on single variable feedback control loops is shown in Fig. 1 and includes control of three of the five state variables:

(1). The product concentration, $C2$, is measured with a continuous in-line refractometer and controlled by manipulating the setpoint of the feed steam flow.

(2). The liquid levels (and thereby the hold-ups $W1$, $W2$) are controlled by manipulating the corresponding output flows ($B1$, $B2$). With conventional controllers, this is done directly, but with DDC the output of the level controllers is cascaded to the setpoints of flow control loops.

(3). The control of other variables such as vacuum in the second effect is relatively straightforward and is not discussed.

The feed system for the evaporator is not shown in Fig. 1. It consists of a two-stream blending system designed to control the total flow, composition, and temperature either at constant values or to introduce programmed disturbances. The system uses ratio control on the two input flows with the ratio reset by a composition measurement from an in-line refractometer, and the individual flows reset by the total flow controller.

A full discussion of the evaporator plus its control system, the relevant literature, and operating experience is available[1,5,8,12] but a couple of points are of particular interest.

First, concentration control can be improved considerably by using "averaging control" (*i.e.* large proportional band and integral constant) on the liquid levels. This permits, for example, the effects of an increase in feed flow-rate to be damped by the liquid capacitance in the first and second effects. However, as discussed by Buckley (Ref. 4, Chapter 18), the system is susceptible to low frequency cycling and there is both an upper and lower limit on the range of proportional gains that will give stability.

Second, the cascade approach to level control is not necessary but is desirable because:

(1). It eliminates the effects of variations, such as changes in operating pressure on liquid level and of non-linearities in the relationship between the control signal and the actual flow-rate. These factors in turn increase the validity of simplified linear process models and thereby improve the performance of all control methods such as feedforward, inferential and optimal schemes which rely on such models.

(2). Perhaps, most important of all for more advanced DDC schemes, the cascade arrangement permits larger sampling intervals to be used for the outer loops. Thus, a "higher level" control program can be run, say every minute, to calculate new setpoints for the flow loops and the flow control can be implemented by a DDC loop operating every one or two seconds.

(3). Finally, since measured values of some of the manipulated and state variables are inputs to the filtering and state estimation algorithm discussed in the previous section, it is desirable to keep the variables as smooth and as close to the setpoint values as possible. This prevents "interactions" from being propagated *via* the estimation algorithm and feedback control law. ("Smooth" variables also facilitate production records and comparative control studies.) Use of non-linear proportional-plus-integral control algorithms[17] in which the control action depends on some power of the error were also evaluated experimentally. They permitted lower gains (which gave "smoother" control action near the setpoint), but still reacted strongly to large deviations and were convenient for "averaging" applications.

Most of the above could be summarised by the obvious statement that for a multivariable interacting system it is necessary to consider more than just single control loops. In anticipation of such multivariable controllers, the control algorithm used in this work was written in matrix form as follows:

$$\mathbf{u}(n) = \boldsymbol{K}_{\mathrm{FB}}\,[\mathbf{y}(n) - \mathbf{y}_{\mathrm{SP}}(n)] + \boldsymbol{K}_{\mathrm{I}} \sum_{i=0}^{n} [\mathbf{y}(i) - \mathbf{y}_{\mathrm{SP}}(i)] + \boldsymbol{K}_{\mathrm{FF}}\mathbf{d}(n) \qquad (5)$$

where $\boldsymbol{K}_{\mathrm{FB}}$, $\boldsymbol{K}_{\mathrm{I}}$, and $\boldsymbol{K}_{\mathrm{FF}}$ are proportional, integral, and feed-forward control matrices respectively. In the work reported in this paper, $\boldsymbol{K}_{\mathrm{FB}}$ and $\boldsymbol{K}_{\mathrm{I}}$ were diagonal and hence, with $\boldsymbol{K}_{\mathrm{FF}} = \mathbf{0}$, equation (5) is equivalent to a set of conventional P. and I. controllers. When digital computers are used to replace conventional controllers, the user normally has a much larger choice of options and parameters to specify. One of the most obvious and most important of these is the sampling or control interval and its effect is discussed in the next section. Multivariable feedforward control is discussed in a later section.

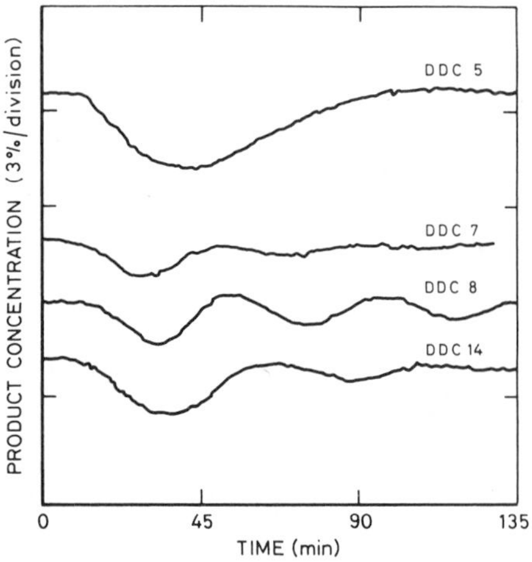


Fig. 2.—*Effect of P. and I. constants and control interval*

*Experimental results*

Computer control of the evaporator is a continuing project and a great deal of operating experience has been gained over the past few years. For each experimental run, the response of the state, load, and manipulated variables were plotted by the computer in a form similar to Fig. 4, and the steady-state conditions and material and energy balance closures were listed as shown in Fig. 5. These results are available[5,8,12] but are not included here.

Figure 2 shows typical responses of the product concentration to a 20% step change in feed flow made using different controller constants and sampling times. The control parameters are summarised in Table II.

A comparison of runs DDC5 and DDC7 shows that increasing the proportional action and decreasing the integral contribution resulted in a significant improvement in control. Other experimental runs confirmed that run DDC7 was typical of the "best" control that could be obtained with the basic feedback control scheme shown in Fig. 1. Attempts to estimate controller constants by analytical means or by simulation using the model defined by equation (1) produced controller gains that were much higher than could be implemented. This was attributed to non-linearities, higher order effects and time delays that are not included in the simplified model. (However, it was noted in later work[12,14] that the same model gave excellent results when used for the design of optimal, multivariable, feedback controllers.)

A comparison of runs DDC7 and DDC8 in Fig. 2 shows the deterioration in the control of the product concentration produced by increasing the sampling interval on the liquid

TABLE II.—*Control Constants for Primary Control Loops*

| Run Number | First-Effect Liquid Level $T$ (s) | $KP$ | $\tau_I$ (s) | Separator Liquid Level $T$ (s) | $KP$ | $\tau_I$ (s) | Product Concentration $T$ (s) | $KP$ | $\tau_I$ (s) |
|---|---|---|---|---|---|---|---|---|---|
| DDC5 | 2 | 1.0 | 2048.0 | 2 | 3.0 | 2048.0 | 64 | 0.31 | 682.0 |
| DDC7 | 2 | 1.0 | 2048.0 | 2 | 3.0 | 2048.0 | 64 | 3.125 | 2728.0 |
| DDC8 | 64 | 1.0 | 2048.0 | 64 | 3.0 | 2048.0 | 64 | 3.125 | 2728.0 |
| DDC13 | 128 | 0.75 | 2048.0 | 128 | 2.5 | 4096.0 | 128 | 1.5 | 2728.0 |
| DDC14 | 64 | 0.75 | 2048.0 | 64 | 2.5 | 4096.0 | 64 | 1.5 | 2728.0 |
| DDC15 | 32 | 0.75 | 2048.0 | 32 | 2.5 | 4096.0 | 32 | 1.5 | 2728.0 |
| INF1 | 60 | 0.75 | 2048.0 | 60 | 2.5 | 4096.0 | 60 | 1.5 | 2728.0 |
| FF1 | 4 | 0.75 | 2048.0 | 4 | 2.5 | 4096.0 | 64 | 1.5 | 2728.0 |
| FF2 | 60 | 0.75 | 2048.0 | 60 | 2.5 | 4096.0 | 60 | 1.5 | 2728.0 |
| FF5 | 60 | 0.75 | 2048.0 | 60 | 2.5 | 4096.0 | 60 | 1.5 | 2728.0 |
| FF3 | 60 | 0.75 | 2048.0 | 60 | 2.5 | 4096.0 | 60 | 1.5 | 2728.0 |

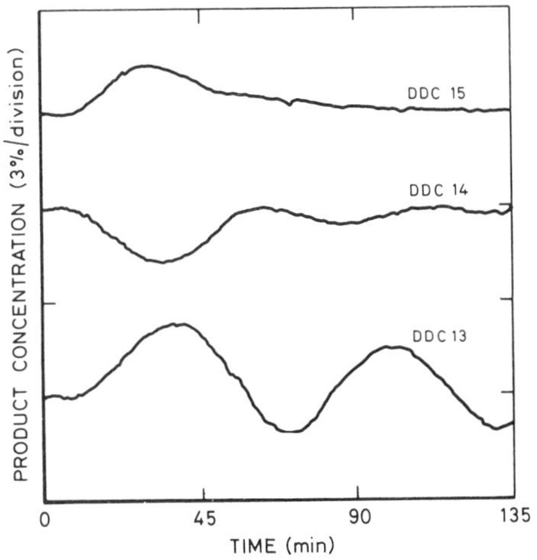


Fig. 3.—*Effect of control interval* (32, 64, 128 *seconds*)

level loops from two to 64 seconds. Figure 3 shows the changes in the evaporator response produced by holding the control parameters constant and using sampling times of 32, 64, and 128 seconds on the level and concentration loops.

The oscillations in run DDC8 were considered excessive and were reduced in run DDC14 by decreasing the controller gains (but at the expense of speed of response). Although "better" control could obviously be obtained by using smaller sampling intervals and higher controller gains, the constants used in run DDC14 were selected as a base case because:

(1). It was thought that the performance of other algorithms, such as feedforward and inferential control, could be more critically evaluated under these conditions.

(2). It was known from other work[8,12] that more sophisticated algorithms would perform well at this sampling interval.

(3). The computer programs used for supervisory control of these variables could be stored on disk and queued for execution every minute without seriously loading the time-shared computer.

An increase in the control interval, $T$, is approximately equivalent to introducing a pure time delay of $T/2$ into the corresponding continuous control loop and therefore degrades the control. It is instructive to relate the sampling interval to the dominant time constants of the system. The time constant of the product concentration, estimated by observing the time required to complete 63% of its response to a 20% increase in feed flow, is in the order of 25 to 30 minutes. Thus, a sampling time of one minute appears conservative and simulated and experimental data support this conclusion. The response time of the liquid levels can be characterised by the volume/throughput ratio. For the first effect ($W1/F$) and second effect ($W2/B1$) these ratios are approximately six and nine minutes. Thus, a one minute sampling time would be expected to be marginal but would probably be acceptable for an isolated control loop. However, there is a strong interaction between the bottoms flow from the first effect, $B1$, and the product concentration. Thus a level disturbance in the first effect changes $B1$, which in turn affects $C2$ faster than it can be controlled by feedback to the feed steam. (The change in steam flow also, affects the first effect level so the circle of interactions is complete.) This points out the need for multivariable design procedures and was the motivation for further work in this area.[12,15]

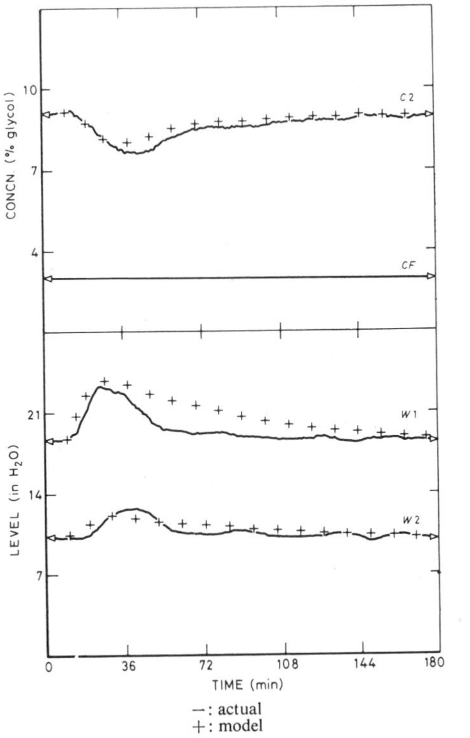


Fig. 4.—*Inferential feedback control of C*2

**Inferential Control**

Figure 4 shows the response of the product concentration to a 20% step increase in feed flow under inferential control. The conditions for this are comparable to those used for run DDC14 (Fig. 3) except, that the value of the product concentration used in the control algorithm was calculated rather than measured directly (*i.e.* $\alpha_{55} = 0$ in equation (3). Theoretically, as the modelling errors are reduced to zero then, inferential control should approach conventional control based on accurate measurements. However, in this example, and in previous work,[5] the inferential control was better than conventional feedback. This was not due to poor measurements but rather to the fact that the response of the model led that of the actual process. This gave the inferential controller an anticipatory action, which for this process, reduced the maximum error and the number of oscillations in the controlled variable. Similar conclusions were reached based on runs using feed concentration as the disturbance variable.[8] Thus, the criterion for a "good" model is not how well it reproduces the open-loop response of the process, but rather how well it performs as part of an (inferential) controller. To some extent the model parameters become controller parameters which can be adjusted empirically, or adaptively, to improve control.

Another interesting point arose due to the integrating nature (zero eigenvalues) of the liquid levels. If, for example, at the final steady-state the measured input to the first effect ($F$) did not equal the sum of measured outputs ($B1 + O1$) then the open-loop model used for state estimation would mathematically integrate this difference and the estimated values of the first effect hold-up would "drift" offscale. This problem was overcome by augmenting the process model so that it included P. and I. feedback control of the liquid levels. The response calculated from this augmented (seventh-order) model are also plotted in Fig. 4. (Part of the discrepancy between the model and actual responses shown for $W1$ is attributed to the difficulty of converting the measured value of the level of a boiling liquid to absolute hold-up.)

The use of a multivariable model (equation (1) in this work gave better results than inferential control based on a simpler second-order transfer function model used in previous work.[5]

**Feedforward Control**

Feedforward control is attractive because in many cases it is capable of compensating for measured input disturbances before they affect the controlled variable. The design of a feedforward controller however, does require quantitative knowledge of the process response (model) and, in practice, it is normally simplified and used in conjunction with feedback action. Feedforward control is discussed in textbooks[4,17] and review articles,[3,11] and has been successfully applied to industrial units including distillation columns and evaporators.[16,18] This section considers the results obtained by adding single variable and multivariable feedforward action to the feedback controllers used with the pilot plant evaporator at the University of Alberta.

*Single variable feedforward control*

Consider an increase in feedflow rate to the evaporator shown in Fig. 1 with the objective of using feedforward control to minimise the effect on product concentration. An analysis of a simplified model of the evaporator leads to the following transfer function relating the inlet steam flow, $S$, and the feed flow, $F$:

$$\frac{S(s)}{F(s)} = \frac{0.40}{5.3\,s+1} \qquad (6)$$

The 0.40 gain term represents the relative effect of steam and feedflow on the product concentration at steady-state. The first order lag approximates the dynamics of the first effect of the evaporator. It appears in equation (6) because the influence of a feedflow disturbance on the product concentration is propagated through both effects of the evaporator whereas the action of the steam passes quickly through the heat transfer surface in the first effect and—to a first approximation—is delayed only by the dynamics of the second hold-up. Feedforward action based on equation (6) has been implemented on the evaporator by Wilson[19] using analog components and by Fehr[5] using DDC. In conjunction with feedback control the results obtained were comparable to those presented in Fig. 6.

Further analysis of the evaporator model suggested that since the dynamic relationship between the flow, $B1$, and the concentration, $C2$, were approximately the same as between the steam flow and $C2$, that the following relationship would also be suitable:

$$S(s) = 0.61\,B1(s) \qquad (7)$$

The results plotted as FF1 in Fig. 6 show that the addition of feed-forward compensation based on equation (7), to the feedback control scheme used for run DDC14, produces dramatic improvements. The 0.61 factor is calculated directly from steady-state gains, but the results from run FF1 suggest that a larger gain would give better results. However, since the evaporator response is non-linear, the gain selected must be appropriate for both increases and decreases in feedflow and for all expected magnitudes. It should also be noted that since $B1$ is an intermediate variable this feedforward scheme produces at least partial compensation for other disturbances which enter the first effect. It also reduces the consequences of the interaction between liquid level control in the first effect and concentration control discussed earlier.

Both feedforward relations can be easily implemented using conventional analog instruments or standard provisions of a DDC monitor program such as the one used at the University of Alberta. The advantage of computer control lies in the convenience afforded the control engineer in evaluating, tuning, and reconfigurating his control system. (Our control students can do this without leaving the process operator's console.) The next two subsections discuss multivariable feedforward controllers which would normally be impractical to implement without a computer.

*Direct multivariable feedforward control*

For the case where the dimensions of the state and control vectors in the process model (equation (1) are equal, and $\mathbf{\Delta}$ is non-singular, the effect of disturbances can be compensated for by feedforward control action based on the following equation:

$$\mathbf{u}_{FF} = \boldsymbol{K}_{FF}\mathbf{d} = -\mathbf{\Delta}^{-1}\boldsymbol{\theta}\,\mathbf{d} \qquad (8)$$

In order to evaluate this approach, the fifth-order process model defined by equation (1) was reduced to the third-order by neglecting variations in the first effect concentration, $C1$, and enthalpy, $H1$. These assumptions are not physically realistic and result in an open-loop model response that is significantly faster than that of the actual process. However, the general form of the response is still correct. The feedforward action calculated from equation (8) using this model is:

$$\begin{bmatrix} B1 \\ B2 \\ S \end{bmatrix} = \begin{bmatrix} 1.0 & 0.25 & 0 \\ 1.0 & 1.00 & 0 \\ 1.0 & -0.49 & 0 \end{bmatrix} \begin{bmatrix} F \\ CF \\ HF \end{bmatrix} \qquad (9)$$

Runs FF2 and FF5 plotted in Fig. 6 were run under the same conditions as DDC14 (Fig. 2) except that $\boldsymbol{K}_{FF}$ in equation (5) was as defined by equation (9) instead of being set to zero. In spite of the model errors, the improvement in the evaporator response is significant. Because of the assumptions made in deriving the process model, there is no feedforward compensation associated with changes in feed enthalpy, $H2$. However, element (3, 3) of $\boldsymbol{K}_{FF}$ in equation (9) could be calculated from the steady-state heat balance on the first effect, or $HF$ could be eliminated.

*"Least squares" multivariable feedforward compensation*

The method described in the previous section does not apply if the dimensions of the state and control vectors are not equal. However, if the feedforward problem is formulated with the objective of minimising the weighted, steady-state contribution of the sum of the last two terms in equation (1) then it follows directly from a least squares analysis that:

$$\mathbf{u}_{FF} = \boldsymbol{K}_{FF}\mathbf{d} = -[\mathbf{\Delta}^{T}\boldsymbol{Q}\,\mathbf{\Delta}]^{-1}\mathbf{\Delta}^{T}\,\boldsymbol{Q}\,\boldsymbol{\theta}\,\mathbf{d} \qquad (10)$$

where $\boldsymbol{Q}$ is a weighting matrix. Equation (10) reduces to equation (9) when $\mathbf{\Delta}$ is square and non-singular.

With $\boldsymbol{Q} = \boldsymbol{I}$ and the model defined by equation (1), it was found that an unacceptably large offset still remained in the product concentration. With the aid of the steady-state model relations it was found that heavier weighting on $C2$, and hence a smaller offset, could be obtained using:

$$\boldsymbol{Q} = \textit{diagonal}\,(10, 100, 1, 10, 1) \qquad (11)$$

The corresponding feedforward action was:

$$\begin{bmatrix} B1 \\ B2 \\ S \end{bmatrix} = \begin{bmatrix} 1.25 & 0.48 & 0 \\ 1.72 & 1.93 & 0 \\ 1.59 & -2.86 & -0.46 \end{bmatrix} \begin{bmatrix} F \\ CF \\ HF \end{bmatrix} \qquad (12)$$

The experimental response of the evaporator using feedforward action based on equation (12) and the same feedback control scheme as run DDC14, is shown as run FF3 in Fig. 6. (Samples taken and analysed off-line suggested that the initial dip in $C2$ was much less than shown and possibly non-existent. It was probably caused by the sharp increase in $B2$ produced by the feedforward action since the in-line refractometer is sensitive to sudden flow changes.) This method is very sensitive to the choice of the weighting matrix, $\boldsymbol{Q}$, and the use of equation (12) led to over-compensation.

It was concluded that all the feedforward—feedback

EXPERIMENT DDC 14

| VARIABLE DESCRIPTION | PROCESS VARIABLE NAME | INITIAL STEADY STATE | FINAL STEADY STATE |
|---|---|---|---|
| Flow rates (lb/min) | | | |
| Total feed flow | *F* 8 | 4·50 | 5·54 |
| Steam flow | *F* 1 | 1·77 | 2·32 |
| First-effect bottoms flow | *F* 2 | 2·91 | 3·61 |
| First-effect overhead flow | *F* 5 | 1·47 | 1·85 |
| Product flow | *F* 6 | 1·42 | 1·85 |
| Second-effect overhead flow | *F* 7 | 1·26 | 1·63 |
| Condenser cooling water flow | *F* 9 | 41·01 | 41·06 |
| Concentrations weight fraction | | | |
| Feed concentration | *C* 1 | 0·030 | 0·030 |
| Product concentration | *C* 6 | 0·090 | 0·090 |
| Temperatures (°F) | | | |
| Feed to first effect | *T* 7 | 187·2 | 186·7 |
| Steam to first effect | *T* 15 | 298·2 | 294·1 |
| Solution in first effect | *T* 19 | 218·3 | 225·9 |
| Vapour in first effect | *T* 2 | 217·0 | 224·2 |
| First-effect steam condensate | *T* 5 | 239·1 | 251·1 |
| Solution to second effect | *T* 4 | 178·2 | 181·4 |
| Steam to second effect | *T* 10 | 217·3 | 224·7 |
| Product | *T* 34 | 154·7 | 159·2 |
| Steam condensate second effect | *T* 28 | 188·2 | 199·8 |
| Separator vapour | *T* 12 | 157·1 | 160·9 |
| Cooling water inlet | *T* 29 | 49·6 | 49·3 |
| Cooling water outlet | *T* 1 | 87·3 | 96·1 |
| Condenser condensate | *T* 11 | 118·0 | 144·9 |
| Pressures (lbf/in$^2$ and in Hg) | | | |
| First-effect pressure | | 5·84 | 10·76 |
| Second-effect pressure | | −13·37 | −13·40 |
| Total mass and component balances (%) | | | |
| Total mass balance | | 4·68 | 2·69 |
| Total component balance | | 1·66 | −0·05 |

Fig. 5.—*Sample of steady-state run documentation produced by the on-line control computer for run DDC*14 (+20% *step in F*)

approaches gave approximately equivalent experimental responses and were significantly better than feedback control used alone.

Further control studies with the same evaporator[12] indicate that *multivariable* feedback controllers give even better results than those described above. The bottom curve in Fig. 6 is a typical concentration response, under multivariable proportional control, to two 20% changes in feed-flow and shows that the concentration remains almost constant.

The feedback control law had the form:

$$\mathbf{u}(nT) = \boldsymbol{K}_{FB}\,\mathbf{x}(nT) \quad . \quad . \quad . \quad (13)$$

when $\boldsymbol{K}_{FB}$ was obtained by solution of the optimal control problem formulated using the process model defined by equation (1) and a generalised quadratic performance index summed over infinite time. Details of the multivariable control experiments are presented elsewhere.[13,20] A comparison of runs DDC14, FF2, and MVC41 in Fig. 6 is typical of the improvement in process control that can be obtained by using more sophisticated control algorithms. When computer control is used these improvements are readily made, often with only a marginal increase in equipment costs.

**Conclusions**

(1). Computer control offers a convenient and flexible means of implementing conventional single variable control loops and facilitates extension to more advanced multivariable control algorithms.

(2). Under direct digital feedback control with constant controller constants the quality of the control deteriorates as the sampling time on the primary control loops is increased. Sampling times cannot be determined on a single loop basis because of process interactions and overall system control objectives.

(3). Inferential control, although sensitive to modelling errors, gave results comparable to control based on direct measurements. It is particularly useful where controlled variables cannot be measured, due to an instrument failure or lack of a suitable measurement transducer.

(4). The addition of feedforward control action to the basic set of feedback control loops significantly improved the evaporator response with only a marginal increase in control complexity.

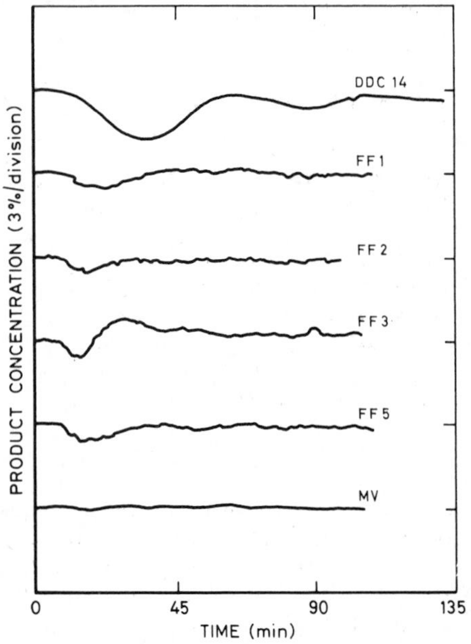


DDC14: feedback control only (base case)
FF1: single variable, equation (7)
FF2: multivariable, equation (8)
FF3: multivariable, equation (10)
FF5: multivariable, equation (8)
MV: optimal multivariable control

Fig. 6.—*Comparison of feedforward—feedback control schemes*

**Symbols Used**

| *Process variables* | | *Steady state* |
|---|---|---|
| **x** = five element state vector | | |
| $W1$ = | holdup in the first effect | 30 lb. |
| $C1$ = | concentration in the first effect | 4.85% glycol |
| $H1$ = | first effect solution enthalpy | 194 Btu./lb. |
| $W2$ = | holdup in the second effect | 35 lb. |
| $C2$ = | concentration in the second effect | 9.64% glycol |
| **u** = three element control vector | | |
| $S$ = | steam flow-rate to the first effect | 1.9 lb./min. |
| $B1$ = | first effect bottoms flow-rate | 3.3 lb/min. |
| $B2$ = | second effect bottoms flow-rate | 1.66 lb/min. |
| **d** = three element load vector | | |
| $F$ = | feed flow-rate | 5.0 lb/min. |
| $CF$ = | feed concentration | 3.0% glycol |
| $HF$ = | feed enthalpy | 162 Btu/lb. |
| *Other process variables* | | |
| $O1$ = | overheads from first effect | 1.7 lb/min. |
| $O2$ = | overheads from second effect | 1.6 lb/min. |
| $P1$ = | pressure in first effect | < 10 lbf/in² gauge |
| $P2$ = | pressure in separator (vacuum) | 15 in. Hg |
| $TF$ = | temperature of feed | 190°F |
| $T1$ = | temperature in first effect | 225°F |
| $T2$ = | temperature in second effect | 160°F |
| *Subscripts* | | |
| FB = | feedback | |
| FF = | feedforward | |
| I = | integral | |

**References**


1 Andre, H. *Ph.D. Thesis*, 1966. University of Alberta, Canada.
2 Andre, H., and Ritter, R. A. *Can. J. chem. Engng*, 1968, **46**, 259.
3 Bertran, D. R., and Chang, K. S. *A.I.Ch.E.Jl.*, 1970, **16**, 897.
4 Buckley, P. S. "*Techniques of Process Control*", 1964. (London: John Wiley and Sons Ltd.).
5 Fehr, M. *M.Sc. Thesis*, 1969. University of Alberta, Canada.
6 Fisher, D. G. *Chem. Eng. Educ.*, 1971, December, p. 24.
7 Fisher, D. G. "*Real-time Computing in the University*", Preprint 38C, 1970, November. (63 Annual Meeting of A.I.Ch.E., Chicago).
8 Jacobson, B. A. *M.Sc. Thesis*, 1970. University of Alberta, Canada.
9 Kalman, R. E. *Trans. Am. Soc. mech. Engrs*, *J. bas. Engng*, 1960, **82**, 35.
10 Luenberger, D. G. *I.E.E.E. Trans*, 1966, **11**, No. 2, p, 190.
11 Miller, J. A., Murrill, P. W., and Smith, C. L. *Hydrocarb. Process Petrol. Refin.*, 1969, July, p. 165.
12 Newell, R. B. *Ph.D. Thesis*, 1971. University of Alberta, Canada.
13 Newell, R. B., and Fisher, D. G. "*Optimal, Multivariable Computer Control of a Pilot Plant Evaporator*", 1971, June. (I.F.A.C. meeting, Helsinki, Finland).
14 Newell, R. B., and Fisher, D. G. *Ind. and Eng. Chem., Process Res. Dev.*, 1972, **11**, 213.
15 Nieman, R. E. *Ph.D. Thesis*, 1971. University of Alberta, Canada.
16 Nisenfeld, A. E., and Hoyle, D. L. *Instrum. Technol.*, 1970, February, p. 49.
17 Shinskey, F. G. "*Process Control Systems*", 1967. (New York: McGraw Hill Book Co. Inc.).
18 Shinskey, F. G. "*Feedforward: A Basic Control Technique*", 1968. (Redhill: Foxboro Co. Publication 170).
19 Wilson, A. H. *M.Sc. Thesis*, 1967. University of Alberta, Canada.
20 Recent publications include:
(*a*) *A.I.Ch.E.Jl*, 1972, **18**, 976.
(*b*) *Automatica*, 1972, **8**, 737.
(*c*) *Trans. Instn chem. Engrs*, 1973, **51**, 132.
(*d*) *A.I.Ch.E.Jl*, 1973, **19**, 901.

# Section 4: MULTIVARIABLE FEEDBACK CONTROL

CONTENTS:



COMMENTS:

In process applications the familiar single-variable, proportional-integral-derivative (PID) controllers probably account for over 90% of the feedback control installations. The objective of the work described in this section was to develop and evaluate analogous multivariable, time-invariant feedback controllers since this appeared to be the logical next step for industrial applications. The discrete (or difference equation) formulation was used because it was assumed that implementation would be via digital control computers.

There are many design approaches that can be applied to multivariable feedback control problems. The papers in this section utilize the "linear-quadratic-Gaussian-optimal" control formulation to produce discrete, time-invariant multivariable feedback controllers which can include:

a) proportional (state) feedback
b) integral feedback
c) feedforward compensation for measured disturbances
d) provision for setpoint changes
e) provision for model-following

Although the multivariable controllers were only slightly more complicated to implement than the corresponding single variable DDC loops, the improvement in control of the evaporator was dramatic. The design technique is essentially a synthesis procedure but does not appear to be overly sensitive to modelling errors, choice of design parameters etc. and in every way appears to be practical and robust enough for use on real processes.

Controllers, such as those discussed above, which require feedback of all the state variables are sometimes more complex than is desired for actual implementation. Papers 4.3 and 4.4 discuss the use of modal analysis techniques to eliminate selected state variables from the feedback control laws so that a simpler "incomplete state feedback", or "output feedback" controller is produced. This approach was effective in the evaporator application and should be useful in many other cases.

Current work involves the design of multivariable controllers using eigenvalue/eigenvector assignment techniques, output feedback techniques, and multivariable frequency domain techniques. No method is clearly superior for all applications. Comparative experimental evaluations are required to determine their scope of application, ease of use, sensitivity to noise and modelling errors, and actual performance under plant conditions that may be more severe than those considered in the design phase.

REFERENCES:

[1] Bryson, A.E. and Y-C. Ho, Applied Optimal Control, Blaisdell, Waltham, Mass. (1969).

[2] Edgar, T.F., J.G. Vermeychuk and L. Lapidus, "The Linear-Quadratic Control Problem: a Review of Theory and Practice", Chemical Engineering Communications 1, 57-76 (1973).

[3] Special Issue on the Linear-Quadratic-Gaussian Problem, IEEE Transactions on Automatic Control, AC-16, December, 1971.

[4] Kwakernaak, H. and R. Sivan, Linear Optimal Control Systems, John Wiley (1972).

# Experimental Evaluation of Optimal, Multivariable Regulatory Controllers with Model-Following Capabilities*

## Evaluation expérimentale des régulateurs optimaux à variables multiples ayant des aptitudes à la simulation

## Experimentelle Ermittlung von optimalen multivariablen Regelungen mit Folgemodellen

## Экспериментальная оценка многокоординатных оптимальных регуляторов способных к слежению модели

R. B. NEWELL† and D. G. FISHER‡

*Discrete, optimal, multivariable controllers are designed which include proportional-plus-integral state feedback, feedforward and model following capabilities. Experimental data from a computer-controlled, pilot plant evaporator demonstrate their practicality and improved performance.*

**Summary**—The formulation of the optimal, linear, time-invariant, multivariable control problem with quadratic performance index is extended so that the solution includes multivariable integral feedback and model-following capabilities in addition to the normal proportional state feedback. The integral action is produced by augmenting the state vector, and eliminates offsets in selected output variables due to constant disturbance inputs. Inclusion of a setpoint vector in the performance index resulted in relatively high controller gains and a fast process response to step changes in setpoints. However, the "real" model-following approach for implementing step changes in setpoints gave greater design flexibility and reliable process responses. The practicality and excellent performance of control systems developed using this approach is demonstrated by experimental data from a computer-controlled pilot-plant evaporator at the University of Alberta.

### 1. INTRODUCTION

THE WORK described in this paper is part of a continuing project at the University of Alberta dealing with the development and application of multivariable control techniques to pilot-plant processes. The specific objective was to develop and test design techniques that would produce multivariable controllers capable of handling both setpoint changes and unmeasured load disturbances and yet simple enough for implementation on industrial processes.

Several design techniques are available for multivariable control systems but most of them result in complex control laws and/or require the real-time solution of systems of non-linear equations. In order to avoid these difficulties the basic approach adopted in this work involved the discrete, optimal, linear control problem with quadratic performance index [1–3]. This approach and the pertinent literature, are reviewed in section 2 and the formulation is extended so that the resulting optimal control law includes multivariable integral feedback and model-following capabilities in addition to the normal proportional state feedback. This is followed, in section 3, by a description of the process equipment and the state space model used in the application of the design technique. Finally, simulated and experimental data from a computer-controlled pilot-plant evaporator are presented, which demonstrate the practicality of this approach and the excellent performance of the "optimal" multivariable controller.

### 2. FORMULATION OF THE OPTIMAL CONTROL PROBLEM

In the formulation of the optimal control problem considered in this paper it is assumed that the process to be controlled can be adequately represented,

* Received 9 June 1971; and in revised form 18 November 1971. This paper is a condensation of two papers presented at the 3rd IFAC/IFIP Conference on Digital Computer Applications to Process Control in Helsinki, Finland, during June 1971, and at the 2nd IFAC Symposium on Multivariable Control in Düsseldorf, Germany, during October 1971. Publication in revised form was recommended by Associate Editor John E. Rijnsdorp.

† Dr. R. B. Newell is with Shell International Petroleum Maatschappij N.V., The Hague 2076, Netherlands.

‡ Dr. D. G. Fisher is a professor in the Department of Chemical and Petroleum Engineering at the University of Alberta, Edmonton 7, Canada.

in the region of normal operation, by the following linear time-invariant model:

$$\dot{\mathbf{x}} = \mathbf{Ax} + \mathbf{Bu} + \mathbf{Dd} \qquad \mathbf{x}(o) = \mathbf{x}_o \tag{1}$$

$$\mathbf{y} = \mathbf{Cx}. \tag{2}$$

The variables are defined such that the origin represents the normal, steady-state operating point of the unforced system and:

$\mathbf{x} = \mathbf{x}(t)$is an $n \times 1$ unconstrained state vector

$\mathbf{u} = \mathbf{u}(t)$ is an $m \times 1$ vector of independent, unconstrained, control variables.

$\mathbf{d} = \mathbf{d}(t)$ is a $p \times 1$ vector of finite disturbances

$\mathbf{y} = \mathbf{y}(t)$ is a $q \times 1$ output vector

**ABCD** are constant coefficient matrices.

The quadratic performance index that is adopted for the optimal regulatory control problem is usually of the form [1–3]

$$J = \tfrac{1}{2} < \mathbf{x}(\tau), \mathbf{Sx}(\tau) > + \tfrac{1}{2}\int_o^\tau < \mathbf{x}, \mathbf{Qx} > + < \mathbf{u}, \mathbf{Ru} > dt \tag{3}$$

where **S**, **Q** and **R** are appropriate weighting matrices and $\tau$ is a specified value of time.

Kalman [1] has solved the general, time-varying optimal control problem (i.e. where **A**, **B**, **C**, **D**, **Q** and **R** in equations (1–3) can be functions of time) and the results are available in the text books such as Athans and Falb [2]. The general *state regulator problem* is concerned with returning the process from its current (initial) condition to its desired operating point (defined as the origin) and/or keeping it near this point in the face of external disturbances. For the case of zero disturbances a globally optimal control exists, is unique, and is given by the linear relationship

$$\mathbf{u}(t) = \mathbf{K}(t)\mathbf{x}(t) \tag{4}$$

where $\mathbf{K}(t)$ is found by solving the appropriate matrix Riccati equation.

If the time varying system is observable then the *output regulator problem* can be formulated by writing the performance index in terms of **y**. However if the quadratic form $< \mathbf{y}, \mathbf{Qy} >$ can be written as $< \mathbf{x}, \mathbf{C}^T\mathbf{QCx} >$ it follows that using $\mathbf{C}^T\mathbf{QC}$ in place of **Q** in the performance index transforms the *state* control problem into the equivalent *output* control problem, and the resulting control law has the same form as equation (4).

The *servo or tracking problem* is similar to the output regulator problem except that the desired output values are specified as $\mathbf{z}(t)$ instead of zero. The performance index has the same form as equation (3) except that **x** is replaced by an error term **e** where $\mathbf{e}(t) = \mathbf{z}(t) - \mathbf{y}(t)$. The resulting control law contains the same state feedback term given by equation (4) but also has a second term which depends on the desired values $\mathbf{z}(t)$ [2, section 9–9].

For time-invariant controllable systems, and with **Q** positive definite, the controller matrix in equation (4) takes on constant values in the limiting case where $\tau \to \infty$. It can also be shown that the system is closed-loop stable so that this formulation produces a simple, practical control scheme. The theory is not as complete for the equivalent *tracking or servo problem* but indicates that for observable and controllable systems with specified constant values for the outputs (i.e. for $\mathbf{z}(t) = \mathbf{y}_{SP}$) the control will be a linear, time-invariant function of the state and the desired outputs, $\mathbf{y}_{SP}$.

Optimal, linear, time-invariant control laws such as those discussed above are very attractive from the point of view of engineering applications and will be pursued further. However, since the work described in this paper is intended for implementation using digital computers a discrete formulation is more convenient and is developed as follows:

If **u** and **d** are constant over the sampling (control) interval $T$, then the following difference equations are equivalent, at the sampling instants, to the system defined by equations (1 and 2),

$$x[(n+1)T] = \boldsymbol{\phi}(T)\mathbf{x}(nT) + \boldsymbol{\Delta}(T)\mathbf{u}(nT) + \boldsymbol{\theta}(T)\mathbf{d}(nT) \tag{5}$$

$$\mathbf{y}(nT) = \mathbf{Cx}(nT) \tag{6}$$

where $\boldsymbol{\phi\Delta}$ and $\boldsymbol{\theta}$ are constant coefficient matrices that can be derived from equation (1) [3, 26].

The discrete quadratic performance index used in this work was:

$$J = \beta^N(\mathbf{x}(NT) - \mathbf{x}_d)^T\mathbf{S}(\mathbf{x}(NT) - \mathbf{x}_d) + \sum_{i=1}^{N-1} \beta^i(\mathbf{x}(iT) - \mathbf{x}_d)^T\mathbf{Q}(\mathbf{x}(iT) - \mathbf{x}_d) + \mathbf{u}((i-1)T)^T\mathbf{Ru}((i-1)T) \tag{7}$$

where $\mathbf{x}_d$ is the desired state (see following discussion) **S** and **Q** are positive, semi-definite, symmetric, constant matrices which penalize the final state offsets and deviations in the state variables, respectively.

**R** is a symmetric matrix of constants which penalize use of control action and must be positive definite to guarantee that the extremum found in the optimization procedure is a minimum. However, it can be made arbitrarily small and in the applications in this paper was actually set to zero.

$N$ is a large integer such that at $t=NT$ the process has normally reached steady state and after $N$ iterations the recursive relations which define the control matrices have converged to constant values. $\beta$ is a scalar parameter which, if given a value greater than one, increases the weighting on state and control deviations as time increases.

The derivation of the discrete optimal control law for the system defined by equations (5–7) makes use of the dynamic programming concept of breaking the system up into subsystems, in this case by time into control intervals. The relevant terms of the performance index in equation (7) can then be optimized with respect to the control vector $\mathbf{u}$ by beginning at the $(N-1)^{st}$ interval and working backwards in time. It is known [3, p. 155] that for the special case of $\mathbf{d}=\mathbf{o}$, $\mathbf{x}_d=\mathbf{o}$, $\mathbf{S}=\mathbf{o}$ and $\beta=1$ that the result is a control law of the form

$$\mathbf{u}(nT)=\mathbf{K}\mathbf{x}(nT) \qquad (8)$$

where the matrix $\mathbf{K}$ is defined by recursive relationships which converge to constant unique values after only a few iterations. The result is analogous to the "infinite time" or time-invariant form of equation (4) discussed earlier, but the controller matrices are considerably easier to calculate and can easily be evaluated on a small process computer. In the following sections the discrete formulation will be extended to add integral feedback to equation (8) and to handle the case where the desired state, $\mathbf{x}_d$, is non-zero. The feedforward control that arises for constant, measurable disturbances is evaluated elsewhere [4].

### 2.1 *Regulator control with proportional-plus-integral feedback*

Multivariable proportional-plus-integral control can, with very little increase in the complexity of the implementation, eliminate the steady-state offsets which arise when proportional control alone is used on systems subject to constant disturbances. It can also compensate for some of the errors which develop when control systems designed on the basis of approximate models are implemented on real systems. However, there were very few papers in the literature, during the period this work was in progress, which dealt with multivariable integral control. SMITH and MURRILL [5] transformed the model equations and defined an integral controller through a solution of the Riccati equation. Their transformation necessitated the assumption of step load changes and they did not present any simulated or experimental results. Their one example was a special case with two control variables and two states. MOORE [6] used a formulation necessitating a term in the criterion involving the control vector derivative. The design is valid for systems with equal numbers of states and control variables but this work has shown that his suggested least squares solution of the general case would not eliminate offsets when $q>m$. Given the proportional feedback matrix, PORTER [7] evaluates integral and derivative control matrices using Liapunov's stability techniques. However, as pointed out by SIMPSON [8], this method is again restricted to the special case of equal numbers of the state and control variables. SIMPSON [9] also presented an alternative, though similar approach to this special case. SHIH [10] obtained optimal integral control for a single control variable and output and this paper extends the formulation to the multivariable case.

To generate integral action in the control law, new variables are defined as

$$z_i=\int_o^t (y_i-y_{SP})\mathrm{d}t \quad \text{or} \quad \dot{z}_i=y_i-y_{SP}. \qquad (9)$$

Since the output and state variables are related by equation (2) a vector $\mathbf{z}$, of dimension $l\leq m$, can be defined as:

$$\dot{\mathbf{z}}=\mathbf{y2}-\mathbf{y2}_{SP}=\mathbf{C2x}-\mathbf{y2}_{SP} \qquad (10)$$

where $\mathbf{C2}$ is defined such that $\mathbf{y2}$ is a subset of $\mathbf{y}$ which contains only the "$l$" output variables to which integral action should be applied. Equations (1 and 10) can then be combined in the form

$$\begin{bmatrix}\dot{\mathbf{x}}\\ \dot{\mathbf{z}}\end{bmatrix}=\begin{bmatrix}\mathbf{A} & \mathbf{O}\\ \mathbf{C2} & \mathbf{O}\end{bmatrix}\begin{bmatrix}\mathbf{x}\\ \mathbf{z}\end{bmatrix}+\begin{bmatrix}\mathbf{B}\\ \mathbf{O}\end{bmatrix}\mathbf{u}+\begin{bmatrix}\mathbf{D} & \mathbf{O}\\ \mathbf{O} & \mathbf{I}\end{bmatrix}\begin{bmatrix}\mathbf{d}\\ -\mathbf{y2}_{SP}\end{bmatrix}. \qquad (11)$$

Equation (11) has the same form as equation (1) and the performance index defined by equation (3) is still applicable if $\mathbf{S}$ and $\mathbf{Q}$ are augmented to be $(n+l)$ square matrices. Thus the equivalent infinite-time, discrete *regulator* problem can be formulated and solved by the same procedure outlined previously [4]. The resulting discrete control law, for the case where the desired state is zero, can be factored to yield

$$\mathbf{u}(nT)=\mathbf{K}_{FB}\mathbf{x}(nT)+\mathbf{K}_I\sum_{i=0}^{n}\mathbf{y2}(iT)+\mathbf{K}_{FF}\mathbf{d}(nT) \qquad (12)$$

which is clearly the discrete form of the desired proportional-plus-integral control law, and with a feedforward term that can be used if the disturbances are measurable. The recursive relations that define $\mathbf{K}_{FB}$, as in equation (8), and $\mathbf{K}_{FF}$ are given in the nomenclature section.

Necessary and sufficient conditions for controllability of the augmented system given by equation (11), with $\mathbf{d}=\mathbf{y}2_{SP}=\mathbf{o}$, are presented by PORTER [11] and YOUNG and WILLIAMS [12]. Following Porter, if $\mathbf{A}$ is non-singular then $(\mathbf{A}, \mathbf{B})$ must be a controllable pair and rank $\mathbf{C}2\mathbf{A}^{-1}\mathbf{B}=l$. If $\mathbf{A}$ is singular, then the augmented system is still controllable if $(\mathbf{A}, \mathbf{B})$ are controllable and a stabilizing matrix $\mathbf{F}$ can be found such that $(\mathbf{A}+\mathbf{BF})$ is non-singular and rank $\mathbf{C}2(\mathbf{A}+\mathbf{BF})^{-1}\mathbf{B}=l$.

It should be noted that if the design were carried out with more "integral" states than control variables, (i.e. $l>m$) it would be found that, in general, the final steady-state offsets are changed but not removed by the integral control action.

The multivariable proportional-plus-integral control problem has more recently been considered by a number of authors working in parallel. PORTER [11] formulates the case for integral state feedback and FOND and FOULARD [13] develop an analogous formulation in terms of the error between the output $\mathbf{y}$ and specified constant values. Both papers discuss the controllability of the augmented system and the latter includes a numerical example. HU [14] in his thesis considers a number of multivariable control algorithms, including the $P+I$ case, and applies them to simulated studies of a distillation problem. DAVISON and SMITH [15] develop the necessary and sufficient conditions for a feedback control system of minimum order to exist such that the outputs of the system, defined by equations (1 and 2) but with $\mathbf{D}=\mathbf{I}$, are driven to zero for all constant external disturbances and such that the closed loop poles are specified. JOHNSON [16] had previously taken a similar approach to the problem assuming a broader class of disturbances but limiting the system to that described by equation (1) with $\mathbf{y}=\mathbf{x}$. LATOUR [17] in a recent presentation showed that for linear time-invariant systems, described in equation (1), subject to constant disturbances, the generalized $P+I$ feedback control law in the infinite-time case is optimal for a performance index of the form defined by equation (3) but with $\mathbf{u}$ replaced by the control rate, (i.e. with $<\dot{\mathbf{u}}, \mathbf{R}\dot{\mathbf{u}}>$.) He considered the infinite time, state regulator problem and the case where the desired value of the state, $\mathbf{x}_d$, was non-zero.

2.2 *Regulator control with provision for setpoint changes*

There are two modifications of the basic optimal control formulation which will generate a regulatory control law which includes provision for driving the process output variables to a specified set of non-zero (setpoint) values. The first uses the *direct approach* of expressing the performance index in terms of the error between the actual and desired values, and the second involves *model-following* techniques.

*Direct approach.* In the direct approach the performance index used is that given by equation (7). However, in most applications to industrial process control the new or desired state, $\mathbf{x}_d$, is an equilibrium point of the process which will become, at least temporarily, the new steady state operating point for the process. Thus all elements of $\mathbf{x}_d$ are not expressed arbitrarily but instead are calculated as a function of the desired output values. In this work it is assumed that $\mathbf{y}$ is a subset of $\mathbf{x}$ such that $q \leq m$ and $\mathbf{x}_d$ in equation (7) can be replaced by:

$$\mathbf{x}_d=\mathbf{C}^T\mathbf{y}_{SP}. \tag{13}$$

Application of the discrete optimization procedure, described above, to the system described by equations (5 and 6) plus the performance index defined by equations (7–13), under the additional assumption that $\mathbf{y}_{SP}$ and $\mathbf{d}$ are constant over the time period of the derivation, leads to a control law that can be put in the form:

$$\mathbf{u}(nT)=\mathbf{K}_{FB}\mathbf{x}(nT)+\mathbf{K}_{SP}\mathbf{y}_{SP}(nT)+\mathbf{K}_{FF}\mathbf{d}(nT). \tag{14}$$

The recursive relations that define the controller matrices are given in the nomenclature section. It should be noted that integral action can be added to the control law given by equation (14) by augmenting the original system model as outlined in equation (11). The augmented $\mathbf{K}_{FB}$ matrix will contain the proportional-plus-integral feedback gains and $\mathbf{K}_{FF}$ will contain the feedforward gains from both the disturbances and the desired values, $\mathbf{y}2_{SP}$.

The control problem associated with driving the process from an initial state, $\mathbf{x}(o)=\mathbf{0}$, to a desired state, $\mathbf{x}_d$, as described above, can be converted to a standard regulator problem by simply transforming the variables so the origin becomes the desired state and the initial conditions are $\mathbf{x}_d$. This is the "regulator" formulation examined by KALMAN and KOEPCKE [18] and LAPIDUS and LUUS [26 chapt. 3]. However this approach is inconvenient for practical applications because the setpoint (desired) values are contained implicitly and require recalculation of the control action for every setpoint change.

The specification of the state feedback matrix is independent of the setpoint matrix [4, 19] so that in the absence of setpoint changes the control is still optimal in the regulator sense.

An examination of the closed-loop, steady-state model shows that, with constant coefficient matrices in the control law, a small offset will exist between the actual and the desired values of the output variables at the final state, such that

$$\mathbf{y}_{ss}=\mathbf{C}\mathbf{x}_{ss}=-\mathbf{C}(\mathbf{A}+\mathbf{B}\mathbf{K}_{FB})^{-1}\mathbf{B}\mathbf{K}_{SP}\mathbf{y}_{SP}=\mathbf{E}\mathbf{y}_{SP}. \tag{15}$$

This offset is, practically speaking, negligible but it can be eliminated by postmultiplication of the setpoint control matrix $\mathbf{K}_{SP}$ by the inverse of matrix **E**.

*Model following approach*

The "*implicit*" model-following schemes presented by TAYLOR [19], MARKLAND [20] and ZINOBER [21] incorporate an unforced model into the performance index and minimize the difference $(\dot{\mathbf{x}} - \dot{\mathbf{x}}_m)$ between the rates of the process response, $\dot{\mathbf{x}}$, and the model response, $\dot{\mathbf{x}}_m$. This problem is shown to be equivalent to the standard *regulator* problem if the performance index includes the appropriate cross product terms between the state and control vectors. The resulting control law therefore has the same form as equation (4), but is intended chiefly to modify the dynamics of the closed-loop plant so that its output behaviour approximates that of the model. As discussed by ERZBERGER [22] this *implicit* model-following scheme is essentially open-loop and can be upset by disturbances. Therefore *real model following* in which the model is actually implemented as part of the physical control system was adopted so that errors arising between model and plant due to uncertainties can be continuously measured and corrected.

Real model-following techniques were examined in some detail by TYLER [19]. The process model considered was homogeneous and the Ricatti equation was used in the solution of the optimal control problem. MARKLAND [20] in a later paper discusses both the real and implicit model-following schemes proposed by Tyler as well as a third synthesis method which minimizes, in an algebraic least-squares sense, the difference between the response of the process model and the desired model. WINSOR and ROY [23] extend the design techniques for linear, model-following problems to include the concepts of "specific optimal control", "trajectory sensitivity" and "perfect model following". ERZBERGER [22] also examines the conditions under which "perfect" model following can be achieved. YORE [24] introduced a model with inputs and considered a configuration with feedback from the process state, and feedforward from the model state and the inputs. The optimal control problem with a summed quadratic error index was solved to give the feedback and model state feedforward control matrices. The input feedforward was evaluated separately and a variety of criteria were mentioned.

In the real model-following formulation a model is selected, in the form defined by equation (16), which exhibits the desired output response, $\mathbf{y}_m(t)$.

$$\dot{\mathbf{y}}_m = \mathbf{H}\mathbf{y}_m + \mathbf{G}\mathbf{y}_{SP}. \tag{16}$$

The optimal control objective is then to have the process output approximate $\mathbf{y}_m(t)$ as closely as possible. For simplicity the model state variables $\dot{\mathbf{y}}_m$ are assumed to be equivalent to a subset of process states as in equation (13). The process model indicated by equation (1), the desired model given by equation (16), and the performance index given by equation (7) with $\mathbf{S}=\mathbf{O}$ and $\beta=1$, can then be rearranged to the following form,

$$\begin{bmatrix} \dot{\mathbf{x}} \\ \dot{\mathbf{y}}_m \end{bmatrix} = \begin{bmatrix} \mathbf{A} & \mathbf{O} \\ \mathbf{O} & \mathbf{H} \end{bmatrix} \begin{bmatrix} \mathbf{x} \\ \mathbf{y}_m \end{bmatrix} + \begin{bmatrix} \mathbf{B} \\ \mathbf{O} \end{bmatrix} \mathbf{u} + \begin{bmatrix} \mathbf{D} & \mathbf{O} \\ \mathbf{O} & \mathbf{G} \end{bmatrix} \begin{bmatrix} \mathbf{d} \\ \mathbf{y}_{SP} \end{bmatrix} \tag{17}$$

$$J = \sum_{n=1}^{N} (\mathbf{x}(nT), \mathbf{y}_m(nT)) \begin{bmatrix} \mathbf{Q} & -\mathbf{QC}^T \\ -\mathbf{CQ} & \mathbf{CQC}^N \end{bmatrix} \begin{bmatrix} \mathbf{x}(nT) \\ \mathbf{y}_m(nT) \end{bmatrix} + \mathbf{u}^T((n-1)T)\mathbf{R}\mathbf{u}((n-1)T). \tag{18}$$

Equation (18) plus the discrete equivalent of equation (17) represent the same problem as the direct approach described in the previous subsection Thus the optimal control law has the same form as equation (14), except that $\mathbf{K}_{SP}=\mathbf{O}$, and the controller matrices are determined from the augmented equivalents of the expressions presented in the nomenclature section for $\mathbf{K}_{FB}$ and $\mathbf{K}_{FF}$. However these controller matrices can be partitioned to give a control law in the following more convenient form:

$$\mathbf{u}(nT) = \mathbf{K}_{FB}\mathbf{x}(nT) + \mathbf{K}_m \mathbf{y}_m(nT) + \mathbf{K}_{SP^2}\, \mathbf{y}_{SP}(nT) + \mathbf{K}_{FF}\mathbf{d}(nT). \tag{19}$$

The implementation of equation (19) is illustrated by Fig. 1. Consideration of equations (14, 16 and 19) will show, since $\mathbf{y}_m(NT) = y_{SP}$, that $\mathbf{K}_{SP}$ in equation (14) is equal to the sum of $\mathbf{K}_m + \mathbf{K}_{SP^2}$ in equation (19). Thus it is possible to consider the *direct approach* as a special case of model following where the response of the assumed model is infinitely fast.

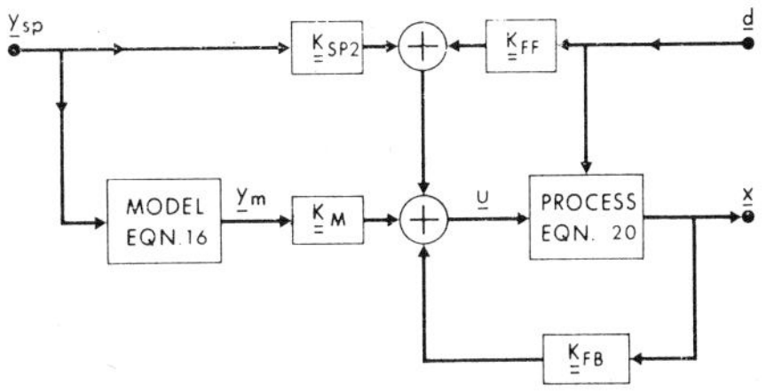


FIG. 1. Schematic representation of equation (19) showing proportional (plus integral) state feedback, feed-forward, and model-following modes of the multivariable controller.

## 3. IMPLEMENTATION

### 3.1 *Process equipment and model*

The multivariable control schemes developed in this paper were evaluated on a pilot-plant double-effect evaporator in the Department of Chemical and Petroleum Engineering at the University of

Alberta. A schematic flow diagram of the evaporator connected for operation in a double-effect, forward-feed mode of operation is presented in Fig. 2,

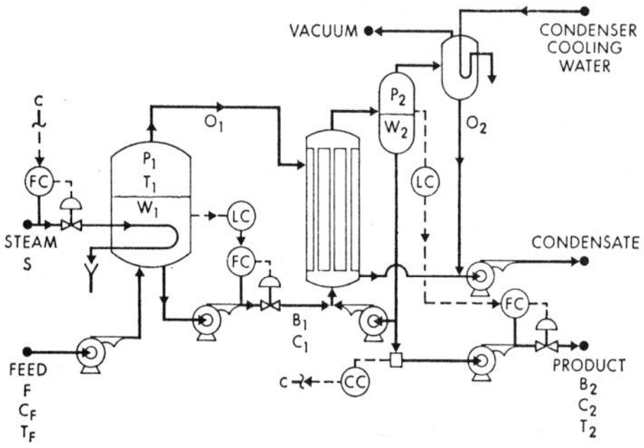


FIG. 2. Flow diagram of computer-controlled, double-effect evaporator used for simulated and experimental studies of multivariable and conventional control schemes. (Symbols are defined in Nomenclature section and state space model is given as equation (20).)

where the symbols and steady-state conditions are defined in the nomenclature section. The first effect is a natural circulation calendria type unit heated with a nominal 0·9 kg/min of fresh steam ($S$) and has a feed stream of 2·25 kg/min of 3 per cent triethylene glycol. The second effect is an externally-forced circulation, long-tube-vertical unit which operates under vacuum and utilizes the first effect overhead vapor ($O_1$) to concentrate the first effect bottoms stream ($B_1$) to a product concentration of about 10 per cent. The unit is constructed largely of glass so that the process response can be observed.

The process is fully instrumented with conventional industrial electronic instruments but is also interfaced to an IBM 1800 digital control computer with 48K of core and three disks. The computer operates under a multiprogramming executive system and provides Direct Digital Control (DDC) capabilities simultaneously with other control programs and off-line jobs. The evaporator can be switched, manually or under computer control, from the analog controllers to supervisory control or to DDC. The principle control loops used for conventional control of the levels and concentration which is measured continuously by in-line process refractometers, are shown in Fig. 2. However over 50 process measurements are interfaced to the computer. These are available for use by computer control algorithms and permit monitoring of the evaporator so it can be run, unattended, 24 hr/day. In addition, shutdown in the event of trouble is initiated by the computer.

The FORTRAN control program used in this work [4] was reloaded from disk every control interval, and it obtained state variable measurements from DDC data acquisition loops. Then it made control variable changes by adjusting the set-points of tightly tuned DDC flow loops ($S$, $B1$, $B2$). It also performed state estimation, for $C1$ since it was not measured, and model calculations. Execution time was only a few seconds.

Several models of the evaporator were developed [4], which ranged from a tenth-order non-linear dynamic model to a second-order transfer function. However the one used for this project is the fifth-order, linear, time-invariant, state-space model given by equation (20). The model is in the form of equations (1 and 2) and the variables are defined so that they represent deviations from the normal operating conditions given in the nomenclature section. The variables were also normalized by dividing by the corresponding steady-state values and are therefore dimensionless. The model is based on material and energy balances and has not been empirically fitted to improve the dynamic response. It is, however, in reasonable agreement with open-loop responses of the evaporator for disturbances of 10–20 per cent of the steady-state values.

$$\begin{bmatrix} \dot{W}1 \\ \dot{C}1 \\ \dot{H}1 \\ \dot{W}2 \\ \dot{C}2 \end{bmatrix} = \begin{bmatrix} 0 & -0{\cdot}00156 & -0{\cdot}1711 & 0 & 0 \\ 0 & -0{\cdot}1419 & +0{\cdot}1711 & 0 & 0 \\ 0 & -0{\cdot}00875 & -1{\cdot}102 & 0 & 0 \\ 0 & -{\cdot}00128 & -0{\cdot}1489 & 0 & +0{\cdot}00013 \\ 0 & +0{\cdot}0605 & +0{\cdot}1489 & 0 & -0{\cdot}0591 \end{bmatrix} \begin{bmatrix} W1 \\ C1 \\ H1 \\ W2 \\ C2 \end{bmatrix} + \begin{bmatrix} 0 & -0{\cdot}143 & 0 \\ 0 & 0 & 0 \\ 0{\cdot}392 & 0 & 0 \\ 0 & +0{\cdot}108 & -0{\cdot}0592 \\ 0 & -0{\cdot}0486 & 0 \end{bmatrix} \begin{bmatrix} S \\ B1 \\ B2 \end{bmatrix}$$

$$+ \begin{bmatrix} +0{\cdot}2174 & 0 & 0 \\ -0{\cdot}074 & 0{\cdot}143 & 0 \\ -0{\cdot}036 & 0 & 0{\cdot}181 \\ 0 & 0 & 0 \\ 0 & 0 & 0 \end{bmatrix} \begin{bmatrix} F \\ CF \\ HF \end{bmatrix} ; \quad \begin{bmatrix} W1 \\ W2 \\ C2 \end{bmatrix} = \begin{bmatrix} 1 & 0 & 0 & 0 & 0 \\ 0 & 0 & 0 & 1 & 0 \\ 0 & 0 & 0 & 0 & 1 \end{bmatrix} \begin{bmatrix} W1 \\ C1 \\ H1 \\ W2 \\ C2 \end{bmatrix} . \qquad (20)$$

A state estimation calculation was included in the implementation of the optimal multivariable control because: it made the implementation more general; it was necessary for those periods when direct measurements of the concentrations were not available due to instrument problems; and it alleviated some of the problems due to measurement noise which, when fedback through the relatively high-gain controllers, produced undesirable variations in the control variables. Current applications are using a Kalman filter and this is the preferred approach. However the experimental runs discussed in this paper employed the simpler approach of using the linear discrete process model to calculate, at each control interval, the current values of the state variables, $\mathbf{x}_{calc}$, based on the estimated value of the state, $\mathbf{x}_{est}$, plus measured values of the control and disturbance variables, from the previous time interval. Normally it was possible to measure all the state variables, $\mathbf{x}_{meas}$, except the concentration $C1$. Therefore the estimated value of the state at each control interval was obtained using the relationship

$$\mathbf{x}_{est} = \boldsymbol{\alpha}\mathbf{x}_{meas} + (\mathbf{I} - \boldsymbol{\alpha})\mathbf{x}_{calc} \qquad (21)$$

where $\boldsymbol{\alpha}$ is a diagonal matrix of "filter constants". It should be noted that, considering the $i^{th}$ element of $\mathbf{x}$, when $\alpha_{ii}=1$ then $x_{est}=x_{meas}$ and when $\alpha_{ii}=0$ then $x_{est}=x_{calc}$. This relationship was successfully used in conjunction with logic in the input subroutine which checked for failure of a measurement transducer. When a measured value of $x_i$ was judged to be unreliable the $\alpha_{ii}$ was changed from unity to zero and control continued. For values of $\boldsymbol{\alpha}$ in the range $\mathbf{O} < \boldsymbol{\alpha} < \mathbf{I}$ then the estimated values, $\mathbf{x}_{est}$, are a weighted combination of measured and calculated values. Individual input measurements were also "lighty" filtered using the standard digital exponential filters within the DDC data acquisition loops. This approach gave excellent results in the regulatory control experiments. However, in the experiments that involved significant changes in the process operating conditions, the errors in $\mathbf{x}_{meas}$, due partially to temperature effects on the refractometers that measured concentration, and the errors in $\mathbf{x}_{calc}$, due mainly to process non-linearities, produced greater uncertainty in the estimated values, $\mathbf{x}_{est}$, and indicated the need for further improvements in this area.

3.2 *Specification of design parameters*

For a given system it is the specification of the weighting factors in the performance index of equation (7) that determine the performance of the closed-loop system. Therefore the parameters $T$, $\beta$, $\mathbf{S}$, $\mathbf{Q}$ and $\mathbf{R}$ will each be discussed in turn.

The control interval, $T$, affects the system indirectly through $\boldsymbol{\phi}(T)$, $\boldsymbol{\Delta}(T)$ and $\boldsymbol{\theta}(T)$ in equation (5). From a mathematical point of view, the control interval should be as small as possible since as $T$ decreases the discrete system given by equation (5) becomes a better approximation to the continuous system given by equation (1) and, in general, control will improve. However from a practical point of view, a compromise must be made between the improvement in control versus the increased amount of computer usage that is required, per unit time. For computer control applications the sampling interval should be as large as possible consistent with the requirement that the discretized system be a "reasonable" representation of the dynamics of the physical system and its inputs $\mathbf{u}$ and $\mathbf{d}$, and this can be conveniently checked by simulation as was demonstrated for the evaporator system [4].

The parameter $\beta$ in equation (7) weights deviations in $\mathbf{x}$ and $\mathbf{u}$ more heavily as time increases and can be used to introduce a "prescribed degree of stability" [27]. In the evaporator study a value of $\beta = 1{\cdot}5$ was found to produce slightly faster response but appeared to do so through increasing the controller gains [4]. For this application, where the process responses were generally overdamped, it was found that similar results could be produced by increasing the elements of $\mathbf{Q}$. Therefore a value of $\beta = 1$ was selected and used for all the experiments reported herein.

The matrix $\mathbf{S}$ in equation (7) which weights the offsets in the state variables at time $(NT)$ has very little significance for large values of $N$, unless the corresponding elements of $\mathbf{Q}$ are zero, and was therefore set to zero for this investigation.

The weighting matrices $\mathbf{Q}$ and $\mathbf{R}$ also in equation (7) determine the penalities to be placed on deviations in the state and control variables respectively, and methods of specifying them have been considered by a number of authors. NICHOLSON [28] mentioned his trials but drew no general conclusions. TYLER and TUTEUR [29] defined a complex procedure for determining the system poles from the elements of diagonal weighting matrices which would enable a trial-and-error design for specified root loci. FALSIDE and SERAJE [30] developed an explicit relationship in the frequency domain between the poles of the optimal, closed-loop system and the weighting matrices $\mathbf{Q}$ and $\mathbf{R}$ that can be used in a two stage design procedure to obtain a specified pole placement. CHANT and LUUS [31] used a "brute force" optimization technique requiring iterative designs and simulations.

Assuming—as is generally done in practice—that $\mathbf{Q}$ is diagonal then a few qualitative generalizations can be made. For example, an increase in the relative magnitude of $q_{ii}$ will "improve" the response of the corresponding state variable but the response of the other states will, in general, not be as good. Also examination of the structure of the system defined by equation (5) may give some guidance since if a control variable acts directly (via $\mathbf{\Delta}$) and indirectly (via $\mathbf{\phi}$) on only one state then that state must be weighted, e.g. $W2$ in the evaporator problem. In the proportional-plus-integral formulation weighting must also be placed on the "integral states" introduced in equation (11). As might be expected, increasing the relative weights on these states results in increased integral action but also slightly increased proportional action.

The weighting matrix **R** cannot be considered as the only weighting on the control variables since the elements of **Q** can produce a significant indirect weighting on control action. For example, a relatively large weight, $q_{ii}$, on a state corresponding to a comparitively large element $\delta_{ik}$ of $\mathbf{\Delta}$ in equation (5) will penalize the use of the corresponding control variable $u_k$. The effect is similar to the direct weighting of $u_k$ by increasing $r_{kk}$ in **R**. In general, as the weighting on $u_k$ are increased the corresponding gains in the controller matrices are decreased, the magnitude of control actions are thereby reduced and, in general, the quality of control on $\mathbf{x}$ decreases. For systems subject to constant disturbances the final steady state, $\mathbf{x}_{ss}$, will not be zero so that $\mathbf{u}_{ss}=\mathbf{K}\mathbf{x}_{ss}$ is non-zero and is influenced by **K**. Therefore the effect of **R** on $\mathbf{u}_{ss}$ could be important in some applications where there are major differences in the "costs" associated with different controls.

It should be noted that it is possible, and in some cases desireable, to use zero weighting on some states. For example, in some of the later studies of the evaporator it was noted that the "best" performance was obtained when $\mathbf{R}=\mathbf{O}$ and the weights on the two states, $C1$ and $H1$, was reduced to zero. This corresponds to a three input, three output problem with unconstrained control variables and hence may approximate the "dead-beat" response that can be obtained with discrete systems [32]. It is significant that the performance of the system was essentially independent of the other elements of **Q** so the generality of the phenomena deserves further study.

There is no explicit procedure relating desired time-domain performance characteristics to the elements of **Q** (or **R**) so that their specification is generally a matter of judgement based on experience. For the evaporator example treated in this paper, the product concentration, $C2$, is of prime interest and hence received the highest weighting. The liquid levels simply had to be controlled within physical limits and hence received only nominal weights. It was found that with all the elements of **R** made arbitrarily small that the control action did not exceed physical constraints and therefore a weighting of $\mathbf{R}=\mathbf{O}$ was used in all experimental runs. The design parameters used in the problem formulation are summarized in Table 1.

TABLE 1. DESIGN BASIS

| | |
|---|---|
| Model: | as defined by equations (5 and 6) with $T=64$ sec |
| Performance index: | as defined by equation (7) with $\beta=1$ |
| | $\mathbf{S}=\mathbf{0}$ |
| | $\mathbf{R}=\mathbf{0}$ |
| | $\mathbf{Q}=$diag. (10, 1, 1, 10, 100, $h$, $h$, $h$) |
| | $h=$o for proportional control and |
| | $h=I$ for $P+I$ control (unless noted) |

## 4. SIMULATED RESULTS: $P+I$ CONTROL

The basic features of the multivariable controller and its sensitivity to different weighting factors can be illustrated best by simulated data. For the system defined in Table 1, Fig. 3 shows the effect on product concentration of varying the weighting on the integral states. When the weighting, $h$, is zero then pure proportional control results with the feedback gain between concentration, $C2$, and steam, $S$, of $-34$. When $h=1$ the proportional gain increases to $-40$ and integral action, with a gain of $-4$, is evident. Increasing $h$ to 10 results in increased controller gains and faster recovery.

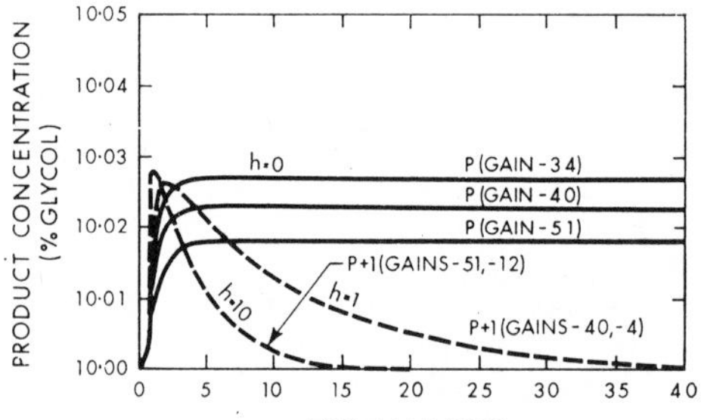


FIG. 3. Simulated responses of evaporator product concentration showing the increase in integral (and proportional) controller gains produced by increasing the weighting on the integral states (see "$h$" in Table 1).

Figure 4 presents a simulated comparison, using controller constants that gave good experimental performance, of the conventional multiloop control scheme shown in Fig. 2 with the multivariable control laws developed in this paper. The transients are the result of a +10 per cent flow disturbance at time zero, and the triangles on the vertical axis indicate the initial conditions. The small deviations and the short transients of the multivariable runs are immediately apparent and it can be seen that

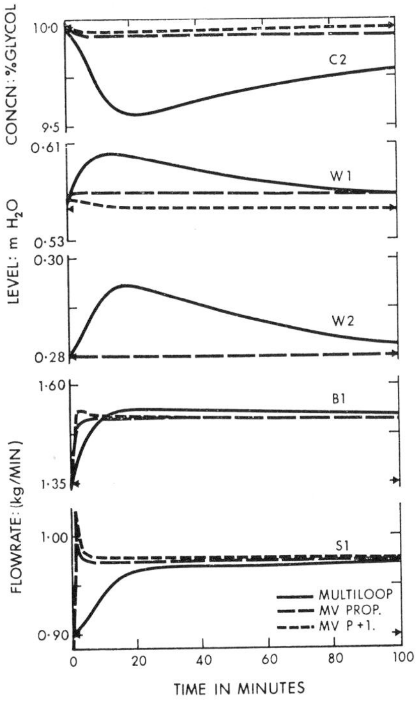


FIG. 4. Comparison of simulated evaporator responses, under multivariable and conventional proportional-plus-integral controllers, to a 10 per cent increase in feed flow. Controller gains are those that give good experimental responses.

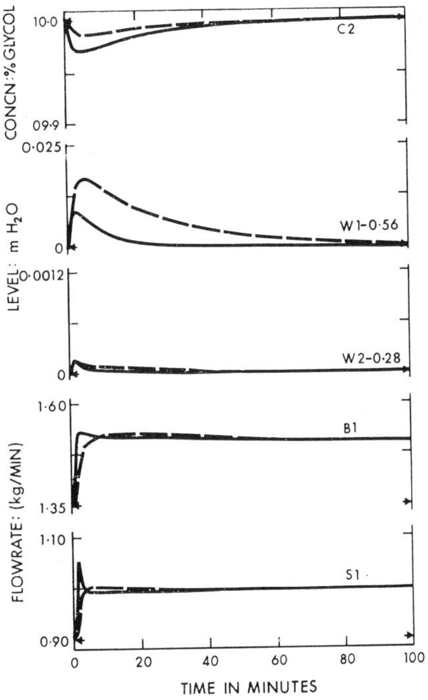


FIG. 5. Simulated effect of decreasing the weighting on the liquid levels from the values given in Table 1 to the values defined in equation (22).

multivariable integral action removes the small offsets present when proportional control is used alone. The excellent performance of the multivariable controllers can be traced to their interacting nature and the high gains of the controllers. The multivariable proportional gains, defined in the nomenclature section, varied from 0 to 19 compared to a range of 2–3 for the single variable controllers, and the integral gains ranged from 0 to 4·3 compared to 0·01–0·04 in the conventional controllers. Note the smooth, bounded response of the manipulated variables and compare this to later experimental runs.

The levels, $W1$ and $W2$, just have to be maintained between limits rather than at a specified steady state, so deviations can be permitted if this leads to improved control of product composition. A simulated run was carried out with reduced weighting on the levels given by:

$$\mathbf{Q} = \text{diag}(1, 1, 1, 1, 100, 0{\cdot}01, 0{\cdot}01, 1). \qquad (22)$$

Figure 5 compares this "averaging" control of levels (dashed lines) with "tight" control (solid lines). The graphs show larger deviations and "slower" control for the levels and improved control for the product concentration. The control variables in the case of averaging control do not exhibit the overshoot produced by the higher values of **Q** in Table 1.

## 5. EXPERIMENTAL RESULTS: $P+I$ CONTROL

A number of experimental runs have been carried out to test the multivariable control schemes developed in this paper and to compare this type of control with conventional multiloop control. Some of the data is reproduced in this paper, but the results from other runs [4] are only summarized in the text. The actual implementation of the "optimal" control laws is of course suboptimal due to the infinite-time design basis, the state estimation step, etc. The purpose of the experimental runs is therefore not so much to verify theoretical or mathematical principles, but rather to evaluate how well the control systems designed on the basis of these principles will perform in practical applications which involve modelling errors, noise, non-linearities, etc. that are not rigorously dealt with by the design procedure.

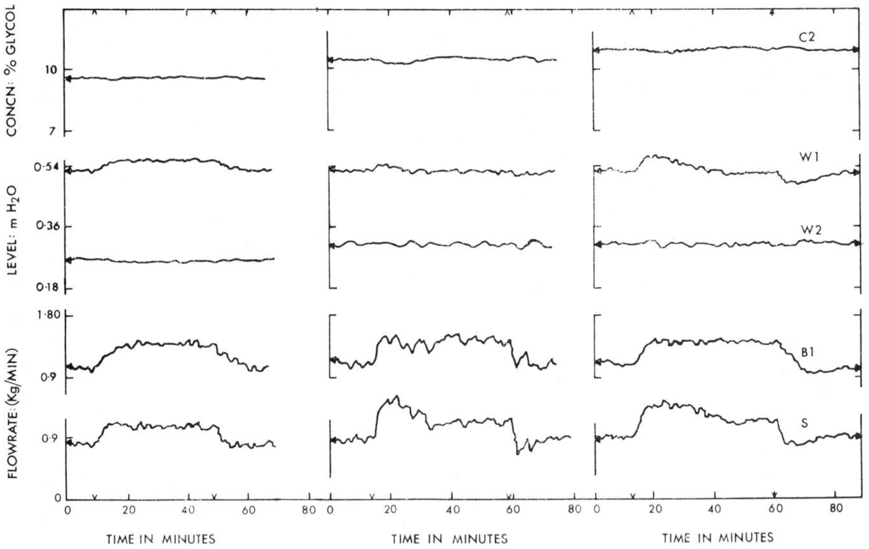


FIG. 6. Experimental evaporator response under multivariable proportional control to a 20 per cent increase (and decrease) in feedflow rate. (Disturbances were introduced at the point shown by a "$v$" mark on time axis.) Design basis as per Table 1 (with $h=o$) and $\mathbf{u}=K_{FB}\mathbf{x}$.

FIG. 7. Experimental evaporator response which, by comparison with Fig. 6, shows the effect of adding integral action. Design basis as per Table 1 (with $h=1$) and $\mathbf{u}=\mathbf{K}_{FB}\mathbf{x}+\mathbf{K}_I\Sigma\mathbf{y}(nT)$.

FIG. 8. Experimental evaporator response which, by comparison with Fig. 6, shows the effect of adding integral action, and by comparison with Fig. 7 the effect of changing the weighting factors in **Q**. Design basis as per Table 1 except that **Q** is given by equation (22) and $\mathbf{u}=\mathbf{K}_{FB}\mathbf{x}+\mathbf{K}_I\Sigma\mathbf{y}(nT)$.

The experimental response of the evaporator to a 20 per cent increase and decrease in feed flowrate while under multivariable proportional and proportional-plus-integral control is shown in Figs. 5–7 where the triangular marks on the time scale indicate the points at which the feed disturbances were introduced. As with the simulated results, the small deviations and short transients are clearly shown.

The effects of noise becomes apparent in the experimental runs. Figures 6 and 7 show 20–40 per cent fluctuations in the control variables resulting from the action of the high gains on noisy measurements. The control of liquid holdup, $W2$, produced by proportional control alone, shown in Fig. 6, is much smoother than the comparable response using multivariable $P+I$ control. This attributed to the higher gains used in the latter case.

Figure 8 presents the results from a multivariable run carried out using the lower weighting factors on the liquid levels given in equation (22). The transient responses of the levels show the expected larger deviations and slower responses. The product concentration response improved as in the simulated run as indicated in Fig. 5.

The experimental response of the evaporator to a 27 per cent change in feed concentration under the "tight" proportional-plus-integral control law was better than its response to 10 per cent feed flowrate changes although these data are not reproduced here.

Figure 9 compares typical transients in the evaporator product concentration, produced by 20 per cent feed disturbances, when controlled by conventional methods, which are shown in Fig. 2, and multivariable methods. The form of the responses can of course be changed significantly by varying the controller gains and in the case of the conventional control by adding feedforward and/or ratio controllers, i.e. by partial implementation of multivariable methods [2].

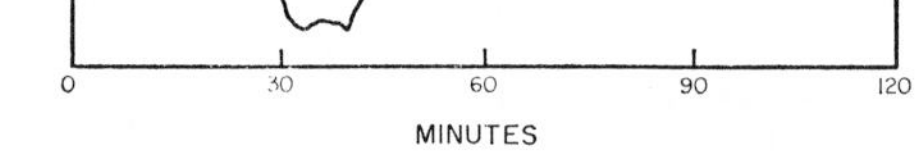

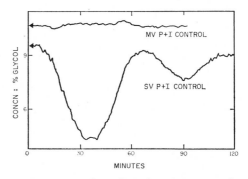

FIG. 9. Comparison of experimental responses of evaporator product concentration using conventional multiloop $P+I$ controllers configurated as in Fig. 2 versus multivariable proportional-plus-integral control.

However, experience gained over the last four years indicates that the results in Fig. 9, and in Fig. 4, present a typical and realistic comparison of single versus multivariable $P+I$ control of the evaporator. It is particularly significant that the "theoretical" model, given by equation (20), could NOT be used to determine satisfactory controller gains for the conventional multiloop controllers, since simulation studies suggested gains that were an order of magnitude larger than could be used experimentally, but it did serve as a basis for developing the multivariable controllers used, for example, in Figs. 6–8.

## 6. EXPERIMENTAL RESULTS: SETPOINT CHANGES

In most process applications, regulatory control performance is much more important than servo or tracking control. Nevertheless it is essential that provision be made to implement setpoint changes, and desirable that the control engineer be able to specify the form of the resulting transient. Note that the desired transient may be dictated by considerations that are outside the boundary of the system defined by the process model, e.g. interactions with upstream and/or downstream units, constraints on utilities and/or physical equipment, safety considerations etc.

For this work the design basis is given by Table 1, where $h=o$ since integral control was not used, and the desired response was defined arbitrarily as that of a non-interacting, unity gain model with first order dynamics, e.g. the output of equation (16) with $\mathbf{H}=-\mathbf{I}$ and $\mathbf{G}=(1/\tau)\mathbf{I}$ where $\tau$ is the desired "time constant" of the response. The matrix $\mathbf{E}$, shown in equation (15), which was used to correct for the small offsets that occur between the desired values and the final steady state, was so close to the identity matrix, indicated in the nomenclature section, that it was ignored. Some of the experimental runs made to evaluate the setpoint control methods are summarized in Figs. 10–14.

Figure 10 compares three setpoint control schemes for a 10 per cent change in product concentration setpoint where the times at which the changes were introduced are indicated by the arrows at the top of the figures. Note:

(a) The $C2$ responses are better for a decrease in setpoint than an increase because of non-linearities in the real process.

(b) The process does not keep up with the model in Fig. 10b indicating that the model time constant ($\tau=1$ min) was smaller than that of the process. As a result the process responds as fast as it can giving comparable results to Fig. 10a, the direct scheme. Note that the direct method can be interpretted as the limiting case of real model following as the time constants of the model approach zero. In Fig. 10c the process follows the "slower" model more closely.

(c) Figure 10c, where the model following operates successfully, shows smaller deviations in levels, $W1$ and $W2$, as a result of the non-interacting model.

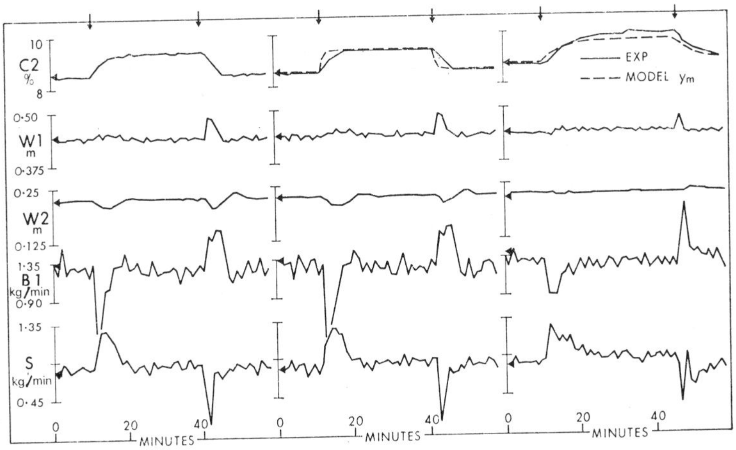


FIG. 10. Experimental implementation of a 10 per cent change in the setpoint (desired value) of the product concentration using the model following technique. (See equation (19).) Fig. 10a is an implementation of the *direct* method but is equivalent to model following with a time constant $\tau=o$. Figures 10b and c show responses following a model with $\tau=1$ and $\tau=5$ min respectively.

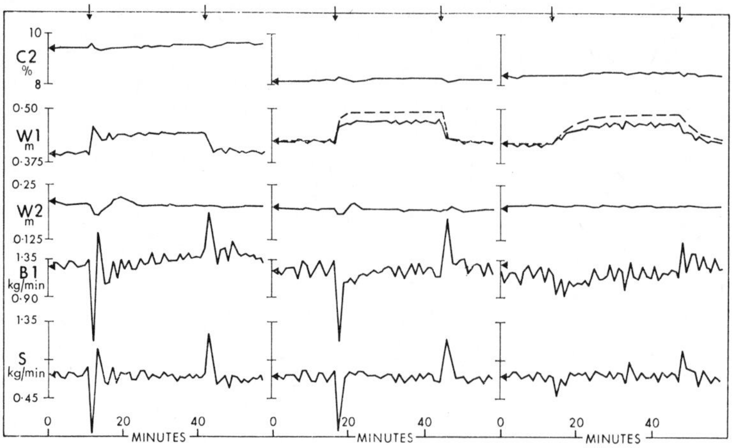


FIG. 11. Evaporator responses obtained under the same conditions as those in Fig. 10 except the setpoint of the first effect level was changed by 15 per cent.

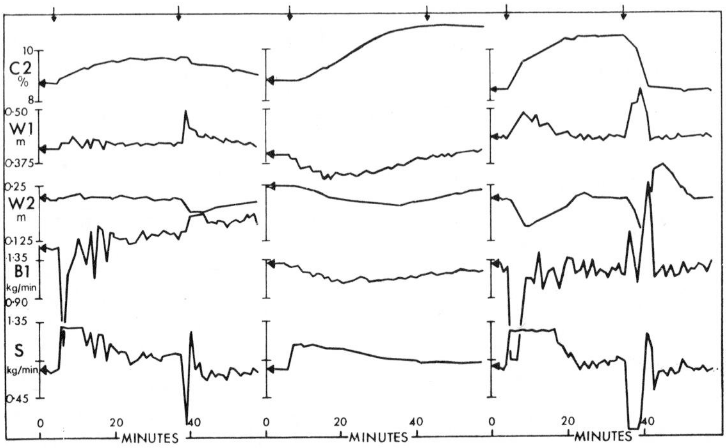


FIG. 12. Evaporator response when a 10 per cent change in feed flow rate is introduced at the same time as a 10 per cent setpoint change. The response is slower than in Fig. 10 and the steam flow reaches its upper limit.

FIG. 13. Evaporator response to a 20 per cent change in the setpoint of the conventional concentration control scheme shown in Fig. 2. Here the controller was tuned for regulatory control.

FIG. 14. Evaporator response to a 20 per cent change in the setpoint of a multivariable controller $\mathbf{u}=\mathbf{K}_{FB}\mathbf{x}+\mathbf{K}_{SP}\mathbf{y}_{SP}$ for comparison with Fig. 13.

(d) The difference, at steady state, between the experimental data and the model response in Fig. 10c is a result of the state estimation procedure used with all runs except those shown in Figs. 10a and b. It involved a weighted average of measured concentrations as defined by equation (21). The concentration, ($x_{\text{calc}}$) calculated from the process model was lower than the value ($x_{meas}$) indicated by the process refractometer. The controller however worked on the basis of the combined value ($x_{\text{est}}$) so that at the final state there was an offset between the model and measured values.

The same three control schemes are compared in Fig. 11 for a 15 per cent change in first effect level. Note that this change is not typical of normal operating practice but was used to investigate interactions between different states. Note also:

(a) The overshoot and oscillation in *W*1 in Fig. 11a, under the direct setpoint scheme, is avoided by the model following control schemes shown in Figs. 10b and c.
(b) The smaller process time constant associated with *W*1 allows the level to keep up with both the one and five minute setpoint models.
(c) The model-following schemes, particularly Fig. 11c, require less control action and produce smaller interactions in *W*2 and *C*2 than the direct approach, as indicated by a comparison of *B*1 and *S*.
(d) The effect of the state estimation procedure and incorrect evaporator model gains is again evident in the plots of *W*1 in Figs. 10b and c.

When setpoint changes were made in the second effect level, *W*2, the same general type of behaviour was observed but, since *W*2 is essentially independent of the other states, since the off-diagonal elements in the *W*2 column of the closed-loop **A** matrix are negligible, very little interaction with *W*1 and *C*1 occurred in any of the three control schemes.

The experimental model following runs showed that a setpoint model with one minute time constants corresponding to the levels and a 5 min constant for *C*2 would have given better overall performance. This conclusion may also have been made by examining the time constants, the eigenvalues, of the closed loop **A** matrix.

Experimental runs were made to evaluate the performance of the systems when load changes were introduced at the same time as the setpoint change. Figure 12 shows the response for a 10 per cent change in setpoint introduced at the same time as a similar 10 per cent change in feed flow and can be compared with Fig. 9. As expected there is a much slower rise to the new setpoint, about 30 min compared to 15 min, when the change in feed flowrate "works against" the setpoint change. However for a decrease in the *C*2 setpoint the "opposing" decrease in feed flow rate had very little effect. This is partially due to the fact that during the increase in setpoint the steam flow reached its maximum limit of approximately 1·35 kg/min, as shown by the flat topped peak in steam flow, so the available control action was not equal for the up and the down change.

A 20 per cent change in the *C*2 setpoint under conventional DDC multiloop control is shown in Fig. 13. The approximately 40 min duration of the transient in *C*2, *W*1 and *W*2 can be compared to the 20 min transient in Fig. 14 where direct multivariable setpoint control is in effect. Note the sharp level disturbances under multivariable control where all three control variables combined to change *C*2 and sacrifice the control of the levels which were not as heavily weighted. In contrast the level disturbances under conventional DDC were a result of interactions with the steam flowrate. The initial fast rate of change in *C*2 shown in Fig. 14 occurred while steam was high and first effect bottoms, *B*1 was low and then the rate of increase of *C*2 slowed when *B*1 acted to control the level, *W*1. The faster reaction of *C*2 to a decrease in setpoint in Fig. 14 was a result of more available control action as already discussed.

## 7. CONCLUSIONS

(1) This work illustrates a successful implementation of "optimal" multivariable control in a process environment. Results are significantly better than with a conventional control scheme despite the presence of significant modelling errors, process nonlinearities, noise etc.
(2) The discrete optimization approach was extended to generate multivariable controllers which include proportional-plus-integral state feedback, feedforward, and model-following capabilities. The controller matrices are time-invariant and thus easily and economically implemented on a digital control computer.
(3) The model-following capability is very flexible and gives the designer control over the form of the setpoint transients, interactions, etc.
(4) Model accuracy and process nonlinearities are more important for setpoint changes than for regulatory control.

## REFERENCES

[1] R. E. Kalman: Contributions to the theory of optimal control. *Pol. Soc. Mat. Mex.* **5**, 102–119 (1960).
[2] M. Athans and P. L. Falb: *Optimal Control.* McGraw-Hill, New York (1966).
[3] A. P. Sage: *Optimum Systems Control.* Prentice-Hall, N.J. (1968).
[4] R. B. Newell: Multivariable computer control of an evaporator. Ph.D. Thesis, Department of Chemical and Petroleum Engineering, University of Alberta (available from University Microfilms, Ann Arbor, Michigan, U.S.A.) (1971).
[5] C. L. Smith and P. W. Murrill: *An Optimal Controller for Multivariable Systems Subject to Disturbance Inputs.* Joint Automatic Control Conference, 10th, Preprints, 469 (1969).
[6] J. B. Moore: Tolerance of nonlinearities in linear optimal regulators with integral feedback. *Electronics Letters* **5**, 409 (1969).
[7] B. Porter: Synthesis of asymptotically stable, linear, time-invariant, closed-loop systems incorporating multivariable 3-term controllers. *Electronics Letters* **5**, 557 (1969).
[8] A. Simpson: Synthesis of asymptotically stable, linear time-invariant, closed-loop systems incorporating multivariable 3-term controllers. *Electronics Letters* **6**, 251 (1970).
[9] A. Simpson: Linear systems with multivariable 3-term controllers: use of the matrix pseudo inverse in their synthesis. *Electronic Letters* **6**, 374 (1970).
[10] Yen-Ping Shih: Integral action in the optimal control of linear systems with quadratic performance index. *Ind. Engng Chem. Fundam.* **9**, 35 (1970).

[11] B. PORTER: Optimal control of multivariable linear systems incorporating integral feedback. *Electronics Letters* **7,** 170–172 (1971).
[12] P. C. YOUNG and J. C. WILLIAMS: An approach to the multivariable servomechanism problem. *Int. J. Control* (to appear).
[13] M. FOND and C. FOULARD: *Linear Multivariable Control in the Presence of Unknown Non-Zero Means Value Perturbations,* paper 2.2.2. 2nd IFAC Symposium on Multivariable Control Systems, Düsseldorf (1971).
[14] Y. C. HU: Application of modern control theory to distillation columns. Ph.D. Thesis, University of Colerado (1971).
[15] E. J. DAVISON and H. W. SMITH: Pole assignment in linear time-invariant multivariable systems with constant disturbances. *Automatica* **7,** 489–498 (1971).
[16] C. D. JOHNSON: Further study of the linear regulator with disturbances—the case of vector disturbances satisfying differential equations. *IEEE Trans. Aut. Control* **AC-15,** 222–228 (1970).
[17] P. R. LATOUR: *Optimal Control of Linear Multivariable Plants with Constant Disturbances.* Joint Automatic Control Conference, St. Louis, Missouri, Paper 8E3, pp. 908–916 (1971).
[18] R. E. KALMAN and R. W. KOEPCKE: *The Role of Digital Computers in the Dynamic Optimization of Chemical Reactors,* p. 107, Proc. Western Joint Computer Conference (1959).
[19] J. S. TYLER: The characteristics of model-following systems as synthesized by automatic control. *IEEE Trans. Aut. Control* **AC-9,** 485 (1964).
[20] C. A. MARKLAND: Optimal model-following control-system synthesis techniques. *Proc. Inst. Elec. Engrs.* **117,** 623–627 (1970).
[21] A. ZINOBER: Discrete-time control using the quadratic performance index, implicit model following and the specific optimal policy. *Electronic Letters* **6,** 841 (1970).
[22] H. ERZBERGER: *Analysis and Design of Model Following Systems by State Space Techniques,* Joint Automatic Control Conference, Ann Arbor, Michigan, U.S.A., pp. 572–581 (1968).
[23] C. A. WINSOR and R. J. ROY: *The Application of Specific Optimal Control to the Design of Desensitized Model Following Control Systems.* Proceeding Joint Auto Control Conference, Atlanta, Georgia, Paper 13A, p. 271 (1970).
[24] E. E. YORE: *Optimal Decoupling Control.* Joint Automatic Control Conference, Ann Arbor, Michigan, U.S.A., Preprints, p. 327 (1968).
[25] B. A. JACOBSON: Multiloop computer control of an evaporator, M.Sc. Thesis, Department of Chemical and Petroleum Engineering, University of Alberta (1970).
[26] L. LAPIDUS and R. LUUS: *Optimal Control of Engineering Processes.* Blaisdell Publishing Company, Waltham, Mass. (1967).
[27] ANDERSON and MOORE: *Linear Optimal Control.* Prentice Hall, Englewood Cliffs, N.J. (1970).
[28] H. NICHOLSON: Dynamic optimization of a boiler. *Proc. Inst. Elec. Engrs.* **111,** 1479 (1964).
[29] J. S. TYLER and F. B. TUTEUR: The use of a quadratic performance index to design multivariable control systems. *IEEE Trans. Aut. Control.* **AC-11,** 84 (1966).
[30] F. FALLSIDE and H. SERAJI: *Decision of Optimal Multivariable Systems by a Frequency Domain Technique.* 2nd IFAC Symposium on Multivariable Control Systems, Düsseldorf, paper 2.2.1 (1971).
[31] V. G. CHANT and R. LUUS: Time suboptimal control of the gas absorber. *Can. J. Chem. Engng* **46,** 376 (1968).
[32] B. C. KUO: *Discrete-Data Control Systems.* Prentice Hall, Englewood Cliffs, N.J. (1970).

## APPENDIX

*Recursive relations for control matrices*

$$\mathbf{K}_{FB}^{N-i} = -(\boldsymbol{\Delta}^T\mathbf{P}^{i-1}\boldsymbol{\Delta}+\mathbf{R})^{-1}\boldsymbol{\Delta}^T\mathbf{P}^{i-1}\boldsymbol{\varphi}$$

$$\mathbf{K}_{FF}^{N-i} = -(\boldsymbol{\Delta}^T\mathbf{P}^{i-1}\boldsymbol{\Delta}+\mathbf{R})^{-1}\boldsymbol{\Delta}^T(\mathbf{P}^{i-1}\boldsymbol{\theta}+\mathbf{O}^{i-1})$$

$$\mathbf{K}_{SP}^{N-i} = -(\boldsymbol{\Delta}^T\mathbf{P}^{i-1}\boldsymbol{\Delta}+\mathbf{R})^{-1}\boldsymbol{\Delta}^T(\mathbf{M}^{i-1}-\mathbf{N}^{i-1}\mathbf{C}^T)$$

where

$$\mathbf{P}^i = \beta(\mathbf{T}^{N-i})^T\mathbf{P}^{i-1}\mathbf{T}^{N-i} + \beta(\mathbf{K}_{FB}^{N-i})^T\mathbf{R}\mathbf{K}_{FB}^{N-i} + \mathbf{O}$$

$$\mathbf{M}^i = \beta(\mathbf{T}^{N-i})^T(\mathbf{P}^{i-1}\boldsymbol{\Delta}\,\mathbf{K}_{SP}^{N-i} - \mathbf{M}^{i-1}) + \beta(\mathbf{K}_{FB}^{N-i})^T\mathbf{R}\mathbf{K}_{SP}^{N-i}$$

$$\mathbf{N}^i = \beta(\mathbf{T}^{N-i})^T\mathbf{N}^{i-1} + \mathbf{Q}$$

$$\mathbf{O}^i = \beta(\mathbf{T}^{N-i})^T(\mathbf{O}^{i-1} + \mathbf{P}^{i-1}(\boldsymbol{\Delta}\mathbf{K}_{FF}^{N-i}+\boldsymbol{\theta})) + \beta(\mathbf{K}_{FB}^{N-i})^T\mathbf{R}\mathbf{K}_{FF}^{N-i}$$

$$\mathbf{T}^{N-i} = \boldsymbol{\varphi} + \boldsymbol{\Delta}\mathbf{K}_{FB}^{N-i}$$

with the counter $i=1, 2, 3, \ldots$ and initial conditions

$$\mathbf{P}_o=\mathbf{Q}+\mathbf{S},\quad \mathbf{M}_o=\mathbf{O},\quad \mathbf{N}_o=\mathbf{Q}+\mathbf{S},\quad \mathbf{O}_o=\mathbf{O}.$$

*Numerical values for control matrices*
*P+I* Controller matrices (equation (12))

$$\mathbf{K}_{FB} = \begin{bmatrix} 6{\cdot}37 & -1{\cdot}48 & -2{\cdot}86 & 0{\cdot}00 & -17{\cdot}04 \\ 4{\cdot}81 & 0{\cdot}35 & 0{\cdot}13 & 0{\cdot}00 & 6{\cdot}98 \\ 6{\cdot}42 & 1{\cdot}17 & -0{\cdot}25 & 18{\cdot}11 & 18{\cdot}95 \end{bmatrix}$$

$$\mathbf{K}_I = \begin{bmatrix} 1{\cdot}31 & 0{\cdot}00 & -1{\cdot}60 \\ 1{\cdot}11 & 0{\cdot}00 & 0{\cdot}67 \\ 1{\cdot}54 & 4{\cdot}28 & 1{\cdot}81 \end{bmatrix}$$

Control matrices (see equations 14 and 19)

$$\mathbf{K}_{FB} = \begin{bmatrix} 5{\cdot}10 & -1{\cdot}47 & -2{\cdot}68 & 0 & -14{\cdot}56 \\ 3{\cdot}95 & 0{\cdot}36 & 0{\cdot}21 & 0 & 7{\cdot}39 \\ 5{\cdot}31 & 1{\cdot}19 & -0{\cdot}11 & 15{\cdot}83 & 18{\cdot}81 \end{bmatrix}$$

$$\mathbf{K}_{FF} = \begin{bmatrix} 2{\cdot}05 & -0{\cdot}136 & -0{\cdot}46 \\ 1{\cdot}02 & 0{\cdot}037 & 0 \\ 1{\cdot}14 & 0{\cdot}116 & 0 \end{bmatrix}$$

Direct setpoint control matrices (see equations 14 and 15)

$$\mathbf{K}_{SP}=\begin{bmatrix} -5{\cdot}10 & 0 & 16{\cdot}08 \\ -3{\cdot}95 & 0 & -7{\cdot}77 \\ -5{\cdot}31 & -15{\cdot}83 & -20{\cdot}06 \end{bmatrix}$$

$$\mathbf{E}=\begin{bmatrix} 1{\cdot}0 & 0{\cdot}9\times10^{-7} & -0{\cdot}4\times10^{-3} \\ 0{\cdot}9\times10^{-4} & 1{\cdot}0 & -0{\cdot}2\times10^{-3} \\ -0{\cdot}3\times10^{-3} & -0{\cdot}5\times10^{-7} & 0{\cdot}999 \end{bmatrix}$$

Model following control matrices (see equation 19)
(a) Model time constant of 1 min ($\mathbf{H}=-\mathbf{I}$, $\mathbf{G}=\mathbf{I}$)

$$\mathbf{K}_{M}=\begin{bmatrix} -1{\cdot}15 & 0 & 3{\cdot}68 \\ -1{\cdot}26 & 0 & -2{\cdot}99 \\ -1{\cdot}86 & -5{\cdot}45 & -6{\cdot}80 \end{bmatrix}$$

$$\mathbf{K}_{SP2}=\begin{bmatrix} -3{\cdot}95 & 0 & 12{\cdot}42 \\ -2{\cdot}70 & 0 & -4{\cdot}77 \\ -3{\cdot}45 & -10{\cdot}38 & -13{\cdot}26 \end{bmatrix}$$

(b) Model time constant of 5 min ($\mathbf{H}=-02{\cdot}\mathbf{I}$, $\mathbf{G}=0{\cdot}2\mathbf{I}$)

$$\mathbf{K}_{M}=\begin{bmatrix} -3{\cdot}67 & 0 & 11{\cdot}61 \\ -3{\cdot}12 & 0 & -6{\cdot}51 \\ -4{\cdot}32 & -12{\cdot}79 & -16{\cdot}13 \end{bmatrix}$$

$$\mathbf{K}_{SP2}=\begin{bmatrix} -1{\cdot}43 & 0 & 4{\cdot}48 \\ -0{\cdot}84 & 0 & -1{\cdot}25 \\ 1-{\cdot}00 & -3{\cdot}04 & -3{\cdot}93 \end{bmatrix}$$

## NOMENCLATURE

| | | | *Normal steady-state value* |
|---|---|---|---|
| *Process variables* | | | |
| State Vector, **x**, (dimension $n$) | $W1$ | first effect holdup | 14·6 kg |
| | $C1$ | first effect concentration | 4·85% |
| | $H1$ | first effect enthalpy | 335 kJ/kg |
| | $W2$ | second effect holdup | 15·8 kg |
| | $C2$ | second effect concentration | 9·64% |
| Control Vector, **u**, (dimension $m$) | $S$ | steam flowrate | 0·86 kg/min |
| | $B1$ | first effect bottoms | 1·5 kg/min |
| | $B2$ | second effect bottoms | 0·77 kg/min |
| Load Vector, **d**, | $F$ | feed flowrate | 2·27 kg/min |
| | $CF$ | feed concentration | 3·2% |
| | $HF$ | feed enthalpy | 3·2 kJ/kg |
| Output Vector, **y**, (dimension $s$) | | equal to $(W1, W2, C2)^T$ | |
| Other Variables | $O1$ | overhead vapor from first effect | 0·77 kg/min |
| | $O2$ | overhead vapor from second effect | 0·73 kg/min |
| | $P1$ | pressure in first effect | 68·9 kN/m² |
| | $P2$ | pressure in second effect | 0·38 m Hg |
| | $TF$ | temperature of feed | 88°C |
| | $T1$ | temperature in first effect | 107°C |
| | $T2$ | temperature in second effect | 71°C |
| Subscripts | $FB$ | feedback | |
| | $FF$ | feedforward | |
| | $I$ | integral | |
| | $m$ | model | |
| | $SP$ | setpoint | |
| | $ss$ | steady state | |
| Superscripts | $T$ | matrix/vector transpose | |
| | $i$ | counter for time intervals | |
| Abbreviations | $CC$ | Concentration Controller | |
| | $LC$ | Level Controller | |
| | $FC$ | Flow Controller | |
| | $DDC$ | Direct Digital Control | |
| | $P+I$ | Proportional-plus-Integral | |
| | $ISE$ | Integral of the Square of the Vector | |

**Résumé**—La formulation du problème de commande optimale linéaire, invariable dans le temps et à variables multiples, avec indice quadratique de performance et généralisée de manière à inclure dans sa solution une réaction intégrale à variables multiples et aptitudes à la simulation, en plus de la reaction proportionnelle d'état habituelle. L'action intégrale est réalisée en augmentant le vecteur d'état et élimine les écarts de grandeurs de sortie choisies, provoqués par des grandeurs perturbatrices constantes. L'inclusion d'un vecteur de consigne dans l'index de performance a eu pour résultat des gains rélativement élévés du régulateur et une reponse rapide du procédé à des variations—échelons des valeurs de consigne. Toutefois, la méthode de simulation "réelle" d'un modèle pour réaliser des variations—échelons des valeurs de consigne a donné une plus grande souplesse de calcul et des reponses fiables du prochèdé. L'intérêt pratique et les excellentes performances des systèmes de commande réalisés en utilisant cette approche sont prouvés pour les résultats expérimentaux en provenance d'une installation-pilote d'évaporation commandée par calculateur à l'Universite d'Alberta.

**Zusammenfassung**—Die Formulierung des optimalen, linearen, zeitinvarianten, multivariablen Steuerproblems mit quadratischem Gütekriterium wird erweitert, so daß die Lösung multivariable integrale Rückführung und Modell-Folge-Tätigkeit zusätzlich zu der normalen proportionalen Zustandsrückführung einschleißt. Die integrale Wirkung wird durch die Zunahme des Zustandsvektors Verreicht und eliminiert die bleibenden Regelabweichungen bei ausgewählten Ausgangsvariablen, die zu konstanten Störeingängen gehören. Der Einschluß eines Sollwertvektors im Gütekriterium resultierte in relativ hoher Reglerverstärkung und einem schnellen Ansprechen auf stufenförmige Änderungen in den Sollwerten. Jedoch gaben die "reale" Folgemodell-Approximation für die Ausführung von stufenförmigen Änderungen von Sollwerten größere Flexibilität beim Entwurf und zuverlässige Prozeßreaktionen. Die Durchführbarkeit und ausgezeichnete Wirkungsweise von unter Benutzung dieser Approximation entwickelten Regelungen wird an experimentellen Daten einer rechnergesteuerten Pilotanlage—eines Entdampfers—an der Universität Alberta gezeigt.

**Резюме**.....Формулировка проблемы многокоординатного линейного оптимального управления, инвариантного по времени, с квадратичным индексом работы обобщена с тем чтобы включить в ее решение многокоординатную интегральную обратную связь и способности к слежению модели, сверх обычной пропорциональной обратной связи состояния. Интегральное действие осуществлено увеличением вектора состояния и устраняет отклонения некоторых избранных выходных координат, вызванные постоянными помехами. Включение вектора заданных значений в индекс работы имело результатом сравнительно высокие коэффициенты усиления регулятора и быстрый ответ продесса на ступенчатые изменения заданных значений. Однако, метод ''действительного'' слежения модели для осуществления ступенчатых изменений заданных значений привел к большей гибкости рассчёта и к надёжным ответам процесса. Практический интерес и отличная работа систем управления осуществленных используя этот подход доказываются экспериментальными результатами подученными на испарительной испытательной установке, управляемой вычислительной машиной, в Университете Альберты.

# Correspondence Item

## Discussion of the Paper—Experimental Evaluation of Optimal, Multivariable Regulatory Controllers with Model-Following Capabilities*†
## by R. B. NEWELL and D. G. FISHER

## Discussion de l'Exposé "Evaluation expérimentale de Contrôleurs de Régulation Multivariable Optimale avec Possibilités de Poursuite du Modèle"

## Diskussion des Berichts—Experimentelle Bewertung optimaler, multiveränderlicher Durchführungsregler mit Modellfolgefähigkeiten von R. B. NEWELL und D. G. FISHER

## Рассмотрение работы "Экспериментальная оценка оптимальных, плюривариантных управляющих регуляторов со следящей способностью"

C. D. JOHNSON‡

**Summary**—In the above paper, the authors proposed a modern state-variable type multi-variable regulatory controller involving "proportional-plus-integral" $(P+I)$ feedback and "disturbance feedforward" terms. This correspondence describes some earlier publications, not included in the author's list of "pertinent literature", which treated these same topics.

THE authors of this paper have presented some convincing evidence that modern state-variable and optimal control theory can indeed be an effective tool for solving realistic, practical control problems.

One of the distinguishing features of the authors proposed controller is the use of a state-variable version of "proportional-plus-integral feedback" $(P+I)$ to eliminate the offset-effect caused by unknown constant external disturbances. In regard to this particular feature, the authors remark [first paragraph of their section 2.1] that: "... there were very few papers in the literature, during the period this work was in progress, which dealt with multivariable integral control". The authors then cite six references, their [5–10], which were published in the period 1969–1970 and which apparently represent those "very few, previously published, papers ... which dealt with multivariable integral control". One of the purposes of this correspondence is to call attention to an earlier reference which Newell and Fisher did not cite but which I think is pertinent because it also dealt with the design of state-variable integral controllers by modern control theory. This earlier reference, "Optimal Control of the Linear Regulator with Constant Disturbances", *IEEE Trans. Aut. Control* **AC-13**, 416–421 (1968), discussed the general problem of coping with multi-dimensional external disturbances in modern multi-dimensional state-variable regulator control formulations and described, for the *first* time, I believe, a scheme for modifying the linear-quadratic optimal regulator theory so as to obtain state variable type "proportional-plus-integral controllers", in the case of constant external disturbances. Although the results contained in this earlier reference are illustrated by considering scalar control, it is clearly stated that those results apply also to multi-variable vector control and, in fact, the one additional consideration required for the vector control case is resolved and explicitly given in footnote No. 4, p. 417, of that paper. Moreover, this earlier reference also showed how to obtain modern state variable "proportional-plus-*multiple* integral controllers" by weighting higher-order time derivatives of **u** in the performance index. This yields a controller that simultaneously accommodates not only constant disturbances but disturbances with constant velocity, constant acceleration, constant jerk, etc. as well. Another pertinent reference, which also was not cited by authors Newell and Fisher, is "Further Comments on 'Optimal Control of the Linear Regulator with Constant Disturbances'"; *IEEE Trans. Aut. Control* **AC-15**, 516–518 (1970). This latter reference presented a new and very general expression [see equation (5)] for a multi-variable "proportional-plus-integral" state feedback controller which can accommodate any multi-variable constant external disturbance. Moreover, that new controller is designed by simple pole assignment methods, thus avoiding the more involved Riccati equations associated with quadratic optimization methods.

Another distinguishing characteristic of the controller proposed by Newell and Fisher is the presence of a disturbance "feedforward term", for the special case when the external disturbances are constant and can be accurately measured on-line, indicated in their equation (12). The subject of previously published literature on "modern state-variable feedforward controllers" is not discussed by the authors, except for a reference to the 1971 Ph.D. thesis of Newell, as indicated in the remarks below their equation (8). In this regard, it perhaps should be mentioned that the subject of multi-variable feedforward controllers for state variable systems with measurable disturbances was studied in detail in the 1970 reference: "Optimal Desaturation of Momentum Exchange Control Systems", *Proc.* 1970 *J.A.C.C.* Atlanta, Ga., pp. 683–694 (1970). This earlier paper, which, I believe, was the *first* study of "multi-variable feedforward controllers for state variable systems with measurable disturbances", considered the linear-quadratic-regulator problem with disturbances and showed that if the external disturbances could be measured on-line then the optimal multi-variable control $\mathbf{u}^*(t)$ could be expressed as the following linear function of the instantaneous plant state $\mathbf{x}(t)$ and instantaneous "disturbance state" $\boldsymbol{\gamma}(t)$

$$\mathbf{u}^*(t) = -\mathbf{R}^{-1}\bar{\mathbf{B}}^T[\bar{\mathbf{K}}_{11}\mathbf{x}(t) + \bar{\mathbf{K}}_{12}\boldsymbol{\gamma}(t)],$$

* Recieved 7 August 1972.
† Published in *Automatica* **8**, 247–262 (1972).
‡ Professor, Electrical Engineering Department, University of Alabama in Huntsville, Huntsville, Alabama 35807.

where $T$ – denotes transpose, as shown in equation (4.13) of that reference. Moreover, this feedforward result was shown to hold not only for constant disturbances $\mathbf{w}(t) \equiv \mathbf{c}$, but for *any* multi-dimensional, $r$-vector, measurable disturbance whose waveform can be described by

$$\mathbf{w}(t) = \mathbf{P}\gamma(t) \qquad \mathbf{P} = r \times p \text{ matrix} \qquad (1)$$

$$\dot{\gamma} = \mathbf{D}\gamma \qquad \mathbf{D} = p \times p \text{ matrix.} \qquad (2)$$

Thus, in addition to constants one can accommodate measurable disturbances $w_i(t)$ such as polynomials in time

$$w_i = C_{io} + C_{i1}t + C_{i2}t^2 + \ldots + C_{im_i}t^{m_i}, \qquad (3)$$

exponentials in time, sinusoids in time, etc. etc. In the case of $r$ independent *constant* measurable disturbances, $\{w_1, \ldots, w_r\}$ one has $r = p$, $\mathbf{P} = \mathbf{I}$, $\mathbf{D} = \mathbf{O}$, in (2) above so that $\mathbf{w}(t) = \gamma(t)$, e.g. the *value* of $\mathbf{w}$ is also the "state" of $\mathbf{w}$. Then, the general feedforward controller (1) above specializes to

$$\mathbf{u}^*(t) = -\mathbf{R}^{-1}\bar{\mathbf{B}}^T[\bar{\mathbf{K}}_{11}\mathbf{x}(t) + \bar{\mathbf{K}}_{12}\mathbf{w}(t)] \qquad (4)$$

which, as far as "feedforward terms" is concerned, is exactly the same as the disturbance feedforward control in equations (12) and (14) of Newell and Fisher's paper. On the other hand, in the case of polynomial disturbances (3) the state $\gamma_i$ of $w_i$ consists of $w_i$ *and* its first $(m_i+1)$ time derivatives—all of which must be "fed forward". It is remarked that some of these special cases of the general disturbance feedforward controller (1), originally presented in the 1970 J.A.C.C. reference cited above, have been re-discovered by other investigators in recent publications.

A third distinguishing feature of the regulator controller proposed by Newell and Fisher is the provision for accommodating a non-zero regulation set-point as discussed in their remarks in their section 2.2. They accomplish this feature by minimizing a performance index involving "set-point error" which leads to the addition of a new controller term linearly proportional to the desired set-point according to their equations (13) and (14). This is the traditional approach. However, there is an alternative method for handling *state vector* set-points which appears to have some advantage over that used by Newell and Fisher. Namely one first observes that the set point error $\mathbf{x}_e = \mathbf{x} - \mathbf{x}_d$ obeys the differential equation

$$\dot{\mathbf{x}}_e = \mathbf{A}\mathbf{x}_e + \mathbf{B}\mathbf{u} + \mathbf{A}\mathbf{x}_d. \qquad (5)$$

Then, one agrees at the outset to split the total control $\mathbf{u}(t)$ into two parts

$$\mathbf{u}(t) = \mathbf{u}_p(t) + \mathbf{u}_R(t) \qquad (6)$$

where $\mathbf{u}_p(t)$ is assigned the task of "counteracting" the bias term $\mathbf{A}\mathbf{x}_d$ in (5) and $\mathbf{u}_R(t)$ is responsible for then regulating (5) to the zero error state $\mathbf{x}_e = 0$. The fundamental necessary and sufficient condition for achieving a given set point $\mathbf{x}_d$ is that rank $[\mathbf{B}|\mathbf{A}\mathbf{x}_d]$ = rank$[\mathbf{B}]$ or, in other words, that

$$\mathbf{A}\mathbf{x}_d = \mathbf{B}\zeta(\mathbf{x}_d) \qquad (7)$$

for some vector $\varsigma(\mathbf{x}_d)$. If $\varsigma$ exists and rank $\mathbf{B}$ = maximum, then $\varsigma$ is given uniquely and explicitly by the simple expression

$$\zeta = (\mathbf{B}^T\mathbf{B})^{-1}\mathbf{B}^T\mathbf{A}\mathbf{x}_d. \qquad (8)$$

Otherwise $\varsigma$ has the same simple linear form as (8) but it is non-unique. In either case, if one chooses $\mathbf{u}_p$ in (6) to be

$$\mathbf{u}_p = -\zeta \qquad (9)$$

then (5) becomes

$$\dot{\mathbf{x}}_e = \mathbf{A}\mathbf{x}_e + \mathbf{B}\mathbf{u}_R. \qquad (10)$$

The remaining component $\mathbf{u}_R = \mathbf{K}\mathbf{x}_e$ can now be designed to achieve $\mathbf{x}_e \to \mathbf{0}$, as usual, by employing either pole-assignment or quadratic regulator theory on (10), where now one uses

$$\mathbf{u}_R^T\mathbf{R}\mathbf{u}_R$$

in the performance index, rather that $\mathbf{u}^T\mathbf{R}\mathbf{u}$. The advantages of this alternative approach are: (i) it *does not* require the traditionally annoying "... re-calculation of the control action for every setpoint change", as described below equation (14) of Newell and Fisher's paper, and (ii) it yields a explicit controller (6), (9) which is exceptionally easy to calculate and which always achieves *exact* set-point regulation $\mathbf{x} = \mathbf{x}_d$ *even* if $\mathbf{x}_d$ is not a natural equilibrium state for the process. The alternative approach just outlined was presented in section VI of the paper "Accommodation of External Disturbances in Linear Regulator and Servo-mechanism Problems", *IEEE Trans. Aut. Control* **AC-16**, 635–644 (1971). This latter reference also contains a variety of schemes for coping with complex external disturbances in multi-variable control problems.

In summary, authors Newell and Fisher have done an impressive job of experimentally evaluating the performance of multi-variable controllers having integral action, feed-forward control, and non-zero set-point capability. We need more application papers like theirs. However, the "formulation extension" claims made in their Summary and Introduction are a little misleading. In particular, the "state variable integral feedback" and "disturbance feed forward" techniques used by the authors are not new and the earlier papers which originally treated those topics are not included in the authors' list of "pertinent literature".

**Résumé**—Dans l'exposé ci-dessus, les auteurs ont proposé un contrôleur de régulation multivariable du type à état variable impliquant des termes de réctroaction "proportionelle plus intégrale" (P+I) et "action advancée de perturbation". Cette correspondance décrit des publications passées, non comprises dans la liste de "littérature pertinente" de l'auteur et qui traitait des mêmes sujets.

**Zusämmenfässung**—In dem obigen Bericht schlugen die Verfasser einen modernen, multiveränderlichen Durchführungsregler des veränderlichen Zustandtyps vor, der "proportionale-plus-integrale" ($P+I$) Rückkoppelung und "Störgrössenaufschaltung" Termen umfasst. Diese Korrespondenz beschreibt einige frühere Veröfffentlichungen, die nicht in der "einschlägigen Literatur" Liste des Verfassers enthalten waren, welche die gleichen Themen behandelte.

**Резюме**—В вышеприведенной работе авторы предлагают на рассмотрение современный плюривариантный управляющий регулятор, включающий "пропорциональную-плюс-интегральную" ($P+I$) обратную связь и "прямую связь возмущения". В настоящей работе описываются более ранние издания не включенные в авторский список "Относящейся литературы" в которой обсуждались подобные же темы.

# Computer Control Using Optimal Multivariable Feedforward-Feedback Algorithms

This paper presents three methods of designing feedforward compensators which can be combined with multivariable feedback controllers in order to minimize or eliminate errors caused by sustained measurable disturbances. The designs are based on a linear, time-invariant state-space model of the process and minimize a quadratic function of the errors and/or constrain selected steady state offsets to zero. Simulated and experimental data from a computer controlled pilot-plant evaporator show that multivariable feedback-plus-feedforward control is relatively simple to implement, is practical, and gives excellent control.

ROBERT B. NEWELL
D. GRANT FISHER
DALE E. SEBORG

Department of Chemical and Petroleum Engineering
University of Alberta
Edmonton, Alberta, Canada

## SCOPE

The justification for improved control of industrial processes normally lies in some combination of increased throughput, improved efficiency, better product quality, safer operation, and reduced pollution. Improved control is generally obtained by adding additional conventional controllers or using more sophisticated designs. Most companies have sufficient experience with single variable control systems so that the question of determining the optimum number is a relatively straightforward task of engineering design and justification. However, the use of more sophisticated multivariable controllers has generally lagged because of the lack of convenient design procedures, difficulties regarding implementation, or lack of convincing demonstrations that the theoretical methods will work on real applications. Fortunately progress is being made in overcoming all these difficulties.

In particular, industry is installing increasing numbers of real-time digital control computers which essentially eliminate the physical restrictions formerly imposed by lack of suitable control hardware. Furthermore, since the central processing unit (CPU) utilization on many of these installations is often as low as 5%, the incremental cost of replacing a system of conventional single variable control algorithms where, for example,

$$\mathbf{u}(t) = \mathbf{K}_{FB}\mathbf{x}(t) \quad \mathbf{K}_{FB} = \text{diagonal} \tag{1}$$

by more sophisticated multivariable controllers is often negligible in terms of equipment required.

The work described in this paper is part of a continuing series of projects at the University of Alberta directed towards developing control techniques that are of interest to industry and evaluating them by experimental implementation on computer controlled pilot plants. Because of their relative simplicity and practicality, attention was first directed towards developing multivariable control laws similar to Equation (1) but without the restriction that $\mathbf{K}_{FB}$ be diagonal. Several methods for determining such a $\mathbf{K}_{FB}$ are referenced later but the method employed in this project was based on a linear, discrete, time-invariant state-space model of the process and a quadratic performance index summed over an infinite interval. The well-known multivariable proportional feedback control law which results from this approach was extended by Newell and Fisher (1972a) to include integral action and provision for making step changes in setpoints or applying model following techniques. This paper describes and evaluates the addition of multivariable feedforward control.

Feedforward control is of considerable interest because, in contrast to feedback control, it can start to compensate for process disturbances before any change occurs in the controlled variables. In certain situations feedforward controllers can completely compensate for disturbances and hence produce perfect regulatory control. However, in practice feedforward controllers are usually combined with feedback action which compensates for modelling errors, controller simplifications, unmeasured disturbances, etc.

Experimental studies by Jacobson (1970) and Newell (1971) showed that multivariable proportional control gave excellent results when applied to a pilot plant evaporator. However, small offsets (deviations) occurred between the actual and desired values of the process variables when constant disturbances were applied to the process. These offsets could be eliminated by integral action, but this increased the dimension of the state vector and the resulting controller. In this study feedforward control is added to the feedback controller as a means of minimizing a quadratic function of the deviations produced by constant, measurable disturbances and/or eliminating a subset of the steady state offsets.

## CONCLUSIONS AND SIGNIFICANCE

This work and related projects show that multivariable optimal control techniques are simple and practical to implement and, when applied to a pilot plant evaporator, gave much better control than was ever achieved using conventional, single variable controllers. The figures in this paper reproduce data taken directly from the process by the computer and show that the concentration of the

Correspondence concerning this paper should be addressed to D. G. Fisher. R. B. Newell is with Shell Internationale Petroleum Maatschappij Postbus 142, The Hague, Netherlands.

AIChE Journal, Vol. 18, No. 5, pp. 976-984 (1972).

**product stream from the evaporator was essentially constant in spite of sustained disturbances in feed flow (20%), feed temperature (16%), or feed concentration (30%). Typical responses of the evaporator product concentration to similar disturbances when the evaporator was controlled by conventional single variable controllers, show maximum deviations of over 20% (of mean value) and oscillations that persist for more than two hours.**

**More specifically it was shown that for a system with *m* inputs subjected to constant measurable disturbances, feedforward control could be combined with a multivariable feedback controller to:**

**1. eliminate steady state offsets in up to *m* variables, and/or**

**2. minimize the steady state offsets in some subset of the state variables, or**

**3. minimize a summed quadratic function of the error between the actual and desired values of the system state variables.**

**It was shown that all three of the above design objectives, plus some of the methods presented in the literature, could be described by a single, generalized interpretation of the optimal control problem for linear, discrete time-invariant processes with quadratic performance indices. This same approach can be used to generate a multivariable control law containing feedback, feedforward, integral, and model following options, and therefore includes all the common options familiar to users of single variable controllers.**

**The simulation results confirmed that the feedforward controllers meet their design objectives. In the experimental runs the improvement due to adding feedforward control was significant, but process noise made it impossible to distinguish between the different feedforward designs. It should be noted that the controllers were designed on the basis of a linear model which contained significant modelling errors, were implemented using only the simplest filtering and state estimation techniques, and were operated in conjunction with an industrial type direct-digital-control (DDC) system, plus other applications, on a multiprogrammed real-time digital control computer. In spite of these factors they gave excellent control and hence would appear to be robust and practical.**

---

## PREVIOUS WORK

During the last decade the design of feedforward controllers for use with conventional single variable control systems has received considerable attention. Miller et al. (1969) in their tutorial article summarize most of the important results and design techniques available at that time. Later investigations include both theoretical and experimental work. Bertran and Chang (1970) studied optimal feedforward control of nonlinear chemical reactors and included distributed parameter terms in their model. Paraskos and McAvoy (1970) used a finite difference approach to develop both transient and steady state feedforward compensation for a class of distributed parameter processes and successfully applied it in experimental studies of a heat exchanger. Feedforward control has also been applied to a laboratory heat exchanger by Corlis and Luus (1969, 1970) and to a stirred tank reactor by West and McGuire (1969b).

By contrast, multivariable feedforward control has received relatively little attention and experimental verification of proposed multivariable control systems has seldom been reported. The transfer function or frequency domain approach was investigated by Bollinger and Lamb (1965), Foster and Stevens (1967), and Luyben (1969). When working with mathematical models it is possible, as shown by Haskins and Sliepcevich (1965) and others, to compensate exactly for load disturbances. However, the required controllers are often complex, sometimes physically unrealizable, and are usually only approximated in physical applications. Greenfield and Ward (1968) also studied feedforward control and point out that the transfer function approach usually requires that the number of control variables must equal the number of outputs. Optimal feedforward-feedback controllers have also been designed using linear state-space process models and minimizing a quadratic performance index of state and control variables. West and McGuire (1969b) used continuous process models, Anderson (1969) used a discrete model, and Solheim and Saethre (1968) considered both types of models. Stochastic disturbances have also been considered by Clifton and McGuire (1971). Heidemann and Esterson (1967) used feedforward action based on the steady state model to reduce offsets. Anderson (1969) presented a design technique based on an error coordinate transformation. Assuming constant disturbances, the system was transformed into the standard optimal regulator problem without disturbances, which could then be solved by existing techniques. The resulting control law, when transformed out of error coordinates, included feedforward action which essentially ensured that selected outputs were at their desired values when the process reached steady state. Latour (1971) has recently shown that for constant disturbances multivariable proportional-integral control is optimal for a quadratic performance index which penalizes the rate of change of the control vector and, of course, eliminates steady state offsets.

Johnson (1968, 1970a, b) has considered the design of disturbance-absorbing controllers. This design approach is based on optimal control theory and seeks to counteract the effects of disturbances by generating a control action which produces a compensating change in the state variables. The disturbance need not be measured but must satisfy a linear differential equation. Sobral and Stefanek (1970) have demonstrated that for certain classes of disturbances, the optimal control formulation reduces to the well-known formulation for systems without disturbances.

This paper considers the problem of designing feedforward controllers that can be added to existing multivariable feedback systems to minimize the effects of measurable, sustained disturbances.

## DESIGN BASIS

Consider the standard state-space model of the form

$$\dot{\mathbf{x}}(t) = \mathbf{A}\,\mathbf{x}(t) + \mathbf{B}\,\mathbf{u}(t) + \mathbf{D}\,\mathbf{d}(t) \quad \mathbf{x}(0) = \mathbf{x}_0 \qquad (2)$$
$$\mathbf{y}(t) = \mathbf{C}\,\mathbf{x}(t)$$

Several of the design techniques mentioned in the previous sections can be used to generate a multivariable control system for a process defined by these equations. In most cases a suitable control law can be formulated in terms of either the $r$ output variables $\mathbf{y}$ or the $n$ state variables $\mathbf{x}$. In this paper the formulation will be in terms of state variables because the optimal control design technique leads naturally to state feedback and because in the evaporator application, to be presented later, the output variables are simply a subset of the state variables. The equivalent formulations in terms of output variables can usually be derived by following the same procedures, and notes to this effect are included in the following sections. Also for simplicity it is assumed that $\mathbf{x}_0 = \mathbf{x}_d = \mathbf{0}$. The case where $\mathbf{x}_d$ is nonzero has been treated by Newell and Fisher (1972a).

Assume that a basic feedback control law with the form given by Equation (3) has already been developed

$$\mathbf{u}(t) = \mathbf{K}_{FB}\,\mathbf{x}(t) \tag{3}$$

This control law will frequently give satisfactory control when used alone. However, consider the case where constant sustained disturbances are applied to a process. If the process is described by Equation (2) and the control action by Equation (3), then when the process finally comes to steady state, $\dot{\mathbf{x}} = \mathbf{0}$ and

$$\mathbf{x}_s = -\ (\mathbf{A} + \mathbf{B}\,\mathbf{K}_{FB})^{-1}\,\mathbf{D}\,\mathbf{d} \tag{4}$$

Since the desired value of the state is zero in this formulation of the regulator control problem, it is obvious from Equation (4 )that constant disturbances will cause an offset in $\mathbf{x}$. In many applications such offsets are undesirable and a design procedure that would eliminate and/or minimize such offsets would be advantageous. Also it is reasonable to expect that if the disturbances are measurable then control could be improved over the entire time period of the transient by adding a feedforward term to Equation (3). The following discussion indicates one way of achieving these design objectives.

Assume that zero offsets are required in $q$ of the state variables. The number of degrees of freedom in a control system is limited by the number of control variables so that in general $q \leq m$. It is then convenient to partition $\mathbf{x}$ and $\mathbf{u}$ as follows

$$\mathbf{x} = \begin{bmatrix} \mathbf{x}_1 \\ \mathbf{x}_2 \end{bmatrix} \qquad \mathbf{u} = \begin{bmatrix} \mathbf{u}_1 \\ \mathbf{u}_2 \end{bmatrix} \tag{5}$$

where $\mathbf{x}_1$ and $\mathbf{u}_1$ both have dimension $q$. Assuming that it is desired to utilize all $m$ elements of $\mathbf{u}$ in the feedback control action implied by Equation (3), then a generalized design procedure can be stated as:

1. Design the $m \times n$ feedback control matrix $\mathbf{K}_{FB}$
2. Utilize $\mathbf{u}_1$ to produce zero offsets in the $q$ elements of $\mathbf{x}_1$ (note that the case where $q = 0$ is allowed)
3. Utilize $\mathbf{u}_2$ to minimize the steady state offsets in some or all of the remaining $(n - q)$ state variables.

Alternatively, in place of the above three steps:

4. Utilize $\mathbf{u}$ to minimize a quadratic function of the error $(\mathbf{x} - \mathbf{x}_d)$ over the entire transient and hence define the feedforward and feedback action simultaneously. (Note, if a subset of $\mathbf{u}$, say $\mathbf{u}_2$ is used in this step then the feedback action will only include $\mathbf{u}_2$ and this is generally not desirable.)

The method used to design the feelback control matrix $\mathbf{K}_{FB}$ in step 1 is arbitrary and could, for example, be based on standard optimal control techniques as presented in textbooks, for example, Ogata (1967) or Lapidus and Luus (1967); design methods for noninteraction, for example, Gilbert (1969); frequency domain design techniques, for example, Rosenbrock (1970); or eigenvalue assignment techniques, for example, Porter (1969) or Wonham (1967). The optimal control approach is illustrated in a later section.

## DESIGN TECHNIQUES FOR FEEDFORWARD CONTROLLERS

The following subsections will follow the general design technique presented in the previous section and develop the design equations for three feedforward controllers, each of which has a different design objective. The designs are all based on a discrete state-space model of the process. If $\mathbf{u}$ and $\mathbf{d}$ are constant during each sampling interval, $iT \leq t \leq (i+1)T$, then the discrete model defined by Equation (6) is equivalent at the sampling instants to the continuous model defined by Equation (2):

$$\mathbf{x}((i+1)T) = \boldsymbol{\varphi}\mathbf{x}(iT) + \boldsymbol{\Delta}\mathbf{u}(iT) + \boldsymbol{\theta}\mathbf{d}(iT), \qquad i = 0, 1, 2, \ldots \tag{6}$$

$$\mathbf{y}(iT) = \mathbf{C}\,\mathbf{x}(iT)$$

where the constant coefficient matrices $\boldsymbol{\varphi}$, $\boldsymbol{\Delta}$, and $\boldsymbol{\theta}$ are evaluated from

$$\boldsymbol{\varphi} = \int_0^T e^{\mathbf{A}t}\,dt \tag{7}$$

$$\boldsymbol{\Delta} = \int_0^T e^{\mathbf{A}(T-t)}\,\mathbf{B}dt \tag{8}$$

$$\boldsymbol{\theta} = \int_0^T e^{\mathbf{A}(T-t)}\,\mathbf{D}dt \tag{9}$$

### Design for Zero Steady State Offsets

The first objective, zero offset in the $q$ elements of $\mathbf{x}$ which make up $\mathbf{x}_1$, can be achieved as follows. Assume that the desired control law has the form

$$\mathbf{u}(iT) = \mathbf{K}_{FB}\,\mathbf{x}(iT) + \mathbf{u}_{FF}(iT) \tag{10}$$

and that $\mathbf{K}_{FB}$ has already been determined. The feedforward control action $\mathbf{u}_{FF}$ which will compensate for the steady state offsets caused by constant disturbances, can then be calculated. Substitution of Equation (10) into Equation (6) gives the closed-loop system response

$$\mathbf{x}((i+1)T) = (\boldsymbol{\varphi} + \boldsymbol{\Delta}\,\mathbf{K}_{FB})\,\mathbf{x}(iT) + \boldsymbol{\Delta}\,\mathbf{u}_{FF}(iT) + \boldsymbol{\theta}\,\mathbf{d}(iT) \tag{11}$$

If the system is at steady state such that $\mathbf{x}((i+1)T) = \mathbf{x}(iT)$ and $\mathbf{T}$ is defined by

$$\mathbf{T} = (\boldsymbol{\varphi} + \boldsymbol{\Delta}\,\mathbf{K}_{FB}) - \mathbf{I} \tag{12}$$

then Equation (11) can be written in partitioned form as defined in Equation (5) to yield

$$\mathbf{0} = [\mathbf{T}_1 \mid \mathbf{T}_2] \begin{bmatrix} \mathbf{x}_1(iT) \\ \mathbf{x}_2(iT) \end{bmatrix} + [\boldsymbol{\Delta}_1, \boldsymbol{\Delta}_2] \begin{bmatrix} \mathbf{u}_{1FF}(iT) \\ \mathbf{u}_{2FF}(iT) \end{bmatrix} + \boldsymbol{\theta}\,\mathbf{d} \tag{13}$$

If $\mathbf{x}_1(iT)$ is constrained to equal the desired values $\mathbf{x}_{1d}$, then Equation (13) can be rearranged to

$$[\boldsymbol{\Delta}_1 \mid \mathbf{T}_2] \begin{bmatrix} \mathbf{u}_{1FF} \\ \mathbf{x}_2 \end{bmatrix} = -\ \boldsymbol{\Delta}_2\,\mathbf{u}_{2FF} - \mathbf{T}_1\,\mathbf{x}_{1d} - \boldsymbol{\theta}\,\mathbf{d} \tag{14}$$

Solving Equation (14) for $\mathbf{u}_{1FF}$ and $\mathbf{x}_2$ gives an expression which can be partitioned and put in the form

$$\mathbf{u}_{1FF}(iT) = \mathbf{K}_1\,\mathbf{u}_{2FF}(iT) + \mathbf{K}_2\,\mathbf{x}_{1d}(iT) + \mathbf{K}_3\,\mathbf{d} \tag{15}$$

In the above derivation it has been assumed that $[\boldsymbol{\Delta}_1, \mathbf{T}_2]^{-1}$ exists and that the disturbance vector $\mathbf{d}$ is constant. However, in practice the resulting feedforward-feedback control law in Equation (15) can be used when the disturbances vary with time providing that the disturbances are measurable. If Equation (15) is substituted into Equation (11), the dynamic response of the controlled system is given by

$$\mathbf{x}((i+1)T) = (\boldsymbol{\varphi} + \boldsymbol{\Delta}\,\mathbf{K}_{FB})\,\mathbf{x}(iT) + (\boldsymbol{\Delta}_1\,\mathbf{K}_1 + \boldsymbol{\Delta}_2)\,\mathbf{u}_{2FF}(iT) + (\boldsymbol{\Delta}_1\,\mathbf{K}_3 + \boldsymbol{\theta})\,\mathbf{d}(iT) + \boldsymbol{\Delta}_1\,\mathbf{K}_2\,\mathbf{x}_{1d} \quad (16)$$

To complete the control system design, $\mathbf{u}_{2FF}$ must be specified. As indicated in the general design procedure, if the dimension of $\mathbf{u}_{2FF}$ is nonzero it can be selected to minimize a function of the offset in $\mathbf{x}_2$. This approach will be examined in the next section.

Thus the control law given in Equation (15) will generate zero offsets in the $q$ state variables forming $\mathbf{x}_1$. It has been shown by Newell (1971) that the "error coordinate" approach of Anderson (1969) is equivalent to the control law given by Equation (15) for the special case where $q = m$ and $\mathbf{K}_{FB}$ is selected to be the optimal feedback matrix obtained using a quadratic performance index.

#### Minimization of Steady State Offsets

An expression for $\mathbf{u}_{FF}$ in Equation (10) or $\mathbf{u}_{2FF}$ in Equation (15) can be obtained by minimizing a weighted sum-of-the-squared-offsets resulting from a disturbance. Define the performance index $J$ as

$$J = \mathbf{e}_s^T\,\mathbf{S}\,\mathbf{e}_s \quad (17)$$

where the steady state error $\mathbf{e}_s$ is defined as $[\mathbf{x}(NT) - \mathbf{x}_d]$. The following expression for $\mathbf{x}(NT)$ is easily derived from Equations (10) and (11) if it is assumed that the disturbance $\mathbf{d}$ is constant

$$\mathbf{x}(NT) = +\,(\mathbf{I} - \boldsymbol{\varphi} - \boldsymbol{\Delta}\,\mathbf{K}_{FB})^{-1}\,(\boldsymbol{\Delta}\,\mathbf{u}_{FF} + \boldsymbol{\theta}\,\mathbf{d}) \quad (18)$$

To determine the value of $\mathbf{u}_{FF}$ which minimizes $J$ set

$$\frac{\partial J}{\partial \mathbf{u}_{FF}} = \frac{\partial}{\partial \mathbf{u}_{FF}}\,(\mathbf{e}^T\,\mathbf{S}\,\mathbf{e}) = 0 \quad (19)$$

Solving for $\mathbf{u}_{FF}$ for the case where $\mathbf{x}_d = \mathbf{0}$ gives

$$\mathbf{u}_{FF} = -\,[(\boldsymbol{\Delta}^T\,\boldsymbol{\varphi}^{\circ T}\,\mathbf{S}\,\boldsymbol{\varphi}^{\circ}\,\boldsymbol{\Delta})^{-1}\,\boldsymbol{\Delta}^T\,\boldsymbol{\varphi}^{\circ T}\,\mathbf{S}\,\boldsymbol{\varphi}^{\circ}\,\boldsymbol{\theta}]\cdot\mathbf{d} \quad (20)$$

where

$$\boldsymbol{\varphi}^{\circ} = +\,(\mathbf{I} - \boldsymbol{\varphi} - \boldsymbol{\Delta}\,\mathbf{K}_{FB})^{-1} \quad (21)$$

For the special case where the state and control vectors have the same dimension (that is, $n = m$) and $\boldsymbol{\Delta}$ is nonsingular then Equation (20) reduces to

$$\mathbf{u}_{FF} = -\,\boldsymbol{\Delta}^{-1}\,\boldsymbol{\theta}\,\mathbf{d} \quad (22)$$

For this special case it is obvious from an inspection of Equation (6) or (18) that $\mathbf{u}_{FF}$ gives perfect dynamic compensation for the effect of the disturbances as well as producing zero offsets.

#### Optimal Feedforward Plus Feedback Controller

The optimal control formulation for determining both the feedforward and feedback control matrices has been studied by Solheim and Saethre (1968), West and McGuire (1969b), and Sobral and Stefanek (1970). In the discrete formulation the process is modeled by a state-space equation in the form of Equation (6) and the quadratic performance index to be minimized is normally of the form

$$J = \mathbf{e}^T(NT)\,\mathbf{S}\,\mathbf{e}(NT) + \sum_{i=1}^{N} \mathbf{e}^T(iT)\,\mathbf{Q}\,\mathbf{e}(iT) + \mathbf{u}^T((i-1)T)\,\mathbf{R}\,\mathbf{u}((i-1)T) \quad (23)$$

Assuming that the disturbances are constant over the interval $0 \leq t \leq NT$ and that the desired operating point is at the origin (that is, that $\mathbf{x}_d = \mathbf{0}$), then the application of discrete dynamic programming techniques results in a control law of the form:

$$\mathbf{u}(iT) = \mathbf{K}_{FB}\,\mathbf{x}(iT) + \mathbf{K}_{FF}\,\mathbf{d}(iT) \quad (24)$$

where the optimal control matrices are defined by the recursive relations given in the Notation section. These recursive relations were found to converge quickly to produce constant control matrices and for the case where the disturbances are neglected (that is, where $\mathbf{d} = \mathbf{0}$), then $\mathbf{K}_{FB}$ is the optimal feedback control matrix. The case where the desired values $\mathbf{x}_d$ (or $\mathbf{y}_d$) are nonzero has been treated by Newell and Fisher (1972a).

A performance index of the same form as Equation (23) can be defined with the error defined in terms of the output variables, (that is, with $\mathbf{e} = \mathbf{y} - \mathbf{y}_d$). However, if $\mathbf{y} = \mathbf{C}\,\mathbf{x}$ then this formulation is the same as that given by Equation (23) with $\mathbf{S}$ and $\mathbf{Q}$ replaced by $\mathbf{C}^T\mathbf{S}\mathbf{C}$ and $\mathbf{C}^T\mathbf{Q}\mathbf{C}$ respectively and the same control law applies. Also it is obvious that if $\mathbf{Q}$ and $\mathbf{R}$ are null matrices that the performance index given by Equation (23) reduces to that given by Equation (17). Thus the design of a feedforward controller to minimize the steady state offsets can be treated as a special case of this more general problem.

Similarly the specification of zero steady state offsets could be considered as simply a constraint on the final values of $q$ of the state variables. Thus all three design objectives treated in this paper could be covered by an optimal control formulation similar to that presented in this section. However, the recursive relations given in this paper for $\mathbf{K}_{FB}$ and $\mathbf{K}_{FF}$ do not include provision for constraints on $\mathbf{x}$, $\mathbf{y}$, or $\mathbf{u}$.

## SIMULATED AND EXPERIMENTAL APPLICATIONS

The following four design objectives, which can all be obtained by adding feedforward control to a multivariable feedback system, were examined in this study.

1. design for zero offsets in $W1$, $W2$, and $C2$
2. design for a zero offset in $C2$ and to minimize offsets in $W1$ and $W2$
3. design to minimize offsets in $W1$, $W2$ and $C2$
4. design both $\mathbf{K}_{FF}$ and $\mathbf{K}_{FB}$ to minimize the quadratic performance index given by Equation (23).

The feedback control matrix used in all four cases was evaluated from the recursive relationships in the notation section with the following parameters:

$$\mathbf{Q} = \text{diag}\,(10,\ 1,\ 1,\ 10,\ 100) \qquad \mathbf{R} = \mathbf{0},\quad \mathbf{S} = \mathbf{0},\quad T = 64 \text{ sec.} \quad (25)$$

Previous work by the authors indicated that these values were a satisfactory combination for this application. The feedforward control matrices evaluated in the four cases are very similar and are shown in Table 1 for cases 1 and 4.

The simulated and experimental applications of the feedforward controllers were carried out using a state-

TABLE 1. CONTROL MATRICES

$$\mathbf{K}_{FB} = \begin{bmatrix} 5.095 & -1.475 & -2.68 & 0 & -14.56 \\ 3.95 & 0.36 & 0.21 & 0 & 7.39 \\ 5.31 & 1.19 & -0.11 & 15.83 & 18.81 \end{bmatrix}$$

$\mathbf{K}_{FF}$ Case 1

$$\begin{bmatrix} 2.055 & -0.090 & -0.463 \\ 1.014 & 0.032 & 0 \\ 1.122 & 0.091 & 0 \end{bmatrix}$$

$\mathbf{K}_{FF}$ Case 4

$$\begin{bmatrix} 2.047 & -0.136 & -0.463 \\ 1.019 & 0.037 & 0 \\ 1.135 & 0.116 & 0 \end{bmatrix}$$

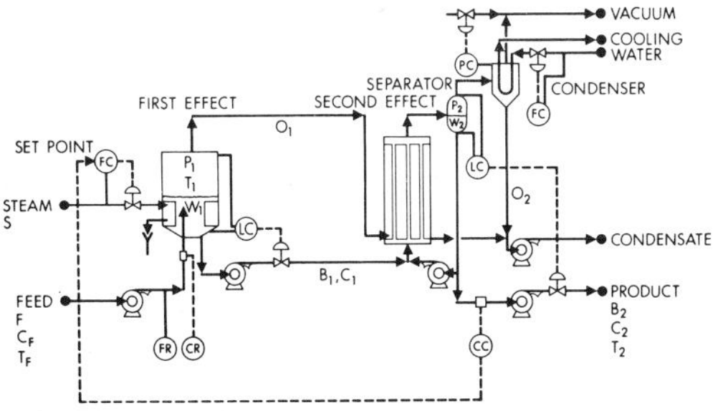


**Fig. 1. Schematic diagram of the double effect, computer controlled, pilot plant evaporator that was used as the basis for all simulated and experimental work. The instrumentation is shown connected as conventional single variable controllers.**

space model of a pilot-plant, double-effect evaporator in the Department of Chemical and Petroleum Engineering at the University of Alberta, which is described in the next section.

**Double Effect Evaporator and Model**

A schematic flow diagram of the evaporator in a double effect "forward feed" mode of operation is shown in Figure 1. The first effect is a calandria type unit with an 8-in. diam. tube bundle. It operates with a nominal feedrate of 2.27 kg/min of 3% aqueous triethylene glycol. The second effect is a long-tube-vertical unit with three $0.0254 \times 1.83$m tubes and is operated with externally forced circulation. The second effect is operated under vacuum and utilizes the vapor from the first effect as the heating medium. The product is about 10% glycol when the steam to the first effect is at its nominal flowrate of 0.9 kg./min.

The state-space model of the evaporator is fifth order and was obtained by linearizing material and energy balances. The process variables in the model are in the normalized perturbation form, for example $(x_i - x_{is})/x_{is}$. The model equations are presented in full in Equation (26). The output vector y is defined to be a subset of the state vector, namely the holdups in the first and second effects $W1$ and $W2$ and the product concentration $C2$. Normal steady state values of the process variable are also given in the notation section.

The multivariable control systems were implemented using an IBM 1800 digital control computer which is interfaced with the pilot-plant double-effect evaporator. The process runs under Direct Digital Control (DDC) and a time-sharing executive system which permits simultaneous execution of off-line jobs.

Multivariable control calculations are carried out by a queued Fortran program which executes every control interval. The basic difference in the computations required for the multivariable versus single variable proportional control algorithms to control the evaporator was 15 multiplications instead of 3. However, the program also does some extra operations such as state estimation and storing desired data on disk. It obtains state variable measurements from DDC data acquisition loops and makes control variable changes by adjusting the set-points of DDC flow control loops.

Additional information on the evaporator, the model, the design procedure, and the implementation are available in the thesis by Newell (1971).

$$\begin{bmatrix} \dot{W1} \\ \dot{C1} \\ \dot{H1} \\ \dot{W2} \\ \dot{C2} \end{bmatrix} = \begin{bmatrix} 0 & -.00156 & -.1711 & 0 & 0 \\ 0 & -.1419 & .1711 & 0 & 0 \\ 0 & -.00875 & -1.102 & 0 & 0 \\ 0 & -.00128 & -.1489 & 0 & .00013 \\ 0 & .0605 & .1489 & 0 & -.0591 \end{bmatrix} \begin{bmatrix} W1 \\ C1 \\ H1 \\ W2 \\ C2 \end{bmatrix} + \begin{bmatrix} 0 & -.143 & 0 \\ 0 & 0 & 0 \\ .392 & 0 & 0 \\ 0 & .108 & -.0592 \\ 0 & -.0486 & 0 \end{bmatrix} \begin{bmatrix} S \\ B1 \\ B2 \end{bmatrix} + \begin{bmatrix} .2174 & 0 & 0 \\ -.074 & .1434 & 0 \\ -.036 & 0 & .1814 \\ 0 & 0 & 0 \\ 0 & 0 & 0 \end{bmatrix} \begin{bmatrix} F \\ CF \\ HF \end{bmatrix} \qquad (26)$$

**Simulation Results**

Simulated results are exemplified by Figure 2 where multivariable feedforward-feedback control (case 4) is compared to multivariable feedback alone for a 10% increase in feed flow rate. The oscillations in $W2$ resulted from simulated roundoff in the measured values of the state variables and the expanded scale on Figure 2. However, it will be noted that in the run with feedforward action, the response of the input variables $B1$ and $S$ was faster and stronger than in the run using only feedback control with the result that the maximum deviations in the controlled variables $C2$, $W1$, and $W2$ were smaller and the steady state offsets negligible.

Comparative results for the four cases are also shown in Table 2 for a 50% increase in feed flow rate. In all cases the response exhibited a very small overshoot in the initial transient. As would be expected, the steady state design technique for zero offset (case 1) gave the smallest average offset, while those for the minimization techniques were larger.

Practically speaking, the resulting feedforward control matrices and transient responses were very similar in the four cases examined. However, the addition of feedforward action resulted in a significant increase in control quality over the feedback only case as illustrated by the results in Figure 2. The zero offset design technique involved the least computation although the dynamic programming design can be carried out in conjunction with the evaluation of the proportional feedback matrix.

A simulation study using a distillation column model was carried out by Newell (1971) and led to essentially the same conclusions.

TABLE 2. COMPARISON OF CONTROL SYSTEMS*
(50% Step Increase in Feed Flow Rate)

| Case | W1 Offset | $W1_{max}$ | W2 Offset | $W2_{max}$ | C2 Offset | $C2_{max}$ |
|---|---|---|---|---|---|---|
| Feedback | 15.5 | 15.5 | 0.11 | 0.25 | −1.45 | −1.45 |
| FF-FB (1) | 0.0002 | 1.22 | 0.0009 | 0.002 | −0.0002 | −0.38 |
| FF-FB (2) | 0.015 | 1.22 | 0.0069 | 0.014 | −0.0093 | −0.38 |
| FF-FB (3) | 0.00003 | 1.19 | 0.00001 | 0.00002 | −0.031 | −0.40 |
| FF-FB (4) | 0.014 | 1.19 | $10^{-5}$ | $10^{-5}$ | −0.028 | −0.40 |

* Note: Offsets and maximum values are expressed as percentage of steady state value.

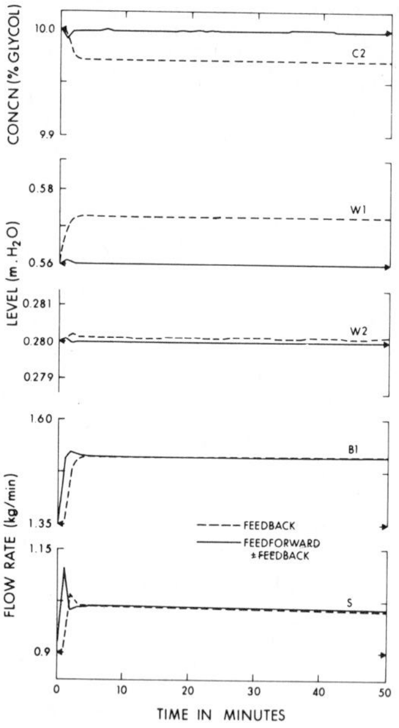


**Fig. 2. Comparison of simulated evaporator responses, for a 10% increase in feed flow, using multivariable feedback and feedback-plus-feedforward controllers (case 4). The arrowheads on the vertical axis define the initial steady state of the evaporator.**

**Experimental Results**

The offsets which result from load changes when multivariable proportional feedback control is implemented are exemplified by the first effect holdup ($W1$) curve in Figure 3. The two 20% feed flowrate changes, an increase and then a decrease, produced significant offsets in first effect holdup ($W1$) and very small offsets in product concentration ($C2$) and second effect holdup ($W2$). However, the essentially constant product concentration C2 represents a dramatic improvement over the best responses obtained from conventional single variable control using experimentally tuned DDC algorithms. Results presented by Jacobson (1970) and Fisher and Jacobson (1971) showed that a typical response to a 20% feed disturbance when the evaporator was controlled by the conventional single variable controllers shown schematically in Figure 1 involved a maximum percentage deviation in $C2$ of approximately 20% and oscillations that persisted for over two hours. [A typical concentration response is also included in the publications by Newell and Fisher (1972a, b).]

The addition of feedforward control (case 4) was found to reduce the offsets due to load changes in feed flow rate (Figure 4), feed concentration (Figure 5), and feed temperature (Figure 6). When feedforward control is used, noise in the controlled inputs **u** could also result from noisy measurements of the disturbances **d**. However, as shown for example by the recording of the flow $F$ in Figure 4, the disturbances in this set of experiments were essentially noise free and hence unlikely to be the cause of the increased variations in $B1$ and $S$. Similarly, since $\mathbf{K}_{FB}$ is independent of $\mathbf{K}_{FF}$ and **d**, it was the same for all runs and hence does not affect the comparison. The increased noise in first effect level, and hence in the control variables in Figures 4 to 7, was eventually traced to a noisy differential pressure transmitter on the first effect liquid level, that is, $W1$. This noise produced a band on the recorder of about 2% of the span (about 0.6 in. of

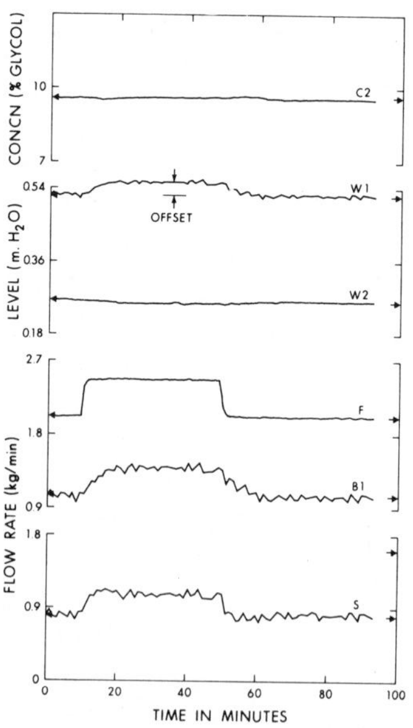


**Fig. 3. Experimental response to two 20% changes in feedflow using optimal feedback control only. Comparison with Figures 4 and 7 shows the effect of adding feedforward control.**

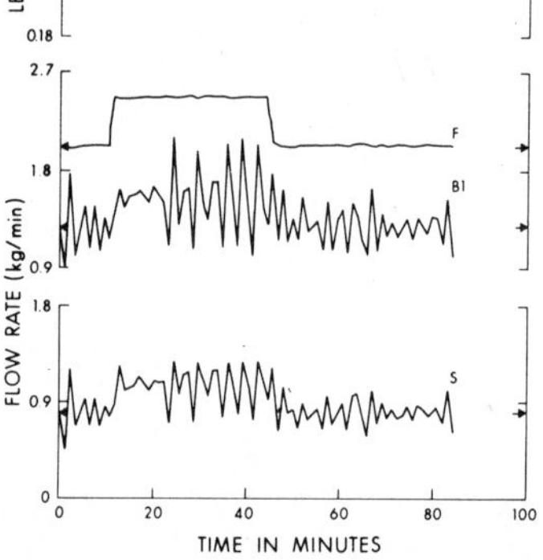

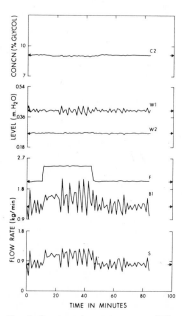

**Fig. 4. Experimental response to two 20% changes in feedflow using the optimal feedback-plus-feedforward control (case 4).**

level). Although due to an instrument fault, these results are a good illustration of what can occur when high gain feedback control is used with noisy measurements to manipulate control variables.

Feedforward control matrices were also calculated using the zero offset technique (case 1) and by minimization of offsets (case 3) and evaluated for a 20% feed flow rate change. A comparison of Figure 7 versus Figure 4 shows that the experimental responses are very similar and no significant difference in the performance of the three control matrices is apparent from the experimental results.

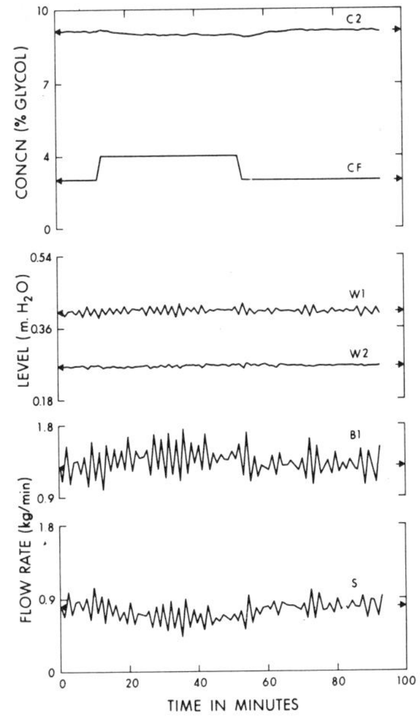


**Fig. 5. Experimental response to two 30% changes in feed composition using the optimal feedback-plus-feedforward control (case 4).**

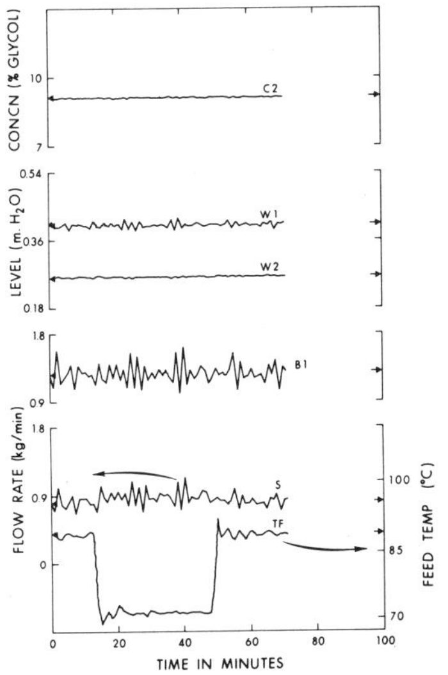


**Fig. 6. Experimental response to two 16% changes in feed temperature, using the optimal feedback-plus-feedforward controller (case 4).**

## ACKNOWLEDGMENT

Financial support from the Canadian government in the form of a Commonwealth Fellowship to R. B. Newell and a National Research Council grant to D. G. Fisher is gratefully acknowledged.

## NOTATION

**A** = constant coefficient matrix, Equation (2)
**B** = constant coefficient matrix, Equation (2)
$\mathbf{B}_1, \mathbf{B}_2$ = partitions of **B**
**C** = constant coefficient matrix, Equation (2)
*CC* = concentration controller
*CR* = concentration recorder
**D** = constant coefficient matrix, Equation (2)
**d** = load vector, $p \times 1$
**e** = error vector, $\mathbf{x} - \mathbf{x}_d$
*FB* = feedback
*FC* = flow controller
*FF* = feedforward
*FR* = flow recorder
**I** = unit matrix
*i* = time interval counter
*J* = performance index
$\mathbf{K}, \mathbf{K}_1, \mathbf{K}_2, \mathbf{K}_3$ = control matrices
*LC* = level controller
*m* = dimension of control vector
*N* = indicator of final time period

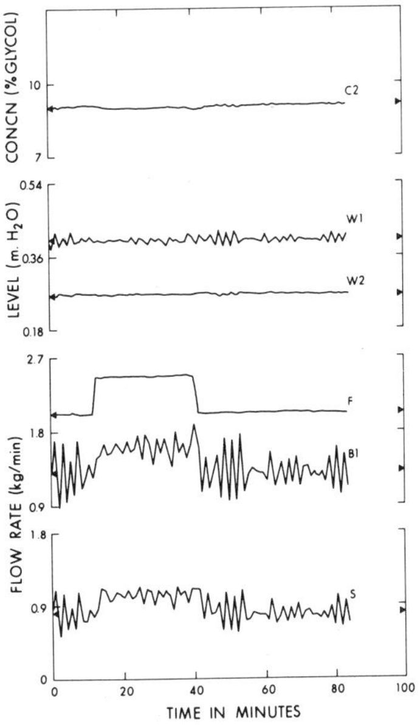


**Fig. 7. Experimental response to two 20% disturbances in feedflow using the optimal feedback controller plus feedforward action designed to minimize the final steady state offsets (case 3). Figure 7 is directly comparable with Figures 3 and 4.**

$n$ = dimension of state vector
$p$ = dimension of disturbance vector
$PC$ = pressure controller
$\mathbf{Q}$ = state weighting matrix, Equation (23)
$q$ = number of states with zero offset
$\mathbf{R}$ = control weighting matrix, Equation (23)
$r$ = dimension of output vector
$\mathbf{S}$ = final state weighting matrix, Equation (23)
$\mathbf{T}$ = closed loop system matrix defined in Equation (12)
$\mathbf{T}_1, \mathbf{T}_2$ = partitions of $\mathbf{T}$
$t$ = time
$T$ = control or sampling interval for discrete systems
$\mathbf{u}$ = control vector, $m \times 1$
$\mathbf{u}_1, \mathbf{u}_2$ = partitions of $\mathbf{u}$
$\mathbf{x}$ = state vector, $n \times 1$
$\mathbf{x}_1, \mathbf{x}_2$ = partitions of $\mathbf{x}$
$\mathbf{y}$ = output vector, $r \times 1$

**Greek Letters**

$\boldsymbol{\Delta}$ = state equation coefficient matrix
$\boldsymbol{\varphi}$ = state equation coefficient matrix
$\boldsymbol{\varphi}^*$ = intermediate matrix defined in Equation (21)
$\boldsymbol{\theta}$ = state transition coefficient matrix

**Subscripts**

$d$ = desired value
$FF$ = feedforward
$FB$ = feedback
$i$ = iteration counter
$s$ = steady state

**Superscripts**

$N - i$ = iteration counter
$T$ = matrix or vector transpose
$-1$ = matrix inverse

**PROCESS VARIABLES (REFERENCE FIGURE 1)**

| State Vector, x | Normal Steady State Value |
|---|---|
| $W1$ = first effect holdup | 14.6 kg |
| $C1$ = first effect concentration | 4.85% |
| $H1$ = first effect enthalpy | 335 kJ/kg |
| $W2$ = second effect holdup | 15.8 kg |
| $C2$ = second effect concentration | 9.64% |
| **Control Vector, u** | |
| $S$ = steam | 0.86 kg/min |
| $B1$ = first effect bottoms | 1.5 kg/min |
| $B2$ = second effect bottoms | 0.75 kg/min |
| **Disturbance Vector, d** | |
| $F$ = feed flowrate | 2.27 kg/min |
| $CF$ = feed concentration | 3.0% |
| $HF$ = feed enthalpy | 382 kJ/kg |

**Other Process Variables (reference Figure 1)**

$O1$ = overhead from first effect
$O2$ = overhead from second effect
$P1$ = pressure in first effect
$P2$ = pressure in second effect
$TF$ = temperature of feed
$T1$ = temperature in first effect
$T2$ = temperature in second effect

**Recursive Relationships for Controller Constants, Equation (24)**

$$\mathbf{K}_{FB}{}^{N-i} = -(\boldsymbol{\Delta}^T \mathbf{P}_{i-1} \boldsymbol{\Delta} + \mathbf{R})^{-1} \boldsymbol{\Delta}^T \mathbf{P}_{i-1} \boldsymbol{\varphi}$$

$$\mathbf{K}_{FF}{}^{N-i} = -(\boldsymbol{\Delta}^T \mathbf{P}_{i-1} \boldsymbol{\Delta} + \mathbf{R})^{-1} (\boldsymbol{\Delta}^T \mathbf{P}_{i-1} \boldsymbol{\theta} + \boldsymbol{\Delta}^T \mathbf{W}_{i-1})$$

where

$$\mathbf{P}_{i-1} = \mathbf{T}^T{}_{N-i+1} \mathbf{P}_{i-2} \mathbf{T}_{N-i+1} + (\mathbf{K}_{FB}{}^{N-i+1})^T \mathbf{R} \mathbf{K}_{FB}{}^{N-i+1} + \mathbf{Q}$$

$$\mathbf{W}_{i-1} = \mathbf{T}^T{}_{N-i+1} \mathbf{W}_{i-2} + \mathbf{T}^T{}_{N-i+1} \mathbf{P}_{i-2} (\boldsymbol{\Delta} \mathbf{K}_{FF}{}^{N-i+1} + \boldsymbol{\theta}) + (\mathbf{K}_{FB}{}^{N-i+1})^T \mathbf{R} \mathbf{K}_{FF}{}^{N-i+1}$$

$$\mathbf{T}_{N-i+1} = \boldsymbol{\varphi} + \boldsymbol{\Delta} \mathbf{K}_{FB}{}^{N-i+1}$$

with initial conditions

$$\mathbf{P}_0 = \mathbf{Q} + \mathbf{S}$$

$$\mathbf{W}_0 = \mathbf{0}$$

**LITERATURE CITED**


Anderson, J. H., "Dynamic Control of a Power Boiler," *Proc. Inst. Elect. Eng.*, **116**, 1257 (1969).

Bertran, D. R., and K. S. Chang, "Optimal Feedforward Control of Concurrent Tubular Reactors", *AIChE J.*, **16**, 897 (1970).

Bollinger, R. E., and D. E. Lamb, "The Design of a Combined Feedforward-Feedback Control System," *Chem. Eng. Progr. Symp. Ser. No. 55*, **61**, 66 (1965).

Clifton, D. M., and M. L. McGuire, "The Feedforward-Feedback Control of Multivariable Stochastic Systems," 1971 Joint Automatic Control Conf., St. Louis (1971).

Corlis, R. G., and R. Luus, "Use of Residuals in the Identification and Control of Two-Input, Single-Output Systems," *Ind. Eng. Chem. Fundamentals*, **8**, 246 (1969).

———, "Compensation for Unknown Input Disturbances in Multiple-Input Identification and Optimal Feedforward Control," paper presented at 20th Can. Chem. Eng. Conf., Sarnia, Ontario (1970).

Fisher, D. G., and B. A. Jacobson, "Computer Control of a Pilot Plant Evaporator," paper presented at 68th National Am. Inst. Chem. Engrs. Meeting, Houston (1971).

Foster, R. D., and W. F. Stevens, "A Method for the Noninteracting Control of a Class of Linear Multivariable Processes," *AIChE J.*, **13**, 340 (1967).

Gilbert, E. G., "The Decoupling of Multivariable Systems by State Feedback," *SIAM J. Control*, **7**, 50 (1969).

Greenfield, G. G., and T. J. Ward, "Feedforward and Dynamic Uncoupling Control of Linear Multivariable Systems," *AIChE J.*, **14**, 783 (1968).

Haskins, D. E., and C. M. Sliepcevich, "The Invariance Principle of Control for Chemical Processes," *Ind. Eng. Chem. Fundamentals*, **4**, 241 (1965).

Heidemann, R. A., and G. L. Esterson, "An Optimal Discrete Controller for a System with Load Changes," 1967 Joint Automatic Control Conference, Philadelphia (1967).

Jacobson, B. A., "Multiloop Computer Control of an Evaporator," M.Sc. thesis, Univ. Alberta, Edmonton, Canada (1970).

Johnson, C. D., "Optimal Control of the Linear Regulator with Constant Disturbances," *I.E.E.E. Trans. Auto. Control*, **AC-13**, 416 (1968).

———, "Further Study of the Linear Regulator with Disturbances—The Case of Vector Disturbances Satisfying a Linear Differential Equation," *ibid.*, **AC-15**, 222 (1970a).

———, "Further Comments on Optimal Control of the Linear Regulator with Constant Disturbances," *ibid.*, 516 (1970b).

Lapidus, L., and R. Luus, "Optimal Control of Engineering Processes," Blaisdell, Waltham, Mass. (1967).

Latour, P. R., "Optimal Control of Linear Multivariable Plants with Constant Disturbances," 1971 Joint Automatic Control Conf., St. Louis (1971).

Luyben, W. L., "Distillation, Feedforward Control with Intermediate Feedback Control Trays," *Chem. Eng. Sci.*, **24**, 997 (1969).

Miller, J. A., P. W. Murrill, and C. L. Smith, "How to Apply Feedforward Control," *Hydrocarb. Process.*, **48**, 7, 165 (1969).

Newell, R. B., "Multivariable Computer Control of an Evaporator," Ph.D. thesis, Univ. Alberta, Edmonton, Canada (1971).
———, and D. G. Fisher, "Experimental Evaluation of Optimal Multivariable Regulatory Controllers with Model Following Capabilities," *Automatica*, **8**, 247 (1972a).
———, "Model Development, Reduction and Experimental Evaluation for an Evaporator," *Ind. Eng. Chem. Design Develop.*, **11**, 213 (1972b).
Ogata, K., "State Space Analysis of Control Systems," Prentice-Hall, Englewood Cliffs, N. J. (1967).
Paraskos, J. A., and T. J. McAvoy, "Feedforward Computer Control of a Class of Distributed Parameter Processes," AIChE J., **16**, 754 (1970).
Porter, B., "Assignment of Closed-Loop Eigenvalues by the Direct Method of Liapunov," *Intern. J. Control*, **10**, 153 (1969).
Rosenbrock, H. H., "State Space and Multivariable Theory," Thomas Nelson, London (1970).
Sobral, M., and R. G. Stefanek, "Optimal Control of the Linear Regulator Subject to Disturbances," *I.E.E.E. Trans. on Auto. Control*, **AC-15**, 498 (1970).
Solheim, O. A., and A. Saethre, "Optimal Control of Continuous and Discrete Multivariable Systems by Feedforward and Feedback," IFAC Conf. Multivariable Control Systems, Dusseldorf (1968).
West, H. H., and M. L. McGuire, "Optimal Feedforward-Feedback Control of Dead Time Systems," *Ind. Eng. Chem. Fundamentals*, **8**, 253 (1969a).
———, "A Limitation of Predictive Controllers," paper presented at the 62nd Ann. Am. Inst. Chem. Engrs. Meeting, Washington, D.C. (1969b).
Wonham, W. M., "On Pole Assignment in Multi-Input Controllable Linear Systems," *I.E.E.E. Trans. Auto. Control*, **AC-12**, 660 (1967).

*Manuscript received September 10, 1971; revision received March 21, 1972; paper accepted March 24, 1972.*

# MODAL APPROACH TO CONTROL LAW REDUCTION

Robert G. Wilson, Dale E. Seborg*

and D. Grant Fisher

Department of Chemical Engineering
The University of Alberta
Edmonton, Alberta, Canada

## ABSTRACT

A method is proposed for reducing the complexity of state feedback control laws for high order systems. Using a modal approach, selected state variables are eliminated to give an incomplete state feedback control law. The proposed method for control law reduction is applied to an optimal control system for a tenth order model of a pilot scale evaporator. Simulation and experimental results show that the modal approach is practical and results in better control than methods based on reduced order models or conventional multiloop design techniques.

## INTRODUCTION

The objective in this investigation is to develop a systematic approach for reducing the complexity of multivariable controllers for high-order systems. It is assumed that the system to be controlled can be adequately described by a linear, time-invariant state space model, and that satisfactory control can be achieved using a multivariable state feedback controller which may also include integral feedback, feedforward, and setpoint control modes. Simplification of the control law is achieved by eliminating selected state variables to give an incomplete state feedback control law. The motivation for this design approach is that in most practical applications, any degradation in system performance will be offset by the practical advantages of fewer measurements (or state estimates) and reduced demands on the online computer.

The first step in the proposed approacn is to select a control law which will give satisfactory control performance when applied to the best available model of the process. In this work, the control law is designed by applying optimal control techniques to a linearized process model. The resulting controller is a linear, time-invariant feedback control law that utilizes the entire state vector. This control law is simplified by eliminating selected state variables from the feedback control terms. The problem of deriving an incomplete state feedback control law from a state feedback control law will be referred to in the study as "control law reduction", and the resulting controller will be referred to as a "reduced-order control law."

## PREVIOUS WORK

The design of incomplete state feedback and output feedback control systems for linear systems has been the subject of a large number of investigations in recent years. Optimal control formulations in which the control configuration is constrained to be a function of the output vector has received considerable attention [2,6,7,16]. However, the calculation of the optimal control policies often requires a significant computational effort, particularly for large systems. Furthermore, the dependence of the optimal control policy on the initial state must be taken into consideration [6,7].

An alternative approach considered by Anderson [1] and Nicholson [12-14] is to derive a low order state-space model by model reduction techniques and then use the optimal control policy for the low order model as a suboptimal control law for the high order system. Rogers and Sworder [15] have proposed that the reduced order model and its optimal control policy be calculated simultaneously with the additional restriction that the low order control law is the best suboptimal controller for the high order system. An iterative solution of nonlinear matrix equations is required in their approach.

None of the above approaches makes direct use of an additional item of information, namely, the optimal control law for the high order system. The strategy adopted in this investigation is to derive an incomplete state feedback law from the high order optimal control law by using modal analysis to eliminate selected state variables.

## CONTROL LAW REDUCTION

Consider the linear, time-invariant, discrete-time model in the form:

$$\underline{x}(j+1) = \underline{\underline{\phi}}\,\underline{x}(j) + \underline{\underline{\Delta}}\,\underline{u}(j) + \underline{\underline{\theta}}\,d(j) \qquad (1)$$

$$\underline{y}(j) = \underline{\underline{C}}^{*}\underline{x}(j) \qquad (2)$$

where $\underline{x}$ is the n-dimensional state vector,
$\underline{u}$ is the m-dimensional control vector,
$\underline{d}$ is the q-dimensional disturbance vector,
$\underline{y}$ is the p-dimensional output vector,
$\underline{\underline{\phi}},\underline{\underline{\Delta}},\underline{\underline{\theta}},\underline{\underline{C}}^*$ are constant coefficient matrices of appropriate dimensions,
j is a counter such that $\underline{x}(j)$ denotes $\underline{x}(t)$ at t=jT, where T is the discrete time interval.

It is assumed that the high-order control law has the form:

$$\underline{u}(j) = \underline{\underline{K}}^{FB}\,\underline{x}(j) + \underline{\underline{K}}^{I}\sum_{i=0}^{j}\underline{y}(i) + \underline{\underline{K}}^{FF}\,\underline{d}(j) + \underline{\underline{K}}^{SP}\,\underline{y}^{SP} \qquad (3)$$

where $\underline{y}^{SP}$ is the constant setpoint vector such that $y_i^{SP}$ is the desired value of $y_i$, i=1,...,p

$\underline{\underline{K}}^{FB}$, $\underline{\underline{K}}^{I}$, $\underline{\underline{K}}^{FF}$, $\underline{\underline{K}}^{SP}$ are the proportional feedback, integral feedback, feedforward and

*Please address all correspondence to Professor D.E. Seborg.

Proc. 1973 Joint Automatic Control Conference, pp. 554-565, IEEE, New York (1973).

setpoint control matrices of appropriate dimensions.

The control law of equation (3) is chosen because it will stabilize the system, and also because it is a practical control law which can be easily applied using an on-line digital computer. The control matrices in equation (3) can be determined by any applicable design technique. In this work they were calculated using the optimal, linear, quadratic (infinite time) approach described by Newell [10,11].

The objective of this investigation is to simplify the control law by reducing the number of state variables to give the following reduced-order control law:

$$\underline{u}(j) = \underline{\underline{K}}_R^{FB}\, \underline{x}_1(j) + \underline{\underline{K}}_R^{I} \sum_{i=0}^{j} \underline{\underline{C}}\, \underline{x}_1(i) + \underline{\underline{K}}_R^{FF}\, \underline{d}(j) + \underline{\underline{K}}_R^{SP}\, \underline{\underline{C}}\, \underline{x}_1^{SP} \qquad (4)$$

where $\underline{x}_1$ is the $\ell$ dimensional subset of $\underline{x}$ which is to be retained in the reduced order control law and where it is assumed that $\underline{x}_1$ can stabilize the system in equations (1) and that $\underline{y}(j)$ in the integral feedback and setpoint modes in equation (3) can be adequately represented by $\underline{\underline{C}}\, \underline{x}_1(j)$. (This restriction can be removed at the expense of including additional terms in equation (4) which involve past values of $\underline{u}$ and $\underline{d}$.) Criteria for selecting the elements to be retained in $\underline{x}_1$ will be discussed later.

## Method Development

If the state vector, $\underline{x}$, is partitioned such that

$$\underline{x} = \begin{bmatrix} \underline{x}_1 \\ \underline{x}_2 \end{bmatrix} \qquad (5)$$

then the state feedback control law of equation (3) can be expanded in terms of $\underline{x}_1$ and $\underline{x}_2$, to give:

$$\underline{u}(j) = \underline{\underline{K}}_1^{FB}\, \underline{x}_1(j) + \underline{\underline{K}}_2^{FB}\, \underline{x}_2(j) + \underline{\underline{K}}^{I} \sum_{i=0}^{j} \underline{\underline{C}}\, \underline{x}_1(i) + \underline{\underline{K}}^{FF}\, \underline{d}(j) + \underline{\underline{K}}^{SP}\, \underline{\underline{C}}\, \underline{x}_1^{SP} \qquad (6)$$

If an approximation for $\underline{x}_2$, in terms of the other variables, can be obtained and substituted into equation (6), the desired form of the reduced-order control law would result. An approximation for $\underline{x}_2$ can be derived using a modal analysis of the original high-order system represented by equations (1) and (2). This approach is similar to the reduction of continuous-time models as originally presented by Marshall [9] and as extended to discrete-time systems by Wilson et. al. [18].

An n x n matrix, $\underline{\underline{M}}$, exists which transforms equation (1) into its Jordan canonical form, as shown by Gantmacher [3]. This similarity transformation preserves the eigenvalues of $\underline{\underline{\phi}}$ and is defined by

$$\underline{x} = \underline{\underline{M}}\, \underline{z} \qquad (7)$$

or in partitioned form,

$$\begin{bmatrix} \underline{x}_1 \\ \underline{x}_2 \end{bmatrix} = \begin{bmatrix} \underline{\underline{M}}_1 & \underline{\underline{M}}_2 \\ \underline{\underline{M}}_3 & \underline{\underline{M}}_4 \end{bmatrix} \begin{bmatrix} \underline{z}_1 \\ \underline{z}_2 \end{bmatrix} \qquad (8)$$

where $\underline{z}$ is the canonical state vector and $\underline{z}_1$ and $\underline{z}_2$ have the same dimensions as $\underline{x}_1$ and $\underline{x}_2$, respectively. When equation (8) is substituted into equation (1), the following Jordan canonical form results (shown here in partitioned form):

$$\begin{bmatrix} \underline{z}_1(j+1) \\ \underline{z}_2(j+1) \end{bmatrix} = \begin{bmatrix} \underline{\underline{\alpha}}_1 & \underline{\underline{0}} \\ \underline{\underline{0}} & \underline{\underline{\alpha}}_2 \end{bmatrix} \begin{bmatrix} \underline{z}_1(j) \\ \underline{z}_2(j) \end{bmatrix} + \begin{bmatrix} \underline{\underline{\delta}}_1 \\ \underline{\underline{\delta}}_2 \end{bmatrix} \underline{u}(j) + \begin{bmatrix} \underline{\underline{\eta}}_1 \\ \underline{\underline{\eta}}_2 \end{bmatrix} \underline{d}(j) \qquad (9)$$

where $\underline{\underline{\alpha}}$, $\underline{\underline{\delta}}$ and $\underline{\underline{\eta}}$ are defined as:

$$\underline{\underline{\alpha}} = \underline{\underline{M}}^{-1}\, \underline{\underline{\phi}}\, \underline{\underline{M}} = \begin{bmatrix} \underline{\underline{\alpha}}_1 & \underline{\underline{0}} \\ \underline{\underline{0}} & \underline{\underline{\alpha}}_2 \end{bmatrix} \qquad (10)$$

$$\underline{\underline{\delta}} = \underline{\underline{M}}^{-1}\, \underline{\underline{\Delta}} = \begin{bmatrix} \underline{\underline{\delta}}_1 \\ \underline{\underline{\delta}}_2 \end{bmatrix} \qquad (11)$$

$$\underline{\underline{\eta}} = \underline{\underline{M}}^{-1}\, \underline{\underline{\theta}} = \begin{bmatrix} \underline{\underline{\eta}}_1 \\ \underline{\underline{\eta}}_2 \end{bmatrix} \qquad (12)$$

Matrix $\underline{\underline{\alpha}}$ is a block diagonal matrix with the eigenvalues of $\underline{\underline{\phi}}$ on its principal diagonal. It then follows directly from equation (9) that each element of $\underline{z}$ is associated with a particular eigenvalue of $\underline{\underline{\phi}}$. Since the columns of $\underline{\underline{M}}$ may be rearranged arbitrarily, it is possible to associate the "dominant" eigenvalues of the system (defined relative to their effect on $\underline{x}_1$) with the elements of $\underline{z}_1$, and this is the key to the success of the proposed design technique. A discussion of the criteria for selecting a particular arrangement of the columns of $\underline{\underline{M}}$ is presented later.

An expression for $\underline{x}_2$ can be derived using equation (8) and matrix $\underline{\underline{V}}$ which is defined as,

$$\underline{\underline{V}} = \begin{bmatrix} \underline{\underline{V}}_1 & \underline{\underline{V}}_2 \\ \underline{\underline{V}}_3 & \underline{\underline{V}}_4 \end{bmatrix} = \underline{\underline{M}}^{-1} \qquad (13)$$

to give

$$\underline{x}_2(j) = \underline{V}_4^{-1} [\underline{z}_2(j) - \underline{V}_3 \underline{x}_1(j)] \quad (14)$$

Equation (14) could be used to eliminate $\underline{x}_2$ from the high-order control law in equation (6). However, the resulting control law would then contain $\underline{z}_2$. As an approximation, $\underline{z}_2(j)$ can be replaced by the steady state value. It will be shown later that this approximation is a reasonable one due to the way that the columns of $\underline{M}$ have been arranged. Thus, assume that $\underline{z}_2$ has reached its final steady state value; then from equation (9):

$$\underline{z}_2(j) = \underline{z}_2(j+1) = (\underline{I}-\underline{\alpha}_2)^{-1} [\underline{\delta}_2 \underline{u}(j) + \underline{\eta}_2 \underline{d}(j)] \quad (15)$$

Equation (15) is then used with equation (14) to give the required approximation of $\underline{x}_2$:

$$\underline{x}_2(j) = \underline{V}_4^{-1} [(\underline{I}-\underline{\alpha}_2)^{-1} \underline{\delta}_2 \underline{u}(j) + (\underline{I}-\underline{\alpha}_2)^{-1} \underline{\eta}_2 \underline{d}(j) - \underline{V}_3 \underline{x}_1(j)] \quad (16)$$

The reduced-order control law can now be obtained by combining equations (6) and (16) such that the coefficient matrices in equation (4) are given by:

$$\begin{aligned} \underline{K}_R^{FB} &= \underline{K}^u (\underline{K}_1^{FB} - \underline{K}_2^{FB} \underline{V}_4^{-1} \underline{V}_3) \\ \underline{K}_R^{I} &= \underline{K}^u \underline{K}^{I} \\ \underline{K}_R^{FF} &= \underline{K}^u (\underline{K}^{FF} + \underline{K}_2^{FB} \underline{V}_4^{-1} (\underline{I}-\underline{\alpha}_2)^{-1} \underline{\eta}_2) \\ \underline{K}_R^{SP} &= \underline{K}^u \underline{K}^{SP} \end{aligned} \quad (17)$$

where

$$\underline{K}^u = (\underline{I}-\underline{K}_2^{FB} \underline{V}_4^{-1} (\underline{I}-\underline{\alpha}_2)^{-1} \underline{\delta}_2)^{-1}$$

Thus, a reduced-order control law has been derived in which the feedback contribution is a function of only a selected subset of the state vector, $\underline{x}_1$, rather than the entire state vector. As indicated earlier, the control system designer also has the freedom to decide which variables are included in the feedforward, integral and setpoint terms.

## Continuous-Time Systems

A similar analysis can be used to obtain a continuous-time, reduced-order control law. Consider the following linear, time-invariant, continuous-time model:

$$\dot{\underline{x}}(t) = \underline{A} \underline{x}(t) + \underline{B} \underline{u}(t) + \underline{D} \underline{d}(t) \quad (18)$$

$$\underline{y}(t) = \underline{C}^* \underline{x}(t) \quad (19)$$

where $\underline{x}$, $\underline{u}$, $\underline{d}$, $\underline{y}$ and $\underline{C}^*$ are as defined earlier and $\underline{A}$, $\underline{B}$, and $\underline{D}$ are constant coefficient matrices of appropriate dimensions. This system can be stabilized by the high-order control law:

$$\underline{u}(t) = \underline{K}^{FB} \underline{x}(t) + \underline{K}^{I} \int_0^t \underline{y}(t)\, dt + \underline{K}^{FF} \underline{d}(t) + \underline{K}^{SP} \underline{y}^{SP} \quad (20)$$

The resulting, reduced continuous-time control system is as follows:

$$\underline{u}(t) = \underline{K}_R^{FB} \underline{x}_1(t) + \underline{K}_R^{I} \int_0^t \underline{C}\, \underline{x}_1(t)\, dt + \underline{K}_R^{FF} \underline{d}(t) + \underline{K}_R^{SP} \underline{C}\, \underline{x}_1^{SP} \quad (21)$$

where

$$\begin{aligned} \underline{K}_R^{FB} &= \underline{K}^u (\underline{K}_1^{FB} - \underline{K}_2^{FB} \underline{V}_4^{-1} \underline{V}_3) \\ \underline{K}_R^{I} &= \underline{K}^u \underline{K}^{I} \\ \underline{K}_R^{FF} &= \underline{K}^u (\underline{K}^{FF} - \underline{K}_2^{FB} \underline{V}_4^{-1} \underline{J}_2^{-1} \underline{H}_2) \\ \underline{K}_R^{SP} &= \underline{K}^u \underline{K}^{SP} \\ \underline{K}^u &= (\underline{I} + \underline{K}_2^{FB} \underline{V}_4^{-1} \underline{J}_2^{-1} \underline{G}_2)^{-1} \end{aligned}$$

and $\underline{J}_2$, $\underline{G}_2$ and $\underline{H}_2$ are defined from the continuous-time canonical system as:

$$\begin{bmatrix} \dot{\underline{z}}_1(t) \\ \dot{\underline{z}}_2(t) \end{bmatrix} = \begin{bmatrix} \underline{J}_1 & \underline{0} \\ \underline{0} & \underline{J}_2 \end{bmatrix} \begin{bmatrix} \underline{z}_1(t) \\ \underline{z}_2(t) \end{bmatrix} + \begin{bmatrix} \underline{G}_1 \\ \underline{G}_2 \end{bmatrix} \underline{u}(t) + \begin{bmatrix} \underline{H}_1 \\ \underline{H}_2 \end{bmatrix} \underline{d}(t) \quad (22)$$

where

$$\underline{J} = \begin{bmatrix} \underline{J}_1 & \underline{0} \\ \underline{0} & \underline{J}_2 \end{bmatrix} = \underline{V}\, \underline{A}\, \underline{M}$$

$$\underline{G} = \begin{bmatrix} \underline{G}_1 \\ \underline{G}_2 \end{bmatrix} = \underline{V}\, \underline{B}$$

$$\underline{\underline{H}} = \begin{bmatrix} \underline{\underline{H}}_1 \\ \underline{\underline{H}}_2 \end{bmatrix} = \underline{\underline{V}}\ \underline{\underline{D}}$$

Matrices $\underline{\underline{M}}$ and $\underline{\underline{V}}$ are the same as those used for the discrete-time system and can be calculated using either the continuous-time or the discrete-time models [18].

This continuous-time control law is subject to all the restrictions which were imposed on the discrete-time control law derived earlier.

### Application Considerations

In applying this method, two important design decisions must be made: first, which subset of $\underline{x}$ should be included in the reduced-order control law as $\underline{x}_1$ and secondly, which subset of the system's eigenvalues should be retained in $\underline{\underline{\alpha}}_1$.

The choice of which state variables to retain in the control law may require a trial and error approach, but a few general guidelines are available. For example, process variables which are to be closely controlled or constrained must be included in the reduced-order control law. Other process variables, which if not controlled would give poor system performance, must also be included in the reduced-order control law. Furthermore, the state variables retained in $\underline{x}_1$ must be able to satisfactorily stabilize the high-order system by means of equation (4). A further consideration in the choice of which state variables to retain in $\underline{x}_1$ is whether a state variable is measurable or not. It is obvious that other considerations aside, those state variables which can be most easily or economically measured are the ones which should be retained. Unmeasured state variables should be included in $\underline{x}_1$ only if they are essential for satisfactory control since a state estimation procedure, such as the Kalman filter [5] or the Luenberger observer [8] would then be required.

After the elements of $\underline{x}_1$ have been selected, the designer must assign the eigenvalues of $\underline{\underline{\phi}}$ to either $\underline{\underline{\alpha}}_1$ or $\underline{\underline{\alpha}}_2$. In general, the eigenvalues which have the most significant effect on the elements of $\underline{x}_1$ should be placed in $\underline{\underline{\alpha}}_1$. The two most important factors to consider in assigning these eigenvalues, as shown by the following analysis, are the magnitude of the eigenvalues of $\underline{\underline{\phi}}$ and the magnitude of the elements of the modal matrix, $\underline{\underline{M}}$. Equation (7) can be expanded as

$$x_i = m_{i1}\, z_1 + m_{i2}\, z_2 + \dots + m_{in}\, z_n, \quad i=1,2,\dots,n \tag{23}$$

where $m_{ij}$ is an element of $\underline{\underline{M}}$. The magnitude of $m_{ij}$ obviously determines the effect of the jth mode on the ith state variable. Similarly, if $\underline{\underline{\phi}}$ has distinct eigenvalues, then the response of the unforced system can be written as (cf. equation (9))

$$z_k(j+1) = \alpha_k\, z_k(j), \quad k=1,2,\dots,n \tag{24}$$

where $\alpha_k$ is an eigenvalue of $\underline{\underline{\phi}}$. For an eigenvalue close to the origin, equation (24) indicates that the corresponding mode decays rapidly and hence its effect on a particular element, $x_i$, of the state vector would be short lived. Equations (23) and (24) imply that the kth mode will most significantly affect a state variable, $x_i$, when $m_{ik}$ is large and $\alpha_k$ is also large. In some situations a trade-off may be necessary, since a large value of $m_{ik}$ may correspond to a small value of $\alpha_k$ or vice versa. After the eigenvalues have been partitioned into $\underline{\underline{\alpha}}_1$ and $\underline{\underline{\alpha}}_2$, the columns of $\underline{\underline{M}}$ must be rearranged accordingly.

If the largest eigenvalues of $\underline{\underline{\phi}}$ are placed in $\underline{\underline{\alpha}}_1$ then $\underline{\underline{\alpha}}_2$ will contain the smaller, or faster, values. In the derivation of equation (16) it was assumed that $\underline{z}_2$ could be represented by steady-state values. This is a reasonable assumption if the eigenvalues in $\underline{\underline{\alpha}}_2$ are small and becomes more accurate as the eigenvalues approach the origin. Thus the magnitude of the eigenvalues in $\underline{\underline{\alpha}}_2$ (plus the associated $m_{ij}$) give the designer a qualitative feel for the accuracy of the reduced-order system.

One additional restriction must be imposed for systems with repeated eigenvalues which do not have linearly independent eigenvectors. The block diagonal matrices, $\underline{\underline{\alpha}}_1$ and $\underline{\underline{\alpha}}_2$, should be assigned so that $\underline{z}_1$ and $\underline{z}_2$ are non-interacting, as was assumed in partitioning equation (9). This will always be the case for systems with n linearly independent eigenvectors, since then $\underline{\underline{\alpha}}$ is a diagonal matrix. However, for systems with less than n linearly independent eigenvectors, the partitioning of the original system must be such that it does not split a Jordan block in $\underline{\underline{\alpha}}$, so that $\underline{z}_1$ and $\underline{z}_2$ will remain non-interacting. This might necessitate increasing or decreasing the size of vector $\underline{x}_1$ slightly. If the original system cannot be satisfactorily partitioned to keep $\underline{z}_1$ and $\underline{z}_2$ non-interacting, then the parameters in the original model could be changed by a small amount so that the eigenvectors would be linearly independent.

## APPLICATION TO THE CONTROL OF AN EVAPORATOR

The proposed method was evaluted by utilizing simulated and experimental response data for a pilot plant evaporator which is described in the Appendix. The simulated responses were generated from a 10th order state space model of the evaporator which was derived by Newell [10]. The following control laws were compared:

(a) a third order control law designed by the proposed method using a tenth order optimal control law as the starting point.
(b) a third order control law which is optimal for a third order state space model and a quadratic performance index. The third order model was derived from the tenth order model using a discrete version [18] of Marshall's model reduction technique [9].
(c) a control law which is optimal for the tenth order model and a quadratic performance index.
(d) a conventional multiloop feedback control scheme using controller constants which were determined experimentally by Jacobson and Fisher [4].

The first step in the application of the proposed method is to determine which states to include in $\underline{x}_1$. An analysis of the evaporator showed that three state variables, product concentration, C2, and the two holdups, W1 and W2, must be retained in $\underline{x}_1$. C2 must be retained since regulation of C2 is specified to be the primary control objective. W1 and W2 must be retained since these are integrating states (corresponding to eigenvalues of 1) which would tend to exceed physical operating limits if they were not controlled. The setpoint and integral control modes were also chosen to be a function of these three state variables thus making $\underline{\underline{C}} = \underline{\underline{I}}$ in equation (4).

The second step in the proposed method is to choose which eigenvalues to retain in $\underline{\alpha}_1$ by an appropriate arrangement of the columns of $\underline{\underline{M}}$ ($\underline{\underline{M}}$ for the evaporator model is shown in the Appendix). If the previous choice of $\underline{x}_1 = [W1, W2, C2]^T$ is made, then the three largest eigenvalues should be retained. This follows from an inspection of matrix $\underline{\underline{M}}$ since the three slowest modes have the most effect on the elements of $\underline{x}_1$.

Once the state vector and eigenvalues have been partitioned and the columns of $\underline{\underline{M}}$ have been arranged, the high-order control law, matrix $\underline{\underline{V}}$ and the canonical state equation are partitioned as shown in equations (6), (13) and (9) respectively. Then, the reduced-order control matrices can be calculated using equation (17).

The third order control laws which were designed by the proposed method are presented in Tables I and II. Table III summarizes the values of the quadratic performance index, J,

$$J = \sum_{j=1}^{N} [(\underline{x}(j) - \underline{x}^{SP})^T \underline{\underline{Q}} (\underline{x}(j) - \underline{x}^{SP}) + \underline{u}^T(j-1) \underline{\underline{R}} \underline{u}(j-1)] \quad (25)$$

TABLE I. CONTROL MATRICES FOR PROPORTIONAL, FEEDFORWARD AND SETPOINT CONTROL

$$\underline{\underline{K}}_R^{FB} = \begin{bmatrix} 2.467 & 0.02163 & 4.128 \\ 4.288 & -1.340 & 9.760 \\ 4.128 & 8.885 & 9.528 \end{bmatrix}$$

$$\underline{\underline{K}}_R^{FF} = \begin{bmatrix} 1.238 & -0.5639 & -0.4138 \\ 0.9815 & 0.2177 & -0.001227 \\ 0.9978 & 0.9877 & -0.001398 \end{bmatrix}$$

$$\underline{\underline{K}}_R^{SP} = \begin{bmatrix} -2.468 & -0.02119 & 5.276 \\ -4.289 & 1.340 & -9.100 \\ -4.128 & -9.763 & -10.52 \end{bmatrix}$$

obtained by controlling the tenth-order model with three of the control laws mentioned above (controllers (a), (b) and (c)). The $\underline{\underline{Q}}$ and $\underline{\underline{R}}$ weighting matrices, the state space models, and the tenth-order optimal control law are presented in the Appendix. The simulation results presented in Table III are for:

(a) Proportional feedback control with a non-zero initial condition (i.e. $\underline{x}(0) \neq \underline{0}$),

TABLE II. CONTROL MATRICES FOR PROPORTIONAL PLUS INTEGRAL CONTROL

$$\underline{\underline{K}}_R^{FB} = \begin{bmatrix} 2.977 & 0.09369 & -5.408 \\ 5.066 & -1.246 & 8.100 \\ 5.686 & 12.29 & 11.58 \end{bmatrix}$$

$$\underline{\underline{K}}_R^{I} = \begin{bmatrix} 0.4117 & 0.02088 & -0.4861 \\ 0.8398 & -0.2887 & 0.8159 \\ 0.8559 & 1.945 & 1.068 \end{bmatrix}$$

(b) Proportional-integral feedback control with a non-zero initial condition,

(c) Proportional feedback-feedforward control with a +20% feed flow disturbance (starting from $\underline{x}(0)=\underline{0}$),

(d) Proportional feedback-setpoint control with a +7% setpoint change in C2.

The results summarized in Table III show that the third order control law calculated by the proposed method was better, as indicated by the lower value of J, than the third-order controller calculated from the reduced-order model. The only exception being the case of proportional feedback-feedforward control where the two third order control laws give almost identical values, differing by only 0.25%. In each case, the optimal control law for the tenth order system gave the lowest value of J, as would be expected.

The transient responses for proportional feedback control in Fig. 1 demonstrate that the reduced order control law calculated by the proposed method gives better control than the control law designed from the third order model. This is especially evident for the primary controlled variable, C2, where a slight oscillation develops when the control law designed from the reduced order model is used. Figure 2 presents a similar comparison for proportional plus integral control law with the same initial state as was used in Fig. 1. Again, the proposed method is the better low order controller but the responses are more oscillatory than for proportional feedback control (cf. Fig. 1). In these and succeeding figures, the arrows along the vertical axes denote the normal steady state values.

Figure 3 and the performance indices in Table III indicate that the two third-order feedback-feedforward control laws are quite similar. All three control laws in Fig. 3 corrected for the +20% feed flow disturbance with no steady-state error. In Fig. 4 the closed-loop responses of three feedback-setpoint controllers are compared for a +7% setpoint change in C2. Again, the control law designed from the third order model results in a more oscillatory response than the control law designed by the proposed method. All three responses attained the desired steady state.

In general, the effectiveness of the reduced-order controller matrices, $\underline{\underline{K}}_R^{FF}$, $\underline{\underline{K}}_R^{SP}$ and $\underline{\underline{K}}_R^{I}$, depends upon the particular $\underline{\underline{K}}_R^{FB}$ with which they are used. This point was investigated for a proportional feedback plus setpoint control law where $\underline{\underline{K}}_R^{FB}$ was calculated from the reduced-order model and used with the $\underline{\underline{K}}^{SP}$ matrix which was optimal for the high order model. For the same +7% setpoint in C2,

TABLE III. CONTROL LAW COMPARISON

| Type of Control Law | Forcing Function or Initial Conditions | Performance Index, J | | | Transient Response Shown in Fig. # |
|---|---|---|---|---|---|
| | | Optimal Controller for 10th Order System | Reduced Control Law (by proposed method) | Optimal Control Law for 3rd Order Model | |
| Proportional feedback | all $x_i(0) = 0$ except W1(0) = 0.2, C2(0) = 0.15 | 3.779 | 3.817 | 4.102 | 1 |
| Proportional plus integral feedback | all $x_i(0) = 0$ except W1(0) = 0.2, C2(0) = 0.15 | 4.657 | 4.789 | 4.969 | 2 |
| Proportional feedback plus feedforward | +20% step in F | 1.018 | 1.226 | 1.223 | 3 |
| Proportional feedback plus setpoint | +7% setpoint change in C2 | 0.9477 | 1.198 | 1.547 | 4 |

this approach gave J = 37.89. This large value was due mainly to the resulting steady state C2 value of +11.5% instead of the desired value of +7%.

In process control applications, feedback control systems are primarily used to correct for unmeasured disturbances which enter the system. Thus, it was of interest to examine the effectiveness of a controller designed by the proposed method in such situations. Figure 5 shows the response of the high-order evaporator model when upset by a +20% feedflow disturbance and controlled by either the optimal tenth order proportional feedback controller or by the third order proportional feedback control law designed by the proposed method. The third order control law results in satisfactory control but is of course, not as good as the optimal control law.

The above examples demonstrate that the reduced order control laws designed by the proposed method give excellent results when applied to the control of the high-order model. The experimental response shown in Fig. 6 indicates that the resulting control law also performs well when controlling the actual pilot plant evaporator. The proportional control scheme results in satisfactory control of W1, W2 and C2 despite two 20% step changes in feed flow. (The control matrices for Fig. 6 are shown in Table I.)

For purposes of comparison, Fig. 7 shows the response of the evaporator when controlled by the "standard" multi-loop proportional plus integral control scheme used by Jacobson and Fisher [4]. Here C2 is controlled by S, W1 is controlled by B1, and W2 is controlled by B2. The control interval and controller constants were those used by Jacobson and Fisher [4]. Notice the much larger deviations and the longer response times in Fig. 7 as compared to the other figures.

In the development of this new approach for designing a reduced order control law, it appeared that a possible application might be for "implied" control. Suppose, for example, there is a key state variable, $x_k$, which cannot be measured, but must be controlled. It appeared that if the optimal, high-order control law for the system was designed with heavy weighting on $x_k$, then the proposed method could be applied to obtain a reduced order control law which was not a function of $x_k$, but which would still control it satisfactorily. This was attempted in the evaporator simulation studies. The variable to be controlled was taken as C2 and several feedback control laws were tried, all without success. It is possible that, with further study into the structure, controllability and/or observability of the system, a satisfactory "implied" control scheme could be devised. However, this was not pursued further as part of this study.

## CONCLUSIONS

A method has been proposed for deriving reduced order control laws which eliminate selected state variables from high order control laws. In effect, a modal analysis is used to derive an incomplete state feedback control law from a high order state feedback control law. Both the high order and low order control laws may contain feedforward, integral and setpoint modes. The feasibility of the proposed method has been demonstrated in simulation and experimental studies involving a pilot scale evaporator. The low order control laws designed using the proposed method gave better results than controllers designed using model-reduction techniques.

ACKNOWLEDGEMENT

Financial support from the National Research Council of Canada is gratefully acknowledged.

REFERENCES

1. ANDERSON, J.H., "Control of a Power Boiler Dynamic Model in the Presence of Measurement Noise and Random Input Disturbances", Industrial Applications of Dynamic Modelling, IEE Conference Publication No. 57, Univ. of Durham, 42 (Sept. 1969).
2. DAVISON, E.J., "The Systematic Design of Control Systems for Large Multivariable Linear Time-Invariant Systems", Preprints of the IFAC 5th World Congress, Part 4A, Paper 29.2, Paris, France (1972).
3. GANTMACHER, F.R., The Theory of Matrices, Chelsea Publishing Co., New York (1959).
4. JACOBSON, B.A. and FISHER, D.G., "Computer Control of a Pilot Plant Evaporator", The Chemical Engineer (UK), in press.
5. KALMAN, R.E. and BUCY, R.S., "New Results in Linear Filtering and Prediction Theory", ASME Trans., J. Basic Eng., Vol. 83, 95 (1961).
6. KOSUT, R.L., "Suboptimal Control of Linear Time-Invariant Systems Subject to Control Structure Constraints", IEEE Trans. on Auto. Control, Vol. AC-15, No. 5, 557 (1970).
7. LEVINE, W.S. and ATHANS, M., "On the Determination of the Optimal Constant Output Feedback Gains for Linear Multivariable Systems", IEEE Trans. on Auto. Control, Vol. AC-15, No. 1, 44 (1970).
8. LUENBERGER, D.G., "An Introduction to Observers", IEEE Trans. on Auto. Control, Vol. AC-16, No. 6, 596 (1971).
9. MARSHALL, S.A., "An Approximate Method for Reducing the Order of a Linear System", Control, Vol. 10, No. 102, 642 (1966).
10. NEWELL, R.B., "Multivariable Computer Control of an Evaporator", Ph.D. Thesis, Dept. of Chem. and Pet. Eng., University of Alberta, Edmonton (1971).
11. NEWELL, R.B. and FISHER, D.G., "Experimental Evaluation of Optimal Multivariable Regulatory Controllers with Model-Following Capabilities", Automatica, Vol. 8, 247 (1972).
12. NICHOLSON, H., "Integrated Control of a Nonlinear Boiler Model", Proc. IEE, Vol. 114, No. 10, 1569 (1971).
13. NICHOLSON, H., "Dual-Mode Control of a Time-Varying Boiler Model with Parameter and State Estimation", Proc. IEE, Vol. 112, No. 2, 383 (1965).
14. NICHOLSON, H., "Dynamic Optimization of a Boiler", Proc. IEEE, Vol. 111, No. 8, 1479 (1964).
15. ROGERS, R.O. and SWORDER, D.D., "Suboptimal Control of Linear Systems Derived from Models of Lower Dimension", Preprints, 1971 Joint Automatic Control Conference, St. Louis, 926.
16. SIMS, C.S., and MELSA, J.L., "A Survey of Specific Optimal Techniques in Control and Estimation", Int. J. Control, Vol. 14, No. 2, 299 (1971).
17. SMITH, H.W. and DAVISON, E.J., "Design of Industrial Regulators: Integral Feedback and Feedforward Control", Proc. IEE, Vol. 119, No. 8, 1210 (1972).
18. WILSON, R.G., FISHER, D.G. and SEBORG, D.E., "Model Reduction for Discrete-Time Dynamic Systems", Int. J. Control, Vol. 16, No. 3, 549 (1972).

APPENDIX

(a) Evaporator Model

The state space model used in this work is that of the double effect evaporator described by Newell [10]. The model is in the form of equations (1) and (2) with the elements of the vectors $\underline{x}$, $\underline{u}$, $\underline{d}$, and $\underline{y}$ defined as normalized perturbation variables; e.g.

$$x_3 = \frac{W1 - W1_{ss}}{W1_{ss}}$$

where $W1_{ss}$ is the normal steady state value of W1. The vectors $\underline{x}$, $\underline{u}$, $\underline{d}$, and $\underline{y}$ are defined as follows:

| State Vector, $\underline{x}$: | | Normal Steady State Value |
|---|---|---|
| TS | Steam Temperature | 118°C |
| TW1 | First effect tube wall temperature | 108°C |
| W1 | First effect holdup | 20.8 kg. |
| C1 | First effect concentration | 4.59% glycol |
| H1 | First effect enthalpy | 441 kJ/kg. |
| TW2 | Second effect tube wall temperature | 83°C |
| W2 | Second effect holdup | 19 kg. |
| C2 | Second effect concentration | 10.11% glycol |
| H2 | Second effect enthalpy | 312 kJ/kg. |
| TW3 | Condenser tube wall temperature | 42°C |

Control Vector, $\underline{u}$:

| | | |
|---|---|---|
| S | Steam flow | 0.91 kg/min. |
| B1 | First effect bottoms flow | 1.58 kg/min. |
| B2 | Second effect bottoms flow | 0.72 kg/min. |

Disturbance Vector, $\underline{d}$:

| | | |
|---|---|---|
| F | Feed flow | 2.26 kg/min. |
| CF | Feed concentration | 3.2% glycol |
| HF | Feed enthalpy | 365 kJ/kg. |

Output Vector, $\underline{y}$:

$\underline{y}^T = (W1, W2, C2)$

(b) Tenth Order Model and Optimal Control Law

The coefficient matrices of equation (1) are shown in Table IV along with the modal matrix $\underline{\underline{M}}$ and the corresponding eigenvalues. Table V shows the high-order, optimal control matrices for the control law of equation (3) and the tenth order state space model. These control laws were calculated with

$$\underline{\underline{Q}} = \text{diag.}[0,0,10,0,0,0,10,100,0,0] \ ,$$

$$\underline{\underline{R}} = \text{diag.}[0.05,0.05,0.05]$$

for the proportional feedback, setpoint case, and with

$$\underline{\underline{Q}} = \text{diag.}[0,0,10,0,0,0,10,100,0,0,0.5,0.5,1.] \ ,$$

$$\underline{\underline{R}} = \text{diag.}[0.05,0.05,0.05]$$

for the proportional plus integral case.

(c) Third Order Model and Optimal Control Law

The third order model was derived from the tenth-order model by model reduction [9] and is in the form of equation (1) with

$$\underline{\underline{\Phi}} = \begin{bmatrix} 0.9998E\ 00 & -0.3795E-07 & -0.2348E-06 \\ 0.8633E-07 & 0.9998E\ 00 & 0.1662E-05 \\ -0.9408E-05 & -0.3070E-04 & 0.9600E\ 00 \end{bmatrix}$$

$$\underline{\underline{\Delta}} = \begin{bmatrix} -0.3250E-01 & -0.8108E-01 & 0.8226E-08 \\ -0.3770E-01 & 0.8539E-01 & -0.4063E-01 \\ 0.5272E-01 & -0.4413E-01 & 0.6023E-06 \end{bmatrix}$$

$$\underline{\underline{\Theta}} = \begin{bmatrix} 0.1200E\ 00 & -0.0780E-04 & -0.1345E-01 \\ 0.3249E-02 & -0.7866E-04 & -0.1561E-01 \\ -0.2177E-01 & 0.3979E-01 & 0.2183E-01 \end{bmatrix}$$

and using the state vector $\underline{x}_1^T = [W1,W2,C2]$.

The optimal control laws for this model were calculated using the quadratic performance index of equation (25) with the following weighting matrices:

$$\underline{\underline{Q}} = \text{diag.}[10,10,100] \ ,$$

$$\underline{\underline{R}} = \text{diag.}[0.05,0.05,0.05]$$

for the proportional feedback, setpoint case and with

$$\underline{\underline{Q}} = \text{diag.}[10,10,100,0.5,0.5,1.0] \ ,$$

$$\underline{\underline{R}} = \text{diag.}[0.05,0.05,0.05]$$

for the proportional plus integral case.

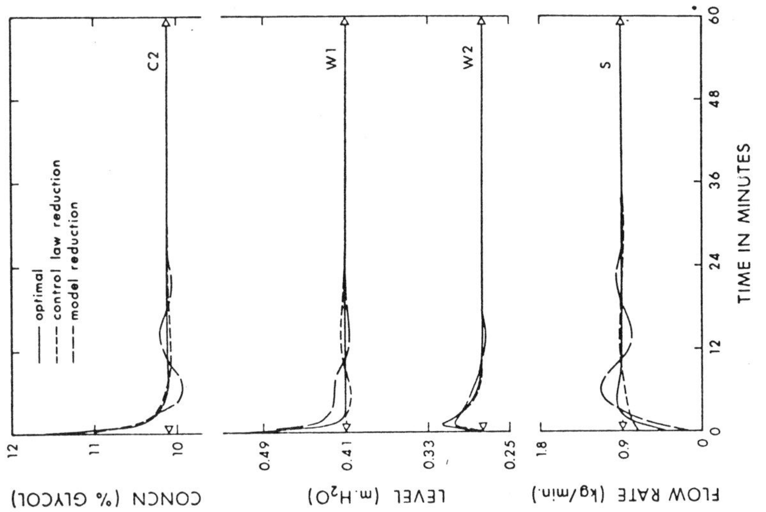


1) Simulated responses for two third order control systems and for the optimal 10th order control system using proportional feedback control.

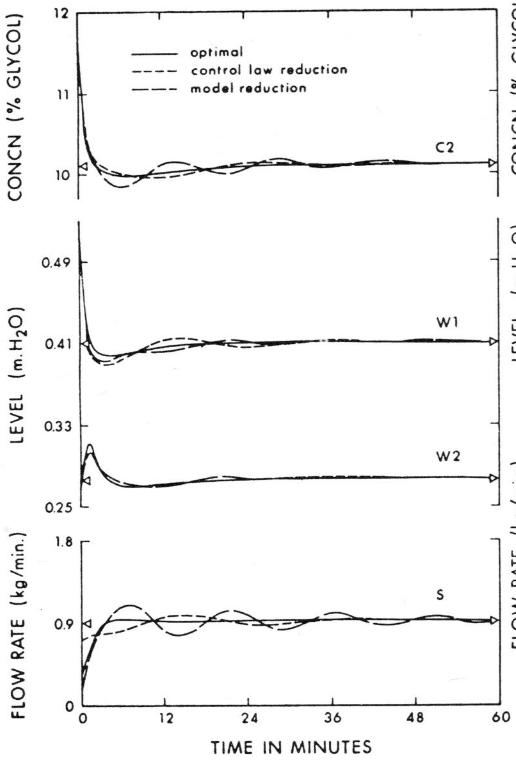


2) Simulated responses for proportional-integral feedback control.

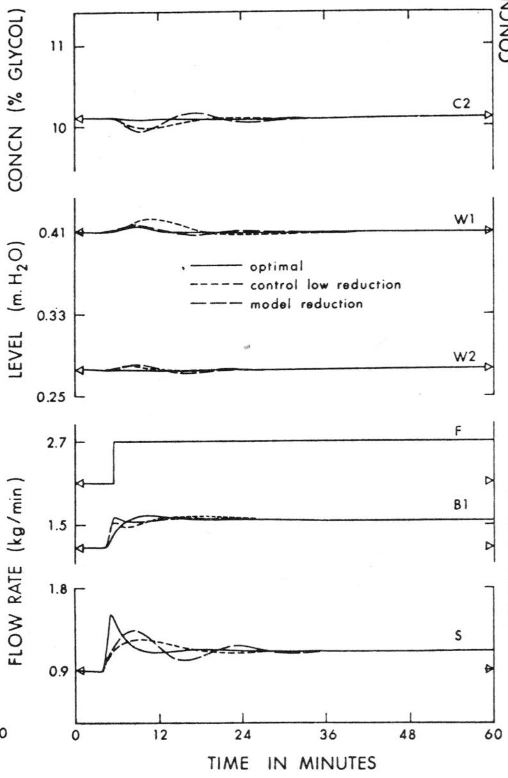


3) Simulated responses for proportional feedback-feedforward control and a +20% step disturbance in feed flow.

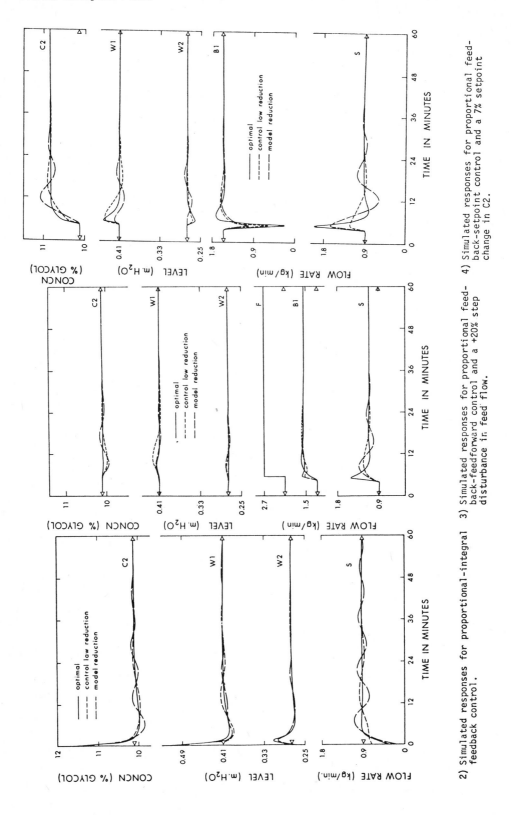

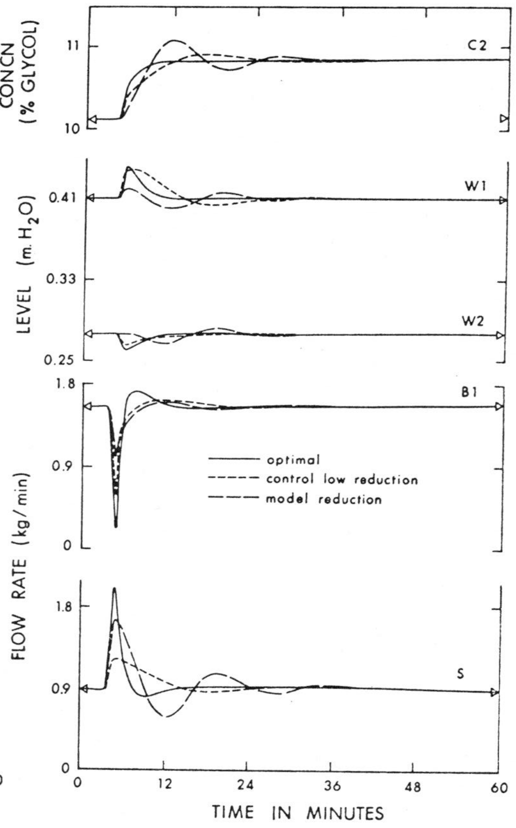


4) Simulated responses for proportional feedback-setpoint control and a 7% setpoint change in C2.

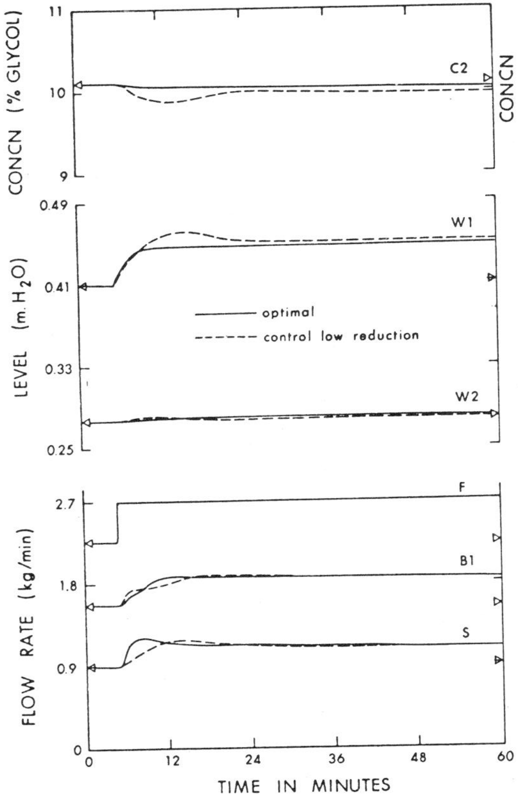


5) Simulated responses for proportional feed-back control and a +20% step disturbance in feed flow.

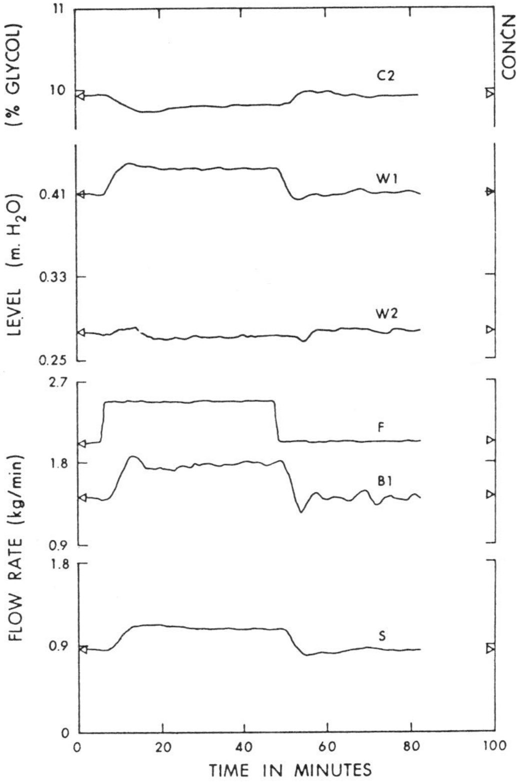


6) Experimental response to two 20% step disturbances in feed flow when the third order proportional feedback control law designed by the proposed "control law reduction" approach is useα.

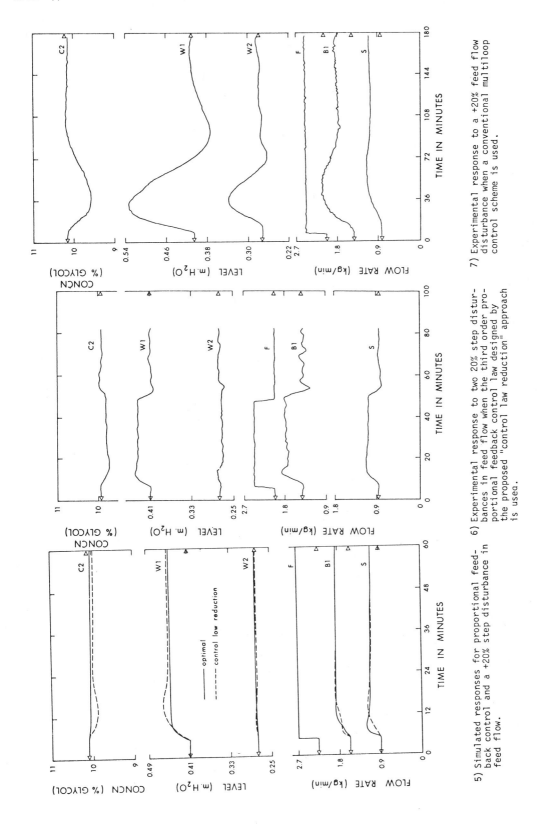

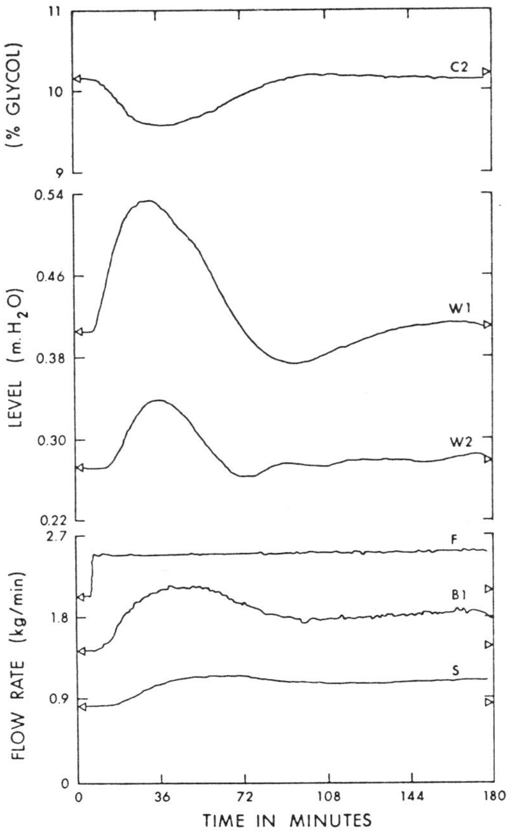


7) Experimental response to a +20% feed flow disturbance when a conventional multiloop control scheme is used.

## TABLE IV. TENTH ORDER EVAPORATOR MODEL

$\underline{\Phi}$ =

| | | | | | | | | | |
|---|---|---|---|---|---|---|---|---|---|
| 0.3836E-02 | 0.1092E 00 | -0.2568E-06 | 0.3219E-02 | 0.3757E 00 | 0.1501E-01 | -0.1414E-06 | 0.1911E-02 | 0.9376E-01 | 0.8400E-02 |
| 0.4188E-02 | 0.1193E 00 | -0.2814E-06 | 0.3515E-02 | 0.4102E 00 | 0.1641E-01 | -0.1553E-06 | 0.2090E-02 | 0.1024E 00 | 0.9192E-02 |
| -0.6923E-03 | -0.1978E-01 | 0.9998E 00 | -0.5016E-03 | -0.5753E-01 | 0.5370E-02 | -0.4962E-07 | 0.6722E-03 | 0.3296E-01 | 0.2933E-02 |
| 0.6573E-03 | 0.1878E-01 | -0.1835E-04 | 0.9217E 00 | 0.5487E-01 | -0.5013E-02 | 0.4812E-07 | -0.6369E-03 | -0.3123E-01 | -0.2799E-02 |
| 0.4961E-02 | 0.1413E 00 | -0.1851E-06 | -0.3800E-02 | 0.4860E 00 | 0.1967E-01 | -0.1934E-06 | 0.2518E-02 | 0.1234E 00 | 0.1110E-01 |
| 0.2950E-02 | 0.8418E-01 | -0.1406E-06 | 0.2438E-02 | 0.2919E 00 | 0.2297E-01 | -0.7174E-06 | 0.3745E-02 | 0.1837E 00 | 0.1900E-01 |
| -0.2556E-03 | -0.7329E-02 | 0.5755E-08 | -0.2180E-03 | -0.2633E-01 | -0.6145E-02 | 0.9998E 00 | -0.1168E-02 | -0.5731E-01 | 0.7482E-02 |
| 0.2687E-03 | 0.7704E-02 | -0.3787E-06 | 0.3846E-01 | 0.2721E-01 | 0.5823E-02 | -0.3054E-04 | 0.9610E 00 | 0.5503E-01 | -0.7216E-02 |
| 0.3158E-02 | 0.9017E-01 | -0.1113E-06 | 0.1806E-02 | 0.3140E 00 | 0.3064E-01 | -0.5159E-06 | -0.1436E-01 | 0.2560E 00 | 0.2725E-01 |
| 0.7696E-03 | 0.2197E-01 | -0.2511E-07 | 0.6338E-03 | 0.7682E-01 | 0.8611E-02 | -0.2671E-06 | 0.1498E-02 | 0.7357E-01 | 0.7963E-02 |

$\underline{\Delta}$ =

| | | |
|---|---|---|
| 0.1857E 00 | 0.2200E-02 | 0.2128E-08 |
| 0.1280E 00 | 0.2417E-02 | 0.2346E-08 |
| -0.9644E-02 | -0.8090E-01 | 0.7565E-09 |
| 0.9347E-02 | -0.7480E-03 | -0.7402E-09 |
| 0.1338E 00 | 0.3009E-02 | 0.3003E-08 |
| 0.5457E-01 | 0.1116E-01 | 0.1688E-07 |
| -0.2161E-02 | 0.8823E-01 | -0.4063E-01 |
| 0.2283E-02 | -0.4662E-01 | 0.6155E-06 |
| 0.4601E-01 | 0.1867E-01 | 0.1546E-07 |
| 0.9651E-02 | 0.4158E-02 | 0.5803E-08 |

$\underline{\Theta}$ =

| | | |
|---|---|---|
| -0.8858E-02 | 0.3069E-03 | 0.4238E-01 |
| -0.9710E-02 | 0.3364E-03 | 0.4645E-01 |
| 0.1180E 00 | -0.3028E-04 | -0.4170E-02 |
| -0.3491E-01 | 0.7847E-01 | 0.4038E-02 |
| -0.1155E-01 | -0.2673E-03 | 0.5671E-01 |
| -0.4878E-02 | 0.1656E-03 | 0.2334E-01 |
| 0.2005E-03 | -0.6683E-05 | -0.9603E-03 |
| -0.9045E-03 | 0.1601E-02 | 0.1013E-02 |
| -0.4134E-02 | 0.1053E-03 | 0.1985E-01 |
| -0.8757E-03 | 0.2913E-04 | 0.4192E-02 |

$\underline{M}$ =

| | | | | | | | | | |
|---|---|---|---|---|---|---|---|---|---|
| -0.3410E-07 | -0.1097E-06 | 0.2825E-05 | -0.1559E-03 | -0.3852E 00 | -0.1901E 00 | -0.2514E-01 | -0.8693E 00 | 0.8483E 00 | 0.9985E 00 |
| -0.3253E-07 | -0.1154E-06 | 0.6100E-05 | -0.1711E-03 | -0.4207E 00 | -0.2070E 00 | 0.3171E-01 | 0.4336E 00 | -0.1778E 00 | 0.1907E-01 |
| -0.7428E 00 | -0.7034E 00 | -0.1962E-02 | -0.3791E-04 | -0.7805E-01 | -0.5549E-01 | 0.1348E-02 | 0.5399E-03 | -0.1810E-02 | 0.2844E-03 |
| 0.1741E-03 | 0.1650E-03 | -0.9696E-05 | 0.7121E 00 | 0.1056E 00 | 0.5815E-01 | -0.1383E-02 | -0.5311E-03 | 0.1826E-02 | -0.2860E-03 |
| -0.1552E-05 | -0.1571E-05 | 0.9881E-05 | -0.6361E-02 | -0.5009E 00 | -0.2372E 00 | -0.2371E-02 | -0.1209E 00 | 0.3609E-01 | -0.1512E-01 |
| -0.6031E-06 | -0.6979E-06 | 0.1005E-04 | -0.2021E-03 | -0.3732E 00 | 0.4339E 00 | -0.6787E-01 | 0.2029E 00 | 0.4948E 00 | -0.4950E-01 |
| 0.6696E 00 | 0.7108E 00 | 0.3921E-03 | -0.4630E-03 | -0.1628E 00 | 0.4483E-01 | -0.1363E-01 | -0.2309E-03 | 0.7180E-04 | -0.1180E-04 |
| -0.3404E-03 | -0.3814E-03 | -0.9998E 00 | -0.7019E 00 | 0.1717E 00 | -0.4719E-01 | 0.1372E-01 | 0.1496E-03 | -0.4512E-04 | 0.7214E-05 |
| 0.5949E-05 | 0.6661E-05 | 0.2041E-01 | 0.1406E-01 | -0.4450E 00 | 0.7779E 00 | -0.1046E 00 | -0.1905E-01 | -0.5072E-01 | 0.5352E-02 |
| -0.2670E-06 | -0.2985E-06 | 0.3026E-05 | -0.6663E-04 | -0.1150E 00 | 0.2465E 00 | 0.9912E 00 | 0.1018E-01 | -0.5961E-02 | 0.8351E-03 |

**Eigenvalues**

| | | | | | | | | | |
|---|---|---|---|---|---|---|---|---|---|
| 0.9999E 00 | 0.9999E 00 | 0.9600E 00 | 0.9212E 00 | 0.7354E 00 | 0.1622E 00 | 0.1271E-03 | 0.2166E-04 | 0.6938E-05 | -0.3728E-06 |

TABLE V. OPTIMAL CONTROL MATRICES FOR TENTH ORDER MODEL

Proportional feedback, feedforward, setpoint control:

$$\underline{K}^{FB} = \begin{bmatrix} -0.2438E{-}01 & -0.6964E\ 00 & 0.8790E\ 01 & -0.1711E\ 01 & -0.2304E\ 01 & -0.1072E\ 00 & -0.4185E{-}01 & -0.1530E\ 02 & -0.1027E\ 01 & 0.9727E{-}01 \\ 0.2031E{-}02 & 0.5833E{-}01 & 0.2547E\ 01 & 0.5391E\ 00 & 0.2371E\ 00 & 0.9395E{-}01 & -0.1307E\ 01 & 0.1163E\ 02 & 0.8415E\ 00 & -0.8783E{-}01 \\ 0.2804E{-}03 & 0.8053E{-}02 & 0.2722E\ 01 & 0.1117E\ 01 & 0.4854E{-}01 & 0.3186E{-}01 & 0.9751E\ 01 & 0.1215E\ 02 & 0.2624E\ 00 & -0.6441E{-}02 \end{bmatrix}$$

$$\underline{K}^{FF} = \begin{bmatrix} 0.3370E\ 01 & -0.2147E\ 00 & -0.4188E\ 00 \\ 0.4360E\ 00 & 0.3824E{-}01 & -0.9721E{-}03 \\ 0.8283E\ 00 & 0.1907E\ 00 & -0.1591E{-}02 \end{bmatrix} \qquad \underline{K}^{SP} = \begin{bmatrix} -0.8791E\ 01 & 0.4340E{-}01 & 0.1721E\ 02 \\ -0.2547E\ 01 & 0.1307E\ 01 & -0.1219E\ 02 \\ -0.2722E\ 01 & -0.9755E\ 01 & -0.1346E\ 02 \end{bmatrix}$$

Proportional-integral feedback control:

$$\underline{K}^{FB} = \begin{bmatrix} -0.2973E{-}01 & -0.8492E\ 00 & 0.1213E\ 02 & -0.1983E\ 01 & -0.2798E\ 01 & -0.1281E\ 00 & 0.2329E\ 00 & -0.2024E\ 02 & -0.1268E\ 01 & 0.1411E\ 00 \\ 0.1333E{-}02 & 0.3841E{-}01 & 0.3133E\ 01 & 0.5214E\ 00 & 0.1743E\ 00 & 0.9276E{-}01 & -0.1261E\ 01 & 0.1108E\ 02 & 0.8215E\ 00 & -0.8202E{-}01 \\ -0.3982E{-}03 & -0.1133E{-}01 & 0.3905E\ 01 & 0.1225E\ 01 & -0.6425E{-}02 & 0.3968E{-}01 & 0.1223E\ 02 & 0.1476E\ 02 & 0.3168E\ 00 & -0.2946E{-}02 \end{bmatrix}$$

$$\underline{K}^{I} = \begin{bmatrix} 0.1693E\ 01 & 0.5076E{-}01 & -0.1807E\ 01 \\ 0.5681E\ 00 & -0.2919E\ 00 & 0.1079E\ 01 \\ 0.6092E\ 00 & 0.1932E\ 01 & 0.1352E\ 01 \end{bmatrix}$$

# Model Reduction and the Design of Reduced-Order Control Laws

This paper is concerned with the problem of designing satisfactory low-order (incomplete state feedback) controllers starting with a high-order, state-space model. Two design approaches are considered: the control law reduction technique, recently developed by the authors (Wilson et al., 1973), and the well-known model reduction approach. These two design techniques are used to develop a variety of low-order controllers for a double-effect evaporator starting with a 10th-order model. Experimental and simulated response data from the computer-controlled evaporator demonstrate the superiority of the control law reduction approach in this application. It is also shown that several of the previously published modal approaches to model reduction are basically equivalent since they yield identical reduced-order models.

ROBERT G. WILSON
D. GRANT FISHER
and
DALE E. SEBORG

Department of Chemical Engineering
The University of Alberta
Edmonton, Alberta, Canada T6G 2G6

## SCOPE

In many practical applications of modern control theory the most difficult problem is to obtain a suitable process model. An analytical approach using basic chemical engineering principles often results in a dynamic model which consists of a large number of nonlinear differential equations. The model is usually too complicated for use in controller design or for implementation as part of the actual control system. Consequently, for purposes of control system design it is common practice to linearize the model and assume time-invariant behavior. Fortunately, the resulting linear "state-space model" (that is, a set of first-order differential equations) will usually adequately describe the process transients in the region of normal operation and provide a suitable basis for control system design. However, many of the multivariable control design techniques (Gould, 1969) result in a control law that requires the availability of all elements of the state vector. Such control laws are frequently impractical because of their complexity and because in many applications it is not practical to measure or estimate all of the state variables. Hence there has been widespread interest in developing control laws that require only a subset of the state vector to be available. The resulting controllers are usually referred to as *incomplete state feedback* or *low-order controllers.*

This investigation is concerned with the problem of designing a satisfactory low-order, multivariable controller starting with a high-order, state-space model of a process. To be judged satisfactory, the low-order controller should perform almost as well as more complicated high-order (state-feedback) controllers. Emphasis is placed on discrete process models and discrete controllers since they are more convenient than their continuous counterparts for on-line implementation via digital process computers. Two basic approaches were used to design low-order controllers:

1. Model Reduction Approach. The original high-order model is simplified using a modal analysis to eliminate selected state variables, that is, to reduce the order of the model. The resulting low-order model then serves as a basis for designing a low-order controller using standard state-feedback design methods (for example, optimal control).

2. Control Law Reduction Approach. The high-order model is first used to design a high-order controller using standard state-feedback design techniques. The high-order controller is then simplified by eliminating selected state variables via a modal analysis of the high-order model. A low-order controller results. In both approaches, the high-order and low-order controllers can include integral feedback, feedforward and setpoint terms, in addition to proportional feedback.

The first approach, model reduction, has been applied in many previous investigations (for example, Nicholson, 1964, 1967; Anderson, 1969). The second approach, control law reduction, has recently been developed by the present authors (Wilson et al., 1973; Wilson, 1974). An objective of this investigation is to critically evaluate the performance of low-order, multivariable controllers designed using these two approaches. The evaluation includes both experimental and simulated data for a double effect evaporator.

## CONCLUSIONS AND SIGNIFICANCE

The two design approaches mentioned above were used to design reduced-order controllers for a computer-controlled, double effect evaporator at the University of Alberta. The starting point in the design was a theoretical 10th-order, state-space model of the evaporator derived from linearized material and energy balances. A variety of model reduction procedures were then employed to generate 3rd, 4th, and 5th-order models as well as 3rd and 5th-order controllers. Additional low-order controllers were designed via the control law reduction approach.

Correspondence concerning this paper should be addressed to D. G. Fisher or D. E. Seborg. R. G. Wilson is with Imperial Oil Enterprises, Ltd., Edmonton, Alberta.

AIChE Journal, Vol. 20, No. 6, pp. 1131-1140 (1974).

**These low-order controllers were then evaluated in simulation and experimental studies.**

**The evaluation of the low-order controllers included the following items: several model reduction techniques; different sequences of design steps to arrive at a low-order controller of specified order (see Figure 1); various multivariable controllers which included proportional feedback, integral feedback, feedforward and setpoint control modes; transient responses to a variety of disturbances.**

**The control law reduction approach developed by the authors (Wilson et al., 1973; Wilson, 1974) produced low-order controllers that were more reliable, smoother-acting and more practical than those resulting from the model reduction approach. Individual elements in the controller matrices produced by the two approaches differed by as much as a factor of three. The controller performance also depended on the modes selected, for example, proportional plus integral control vs. proportional feedback plus feedforward control.**

**In applying the model reduction approach, it was possible to design satisfactory 3rd-order controllers even though the 3rd-order process models showed significant errors in their open-loop responses. The resulting controllers were practical, easy to implement, and performed almost as well as 5th-order controllers. Furthermore, the 3rd-order controllers were superior to a conventional multi-loop control scheme consisting of three single variable controllers.**

**Of the various model reduction techniques, those techniques that guarantee agreement between the high-order and the reduced-order models at both the initial conditions and at steady state appear more desirable for control applications. Also the choice of which eigenvalues of the high-order model to retain in the low-order model can be a very important decision (compare Figure 3).**

**A further conclusion of this study is that several of the previously published modal approaches to model reduction are essentially equivalent since they produce identical reduced-order models. Apparently, the equivalence of these independently developed techniques has not been previously reported.**

**The design methods and practical experience resulting from this investigation should be of interest to those planning industrial applications of multivariable control techniques.**

---

Multivariable control systems designed using state-space models often require that all of the system state variables are available. However, in most process control problems this is not the actual situation, and consequently there has been considerable interest in the development of control algorithms which require the availability of only some, rather than all, state variables. Such controllers are referred to in this paper as *low-order* or *reduced-order* controllers.

Figure 1 illustrates the various approaches that can be used to design a discrete reduced-order controller starting from a high-order, continuous model. In the model reduction approach, a low-order model is first derived from the high-order model and a controller is then designed for the low-order model using state feedback design methods (for example, Anderson, 1969; Nicholson, 1964, 1967). This controller can then be used as a reduced-order controller for the high-order system. This approach consists of steps 1, 3, and 7 or steps 2, 4, and 7 depending on whether the model reduction step precedes the discretization step or vice versa. Rogers and Sworder (1971) have suggested that the model reduction and controller design steps be performed simultaneously, subject to the additional restriction that the resulting reduced-order controller be the best suboptimal controller for the high-order system.

An alternative approach is to first design a satisfactory state feedback controller for the high-order model and then use it and the high-order model to design the low-order controller (that is, steps 2, 5, and 8 in Figure 1). This approach, control law reduction, has been developed by the authors (Wilson et al., 1973) and does not require the calculation of a low-order model.

A third design approach is to generate a reduced-order controller directly from the high-order model without calculating a reduced-order model or a high-order controller

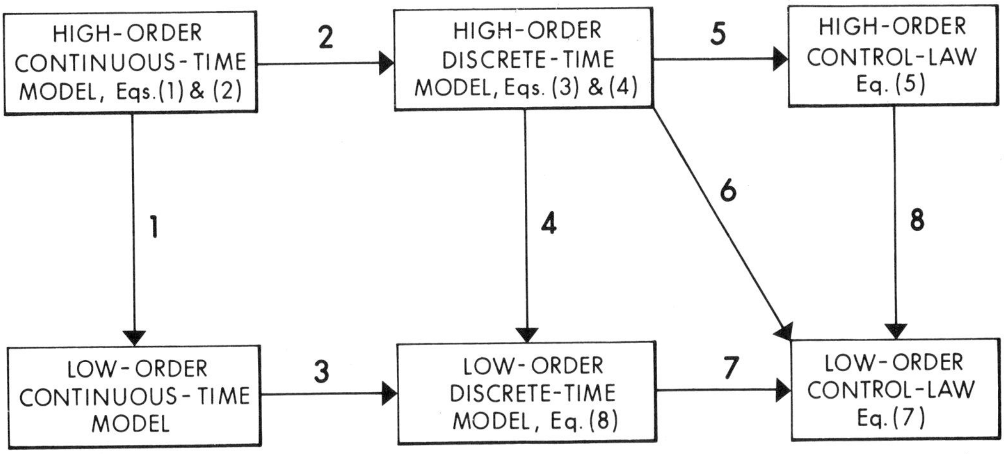


**Fig. 1. Diagrammatic representation of the different steps and paths for proceeding from a high-order continuous process model to a low-order discrete control law.**

(that is, steps 2 and 6). This approach includes a wide variety of design methods such as modal control (Gould, 1969), specific optimal control, and eigenvalue assignment techniques (Davison, 1972). These design methods are the subject of current investigations by the authors and co-workers and will not be considered further in this paper.

## PROBLEM FORMULATION

Consider the linear, time-invariant, continuous-time, state-space model

$$\dot{\mathbf{x}}(t) = \mathbf{A}\,\mathbf{x}(t) + \mathbf{B}\,\mathbf{u}(t) + \mathbf{D}\,\mathbf{d}(t) \quad (1)$$

$$\mathbf{y}(t) = \mathbf{C}\,\mathbf{x}(t) \quad (2)$$

where the state vector $\mathbf{x}$, input vector $\mathbf{u}$, disturbance vector $\mathbf{d}$, and output vector $\mathbf{y}$ are column vectors of dimensions $n$, $m$, $q$ and $p$, respectively. Matrices $\mathbf{A}$, $\mathbf{B}$, $\mathbf{C}$, and $\mathbf{D}$ are constant matrices of the appropriate dimensions. The corresponding discrete-time model can be derived from Equations (1) and (2) using well-known techniques (for example, Lapidus and Luus, 1967; Ogata, 1967) and is written

$$\mathbf{x}(j+1) = \boldsymbol{\varphi}\,\mathbf{x}(j) + \boldsymbol{\Delta}\,\mathbf{u}(j) + \boldsymbol{\theta}\,\mathbf{d}(j) \quad (3)$$

$$\mathbf{y}(j) = \mathbf{C}\,\mathbf{x}(j) \quad j = 0, 1, 2, \ldots \quad (4)$$

A general multivariable controller (Newell et al., 1972a, c) for the system in Equations (3) and (4) is given by Equation (5)

$$\mathbf{u}(j) = \mathbf{K}_1^{FB}\,\mathbf{x}_1(j) + \mathbf{K}_2^{FB}\,\mathbf{x}_2(j) + \mathbf{K}^{FF}\,\mathbf{d}(j) + \mathbf{K}^{SP}\,\mathbf{y}^{SP}(j) + \mathbf{K}^{I}\sum_{i=0}^{j}[\mathbf{y}(i) - \mathbf{y}^{SP}(i)] \quad (5)$$

where $\mathbf{x}$ has been partitioned into a $l$ vector $\mathbf{x}_1$ of variables which are available for purposes of control and a $(n-l)$ vector $\mathbf{x}_2$:

$$\mathbf{x} = \begin{bmatrix} \mathbf{x}_1 \\ \mathbf{x}_2 \end{bmatrix} \quad (6)$$

Vector $\mathbf{y}^{SP}$ denotes the setpoint of $\mathbf{y}$ and the superscripts on the controller gain matrices (for example, $\mathbf{K}^{FB}$) denote proportional feedback, feedforward, setpoint, and integral feedback control modes, respectively.

The multivariable control law in Equation (5) can only be implemented if a measurement or estimate of each state variable is available. A suitable reduced-order control law which requires only $\mathbf{x}_1$ rather than $\mathbf{x}$ is of the form:

$$\mathbf{u}(j) = \mathbf{K}_R^{FB}\,\mathbf{x}_1(j) + \mathbf{K}_R^{FF}\,\mathbf{d}(j) + \mathbf{K}_R^{SP}\,\mathbf{y}^{SP}(j) + \mathbf{K}_R^{I}\sum_{i=0}^{j}[\mathbf{y}(i) - \mathbf{y}^{SP}(i)] \quad (7)$$

where the subscript $R$ refers to the reduced-order controller. Note that $\mathbf{y}$ is retained in the reduced-order control law since its elements are, by definition, the measured output variables.

The control problem of interest is to design a reduced-order controller of the form of Equation (7) starting with the high-order model in Equations (3) and (4). In the following sections two general design approaches, model reduction and control law reduction, are considered.

## MODEL REDUCTION

One approach for designing a reduced-order controller is to first use a model reduction technique to derive a reduced-order model which contains $\mathbf{x}_1$ but not $\mathbf{x}_2$. The resulting reduced-order model and state feedback design methods can then be used to design a reduced-order controller of the form of Equation (7) (Nicholson, 1964, 1967; Anderson, 1969). In the model reduction step, the objective is to derive the reduced-order model in Equation (8) starting from the high-order model in Equation (3):

$$\mathbf{x}_1(j+1) = \boldsymbol{\varphi}_R\,\mathbf{x}_1(j) + \boldsymbol{\Delta}_R\,\mathbf{u}(j) + \boldsymbol{\theta}_R\,\mathbf{d}(j) \quad (8)$$

Existing techniques for reducing the order of a multi-input, multi-output, state-space model can be divided into two categories: modal approaches and least squares approaches (Sinha and De Bruin, 1973; Wilson, 1974). Various modal approaches are briefly described below; least squares methods have been considered elsewhere (Wilson, 1974).

### Modal Approach to Model Reduction

In this approach the strategy is to retain certain modes (or eigenvalues) of the high-order model in the low-order model. Consequently, the approach is based on the following modal analysis (for example, Gould, 1969; Friedly, 1972) of the high-order model.

For any square matrix $\boldsymbol{\varphi}$, there exists an $n \times n$ matrix $\mathbf{M}$ which transforms Equation (3) into its Jordan canonical form (Gantmacher, 1959; Friedly, 1972):

$$\begin{bmatrix} \mathbf{z}_1(j+1) \\ \mathbf{z}_2(j+1) \end{bmatrix} = \begin{bmatrix} \boldsymbol{\alpha}_1 & \mathbf{0} \\ \mathbf{0} & \boldsymbol{\alpha}_2 \end{bmatrix}\begin{bmatrix} \mathbf{z}_1(j) \\ \mathbf{z}_2(j) \end{bmatrix} + \begin{bmatrix} \boldsymbol{\delta}_1 \\ \boldsymbol{\delta}_2 \end{bmatrix}\mathbf{u}(j) + \begin{bmatrix} \boldsymbol{\eta}_1 \\ \boldsymbol{\eta}_2 \end{bmatrix}\mathbf{d}(j) \quad (9)$$

where $\boldsymbol{\alpha}$, $\boldsymbol{\delta}$ and $\boldsymbol{\eta}$ are defined as

$$\boldsymbol{\alpha} \equiv \mathbf{M}^{-1}\,\boldsymbol{\varphi}\,\mathbf{M} = \begin{bmatrix} \boldsymbol{\alpha}_1 & \mathbf{0} \\ \mathbf{0} & \boldsymbol{\alpha}_2 \end{bmatrix} \quad (10)$$

$$\boldsymbol{\delta} \equiv \mathbf{M}^{-1}\,\boldsymbol{\Delta} = \begin{bmatrix} \boldsymbol{\delta}_1 \\ \boldsymbol{\delta}_2 \end{bmatrix}, \quad \boldsymbol{\eta} \equiv \mathbf{M}^{-1}\,\boldsymbol{\theta} = \begin{bmatrix} \boldsymbol{\eta}_1 \\ \boldsymbol{\eta}_2 \end{bmatrix} \quad (11)$$

and $\boldsymbol{\alpha}_1$ and $\boldsymbol{\alpha}_2$ are block diagonal matrices with the eigenvalues of $\boldsymbol{\varphi}$ as the diagonal elements. Vector $\mathbf{z}$ is the canonical state vector defined by the similarity transformation

$$\mathbf{x} = \mathbf{M}\,\mathbf{z} \quad (12)$$

or in partitioned form as

$$\begin{bmatrix} \mathbf{x}_1 \\ \mathbf{x}_2 \end{bmatrix} = \begin{bmatrix} \mathbf{M}_1 & \mathbf{M}_2 \\ \mathbf{M}_3 & \mathbf{M}_4 \end{bmatrix}\begin{bmatrix} \mathbf{z}_1 \\ \mathbf{z}_2 \end{bmatrix} \quad (13)$$

The partitions of $\mathbf{z}$, namely $\mathbf{z}_1$ and $\mathbf{z}_2$, have the same dimensions as $\mathbf{x}_1$ and $\mathbf{x}_2$, respectively.

At this point in the analysis, various expressions for $\boldsymbol{\varphi}_R$, $\boldsymbol{\Delta}_R$, and $\boldsymbol{\theta}_R$ can be derived depending on the particular assumptions that are made (Marshall, 1966; Davison, 1968a, b; Chidambara and Davison, 1967a, b, c). In the original papers, the model reduction techniques were developed for continuous models; however, the extensions to discrete models are straightforward and have been presented by the authors (Wilson et al., 1972; Wilson, 1974). Also, when a modal approach is used to reduce a high-order continuous model to a low-order discrete model, the same low-order model is obtained regardless whether the continuous model is first discretized and then reduced, or vice versa (Wilson et al., 1972).

### Marshall's Method

In Marshall's model reduction technique (Marshall, 1966), it is assumed that the dominant eigenvalues of the high-order system are located in $\boldsymbol{\alpha}_1$ and the less significant eigenvalues are located in $\boldsymbol{\alpha}_2$. (The dominant eigenvalues

TABLE 1. DISCRETE REDUCED-ORDER MODELS DERIVED FROM THREE MODAL APPROACHES

| Design method | $\boldsymbol{\varphi}_R$ | $\boldsymbol{\Delta}_R$ | $\boldsymbol{\theta}_R$ | $\mathbf{E}_R$ | $\mathbf{F}_R$ |
|---|---|---|---|---|---|
| Marshall | $\boldsymbol{\varphi}_1 - \boldsymbol{\varphi}_2\mathbf{V}_4^{-1}\mathbf{V}_3$ | $\boldsymbol{\Delta}_1 + \boldsymbol{\varphi}_2\mathbf{V}_4^{-1}(\mathbf{I} - \boldsymbol{\alpha}_2)^{-1}\boldsymbol{\delta}_2$ | $\boldsymbol{\theta}_1 + \boldsymbol{\varphi}_2\mathbf{V}_4^{-1}(\mathbf{I} - \boldsymbol{\alpha}_2)^{-1}\boldsymbol{\eta}_2$ | — | — |
| Davison | same | $\mathbf{M}_1\boldsymbol{\delta}_1$ | $\mathbf{M}_1\boldsymbol{\eta}_1$ | — | — |
| Revised Davison | same | $\mathbf{M}_1\boldsymbol{\delta}_1$ | $\mathbf{M}_1\boldsymbol{\eta}_1$ | $\mathbf{M}_2(\mathbf{I} - \boldsymbol{\alpha}_2)^{-1}\boldsymbol{\delta}_2$ | $\mathbf{M}_2(\mathbf{I} - \boldsymbol{\alpha}_2)^{-1}\boldsymbol{\eta}_2$ |

are those which have the largest effect on $\mathbf{x}_1$ and/or have large absolute values and thus correspond to slow modes). This situation can always be realized by an appropriate rearrangement of the elements of $\boldsymbol{\alpha}$ and the columns of $\mathbf{M}$. This partitioning of $\boldsymbol{\alpha}$ justifies the simplifying assumption that $\mathbf{z}_2$ reacts instantaneously to changes in $\mathbf{u}$ or $\mathbf{d}$. Once this approximation is made, the expressions for $\boldsymbol{\varphi}_R$, $\boldsymbol{\Delta}_R$, and $\boldsymbol{\theta}_R$ in Table 1 can be derived (Wilson et al., 1972, 1973). In Table 1, $\boldsymbol{\Delta}_1$, $\boldsymbol{\varphi}_1$ and $\boldsymbol{\varphi}_2$ are partitions of $\boldsymbol{\Delta}$ and $\boldsymbol{\varphi}$, and $\mathbf{V}_3$, and $\mathbf{V}_4$ are partitions of $\mathbf{V}$ where $\mathbf{V} \equiv \mathbf{M}^{-1}$ (Wilson et al., 1972, 1974). An important advantage of Marshall's method is that the reduced-order model has the same steady state as the high-order model for step disturbances.

As noted by Graham (1968), one of Chidambara's proposed methods (Chidambara and Davison, 1967b) is equivalent to Marshall's method in the sense that the resulting reduced-order models are the same.

#### Davison's Method

A second basic model reduction technique has been developed by Davison (1966) and Nicholson (1964). As shown by Wilson (1974), these two formulations generate the same reduced-order model. Since this approach is referred to in the literature as *Davison's method,* the same designation will be used in this paper.

The basis of this approach is to derive a reduced-order model in which the modes are excited in the same relative proportions as they are in the high order model. This objective is achieved but at the expense of a steady state error between the high-order and low-order model responses to a step disturbance. The resulting reduced-order model is shown in Table 1. Several modifications of Davison's method have been reported but many of these are equivalent, as demonstrated in Appendix I.

Davison (1968b) suggested a revision to his earlier method (Davison, 1966) which would ensure steady state agreement for single-input systems. Here each element of $\mathbf{x}_1$ is multiplied by the ratio of the desired and actual steady state values and it is assumed that both $\boldsymbol{\varphi}$ and $\boldsymbol{\varphi}_R$ are nonsingular.

A second modification of Davison's method has been presented in different forms by both Chidambara and Davison (1967b). Chidambara later noted the equivalence of these two formulations (Chidambara and Davison, 1967c). The reduced-order model is shown in Table 1 as the revised Davison method. Fossard (1970) and Graham (1968) have also proposed model reduction techniques, but their approaches yield reduced-order models which are identical to those of the previously reported revision of Davison's method (Wilson, 1974). Apparently, the equivalence of these methods has not been previously reported and hence is included in the Appendix. The resulting reduced-order model can be written as

$$\mathbf{x}_R \equiv \mathbf{x}_l + \mathbf{E}_R\,\mathbf{u} + \mathbf{F}_R\,\mathbf{d} \qquad (14))$$

where the $l$-dimensional vector $\mathbf{x}_l$ is calculated from

$$\mathbf{x}_l(j+1) = \boldsymbol{\varphi}_R\,\mathbf{x}_l(j) + \boldsymbol{\Delta}_R\,\mathbf{u}(j) + \boldsymbol{\theta}_R\,\mathbf{d}(j) \qquad (15)$$

As shown in Table 1, the $\boldsymbol{\varphi}_R$, $\boldsymbol{\theta}_R$ and $\boldsymbol{\Delta}_R$ matrices in Equation (15) are the same as in Davison's original method.

## CONTROL LAW REDUCTION

In many control problems a satisfactory state feedback controller for the high-order model can be designed but is not practical for actual implementation due to the unavailability of a significant number of state variables. The model reduction approach described in the previous section could be used, but it does make use of all the available information, namely, the high-order control law. An alternative strategy is to use both the high-order controller and the high-order model to design the reduced-order controller. This approach will be referred to as control law reduction and consists of steps 5 and 8 in Figure 1. Control law reduction based on a modal analysis of the high-order model has been developed by the authors (Wilson et al., 1973) and is briefly summarized below.

The basis of the method is to use a modal analysis of the high-order model in Equation (3) to derive an approximate expression for $\mathbf{x}_2$ in terms of $\mathbf{x}_1$, $\mathbf{u}$, and $\mathbf{d}$. The resulting expression for $\mathbf{x}_2$ is approximate because of the simplifying assumption that the fast modes of the high-order model (corresponding to $\mathbf{z}_2$) react instantaneously (that is, the same assumption that is made in Marshall's model reduction method). Substitution of this expression into Equation (5) and subsequent rearrangement yields the reduced-order controller of Equation (7) with the following control matrices (Wilson et al., 1973):

$$\begin{aligned}
\mathbf{K}_R^{FB} &= \mathbf{K}^u\,(\mathbf{K}_1^{FB} - \mathbf{K}_2^{FB}\,\mathbf{V}_4^{-1}\,\mathbf{V}_3) \\
\mathbf{K}_R^{I} &= \mathbf{K}^u\,\mathbf{K}^{I} \\
\mathbf{K}_R^{FF} &= \mathbf{K}^u\,(\mathbf{K}^{FF} + \mathbf{K}_2^{FB}\,\mathbf{V}_4^{-1}\,(\mathbf{I} - \boldsymbol{\alpha}_2)^{-1}\,\boldsymbol{\eta}_2) \qquad (16) \\
\mathbf{K}_R^{SP} &= \mathbf{K}^u\,\mathbf{K}^{SP} \\
\mathbf{K}^u &= (\mathbf{I} - \mathbf{K}_2^{FB}\,\mathbf{V}_4^{-1}\,(\mathbf{I} - \boldsymbol{\alpha}_2)^{-1}\,\boldsymbol{\delta}_2)^{-1}
\end{aligned}$$

where

$$\mathbf{V} = \begin{bmatrix} \mathbf{V}_1 & \mathbf{V}_2 \\ \mathbf{V}_3 & \mathbf{V}_4 \end{bmatrix} = \mathbf{M}^{-1} \qquad (17)$$

## APPLICATION TO A PILOT PLANT EVAPORATOR

The double effect evaporator used in this work is located in the Department of Chemical Engineering at the University of Alberta and has been described in previous publications (for example, Newell et al., 1972a, b, c). However, for convenience, a schematic diagram, a list of the process variables, and the normal steady state operating conditions are included in Appendix II.*

The control objective is to maintain the product concentration C2 constant in spite of variations in load variables such as feedflow rate F. It is also necessary to control the liquid holdups in the two effects, W1 and W2, within practical limits. Control is maintained by manipulating the inlet stream flow S, the liquid flow rate (bottoms) B1, from the first effect, and the product flow rate B2. The evaporator feed and product handling units and the computer control system form an important part of the pilot plant, but their description is not necessary for the purpose of this paper.

* Appendix II is available from the authors upon request.

Several of the evaporator models developed over the last eight years are discussed and compared by Newell and Fisher (1972b). The most rigorous and accurate model that has been used consists of ten nonlinear, first-order ordinary differential equations. This model was linearized and put into the standard linear state-space form defined by Equations (3) and (4) and used as the starting point for this work.

## DISCUSSION OF RESULTS

The objective of this investigation was to start with a high-order theoretical model and develop the lowest-order, practical, multivariable, feedback controller that would give satisfactory control of the existing pilot plant evaporator. As discussed in the previous sections, the work divided into two main areas: model reduction (paths 4 and 7 in Figure 1) and control law reduction (paths 5 and 8). However, the task of evaluating the different design methods via simulated and/or experimental runs was complicated by the large number of different factors involved. The most important factors are summarized below:

1. Model Order: The lower-order models referred to in this paper were derived from the linear tenth-order, state-space model using the discrete equivalent (Wilson et al., 1972) of Marshall's model reduction technique (Marshall, 1966). The 10th order plus 5th, 4th, and 3rd-order models were evaluated in open-loop and closed-loop runs.
2. Reduction Techniques: The model reduction and control law reduction steps were carried out using several different techniques. The modal approaches are presented here and the least squares results are described elsewhere (Wilson, 1974).
3. Design Strategy: The major emphasis was placed on comparing model reduction versus control law reduction, that is, paths 4 and 7 versus 5 and 8. For ease of reference the optimal control law for a $\beta$ order model is designated $\beta$th, for example, for a third-order system 3OPT. When the optimal control law for the tenth-order system is reduced to a third-order controller using the procedure described previously, the resulting controller is designated as 3RED10.
4. Type of Control Law: Multivariable feedback controllers were designed for the tenth-order and the reduced-order models using the same optimal, linear, quadratic formulation used in previous work (for example, Newell et al., 1972a). The weighting factors in the quadratic performance index (that is, the elements of the **Q** and **R** matrices) were selected based on the previous results and the same values were used for all runs discussed in this paper. The basic control mode was multivariable proportional feedback control, but runs were also done to evaluate the effect of adding integral feedback, proportional feedforward, and setpoint (servo) control. Note that any other design procedure for multivariable state feedback controllers could be substituted for the optimal control approach used in this work.
5. Type of Disturbance: Since the reduced-order models only approximate the behavior of the actual nonlinear process, the control schemes were evaluated for disturbances in different load variables, for nonzero initial conditions, and for setpoint changes.
6. Multivariable Aspects: The evaporator is a multi-input, multi-output system and hence performance evaluation is complicated. For example, techniques that produce an improved response for one state variable could result in poorer control of other variables and/or excessive manipulation of input variable(s). No single quantitative criterion was found that would satisfactorily characterize the performance of all the runs. Therefore, all of the variables were plotted (Wilson, 1974) and the evaluation made subjectively. In this paper, the most important state variable C2 is plotted in each figure and typical responses of a second state variable and/or manipulated variables are included in some figures.

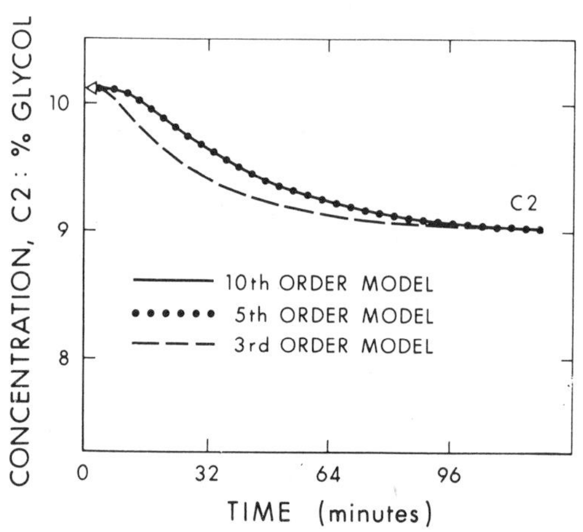


**Fig. 2. Simulated response of different order evaporator models to a +20% step in feedflow to show the effect of model order. (Arrowheads indicate initial steady state values.)**

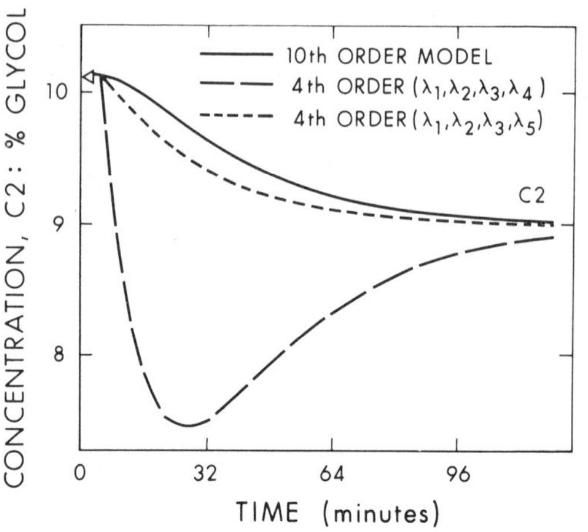


**Fig. 3. Simulated responses of different evaporator models to a +20% step in feedflow to show the effect of retaining different eigenvalues in the reduced-order model.**

The coefficient matrices for the tenth-order model, the tenth-order optimal controller, and the third-order model from the paper by Wilson et al. (1973) are reproduced in Appendix 2. Further details are available from the thesis by Wilson (1974). The following discussion describes some of the results obtained and the conclusions reached.

### Effect of Model Order

The accuracy of a state-space model generally decreases as the model order is reduced. For example, Figure 2 shows that the product concentration response of the fifth-order evaporator model was essentially identical to that of the original tenth-order model, but the third-order model exhibits a significant error. The models were reduced using modal analysis techniques, retaining the largest system

TABLE 2A. CONTROL MATRICES FOR PROPORTIONAL-PLUS-INTEGRAL CONTROL

| Design method | $K_R^{FB}$ | | | $K_R^{I}$ | | |
|---|---|---|---|---|---|---|
| 3RED10 | 2.977 | 0.0937 | −5.408 | 0.4117 | 0.02088 | −0.4861 |
| | 5.066 | −1.246 | 8.10 | 0.8398 | −0.2887 | 0.8159 |
| | 5.686 | 12.29 | 11.58 | 0.8559 | 1.945 | 1.068 |
| 30PT | 5.49 | −0.1903 | −12.00 | 0.9893 | −0.05066 | −1.175 |
| | 6.429 | −1.386 | 4.487 | 1.156 | −0.3255 | 0.4373 |
| | 5.519 | 12.26 | 11.81 | 0.8254 | 1.935 | 1.090 |

TABLE 2B. CONTROL MATRICES FOR PROPORTIONAL FEEDBACK, FEEDFORWARD AND SETPOINT CONTROL

| Design method | $K_R^{FB}$ | | | $K_R^{FF}$ | | | $K_R^{SP}$ | | |
|---|---|---|---|---|---|---|---|---|---|
| 3RED10 | 2.467 | 0.02163 | −4.705 | 1.238 | −0.5639 | −0.4138 | −2.468 | −0.0212 | 5.276 |
| | 4.288 | −1.340 | 8.885 | 0.9815 | 0.2177 | −0.00123 | −4.289 | 1.340 | −9.100 |
| | 4.128 | 9.760 | 9.528 | 0.9978 | 0.9877 | −0.00140 | −4.128 | −9.763 | −10.52 |
| 30PT | 4.904 | −0.4013 | −11.92 | 1.238 | −0.5583 | −0.4128 | −4.904 | 0.4017 | 12.47 |
| | 5.784 | −1.600 | 4.425 | 0.9832 | 0.2231 | 0.00055 | −5.784 | 1.599 | −4.648 |
| | 4.093 | 9.685 | 9.357 | 0.9983 | 0.9937 | −0.00122 | −4.094 | −9.686 | −10.35 |

eigenvalues and with $\mathbf{x}_1$ = [W1, C1, H1, W2, C2]$^T$ and [W1, W2, C2]$^T$, respectively.

The results in Figure 3 show that the assumptions and/or design decisions made during the model reduction step can be more important than how much the model order is reduced, that is, than the final model order. The two fourth-order models used in Figure 3 have the same state vector, $\mathbf{x}_1$ = [W1, H1, W2, C2]$^T$, but differ in the subset of eigenvalues of the tenth-order model that is retained in the reduced model [compare Equation (9) and related discussion]. In this run the objective was to obtain a reduced model that would satisfactorily represent the open-loop response to a step disturbance. Thus it would appear logical to retain the dominant (largest in absolute value) eigenvalues of the original system in the reduced model. However, when the four largest eigenvalues of $\boldsymbol{\varphi}$ were retained in $\boldsymbol{\varphi}_R$, the model response was very poor as illustrated by the C2 response in Figure 3 (long dashes). When the fourth largest eigenvalue (0.9292) was replaced by the fifth largest (0.7354), the response was much better as shown by the curve of short dashes. This difference can be explained (or predicted a priori) by an examination of the modal matrix for the tenth-order model which shows that element H1 of the state vector depends significantly on the fifth eigenvalue but not on the four largest ones (Wilson et al., 1973). In other words, since $\mathbf{x} = \mathbf{M}\,\mathbf{z}$, it is important when considering the response of the $i$th element of $\mathbf{x}$ to inspect the $i$th row of $\mathbf{M}$ as well as at the dynamics of all the elements of $\mathbf{z}$ (which are characterized by the diagonal elements of $\boldsymbol{\alpha}$ which in turn are equal to the eigenvalues of $\boldsymbol{\varphi}$).

**Model Reduction Techniques**

The modal methods applied to the reduction of the evaporator model are those summarized in Table 1.

Figure 4 compares the simulated time domain response to a +20% step change in feedflow using three third-order and the tenth-order open-loop models. The third-order model calculated using Davison's method has considerable steady state error. [It should be pointed out, however, that the fifth-order model calculated using Davison's method had negligible steady state error (Wilson, 1974).] Figure 4 also shows that the third-order models calculated using Marshall's method and the revised Davison method are very similar, lead the tenth-order model slightly, and result in zero steady state error. In order to have zero initial error

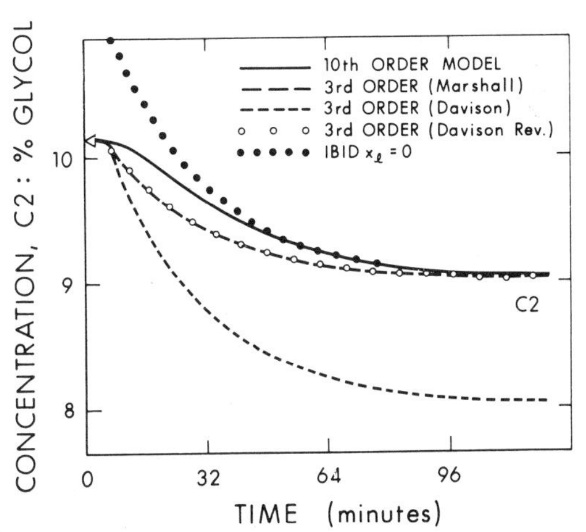


**Fig. 4. Simulated responses of the tenth-order and reduced-order evaporator models to a 20% step in feedflow to compare different modal analysis techniques.**

using the revised Davison method, the initial values of the elements of $\mathbf{x}_l$ were calculated as follows:

$$\mathbf{x}_l(0) = \mathbf{x}_1(0) - \mathbf{E}_R\,\mathbf{u}(0) - \mathbf{F}_R\,\mathbf{d}(0) \qquad (18)$$

If $\mathbf{x}_R(0) \neq \mathbf{x}_1(0)$ is used, as proposed by Graham (1968), then there is a significant error in the initial value of $\mathbf{x}_R$ but, as illustrated by the dotted C2 response in Figure 4, the response converges rapidly to that of the tenth-order model.

**Effect of the Design Sequence**

In the closed-loop control runs, the main objective was to compare the low-order controllers designed using the control law reduction and model reduction approaches. In each design the modal analysis of Marshall was employed to simplify either the controller or the model. Most of the closed-loop experimental runs presented in this paper used third-order controllers with $\mathbf{x}_1$ = [W1, W2, C2]$^T$. The controller matrices are given in Table 2. Typical results using fifth-order controllers are presented in Figures 8 and 9 and in the literature (Newell et al., 1972a, c).

Figure 5 shows typical experimental responses to 20% changes in feedflow when using a proportional feedback plus feedforward controller designed using the control law

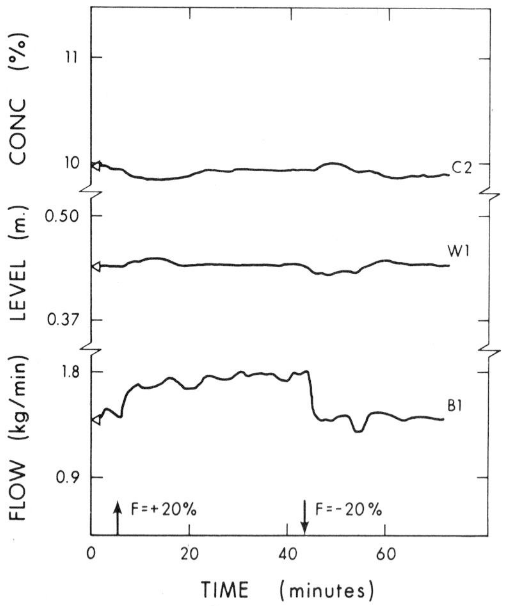


**Fig. 5. Experimental evaporator response to 20% step changes in feedflow with a 3RED10 controller incorporating FB+FF modes. (Arrows on the time axis denote a ±20% disturbance in feedflow.)**

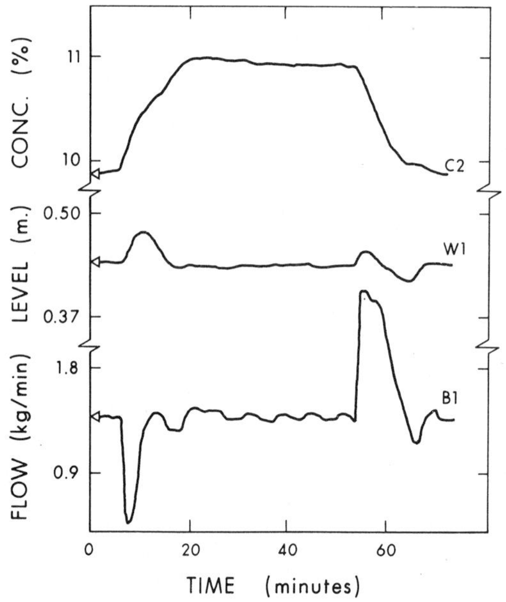


**Fig. 6. Experimental evaporator response to 10% step changes in the product concentration (C2) setpoint with a 3RED10 controller incorporating FB+SP modes.**

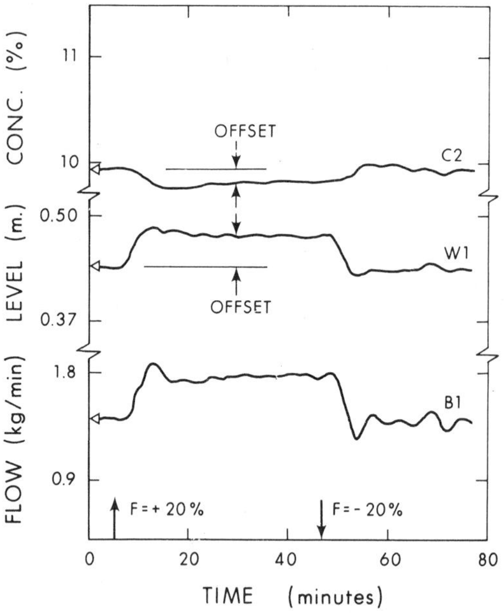


**Fig. 7. Experimental evaporator response to 20% step changes in feedflow with a 3RED10 controller incorporating only multivariable proportional feedback control. (Note the offset in C2 and W1 compared to Figure 5.)**

reduction approach. This 3RED10 controller produced slightly steadier control with less oscillatory responses than a similar 30PT controller designed on the basis of a third-order model (results not shown—see Wilson, 1974, Figure 6.11). Note that some of the corresponding elements in the controller matrices presented in Table 2 differ by over a factor of two. A comparison of Figures 5 and 7 shows that the addition of proportional feedforward control eliminated the steady state offsets in the state variables.

The results in Figure 6 were obtained using a 3RED10 controller designed using the same approach as for Figure 5. However, in this case the controller included proportional feedback plus setpoint control modes and the disturbances were 10% changes in the setpoint of the product concentration C2. Note that since the control law is

$$\mathbf{u} = \mathbf{K_R}^{FB}\, \mathbf{x}_1 + \mathbf{K_R}^{SP}\, \mathbf{y}^{SP} \qquad (19)$$

and the variables represent perturbations about the initial steady state, then when $\mathbf{y}^{SP} = \mathbf{0}$ (as in Figure 6 for $t >$ 55 min.) only the proportional feedback control mode is active. Thus the responses represent a return from nonzero initial conditions. (Figure 7 presents the comparable results for a disturbance in feed flow.) The results in Figure 6 showed significantly less overshoot in C2 and smoother, less oscillatory responses than those obtained using a comparable 30PT controller designed using the model reduction approach. The latter results are available elsewhere (see Wilson 1974, Figure 6.15).

The results in Figure 7 were also obtained by control law reduction and represent the response to 20% step changes in feedflow when using proportional feedback control only. The responses of both the state and manipulated variables were smoother and less oscillatory than any of the multiple mode controllers. The control matrix $\mathbf{K_R}^{FB}$ is given in Table 2. Note that the steady state offsets which are expected when proportional control is used alone are evident in both C2 and W1. These offsets were eliminated when the second feed flow disturbance brought the system back to the original steady state values. The comparable

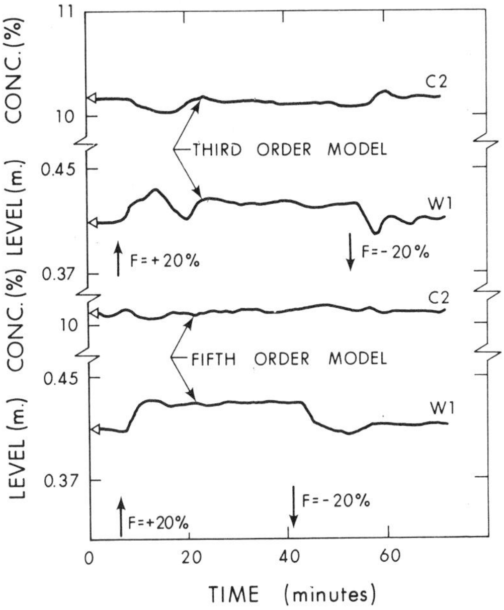


**Fig. 8. Experimental evaporator responses to 20% step changes in feedflow to compare "30PT" (top) and "50PT" (bottom) proportional feedback controllers.**

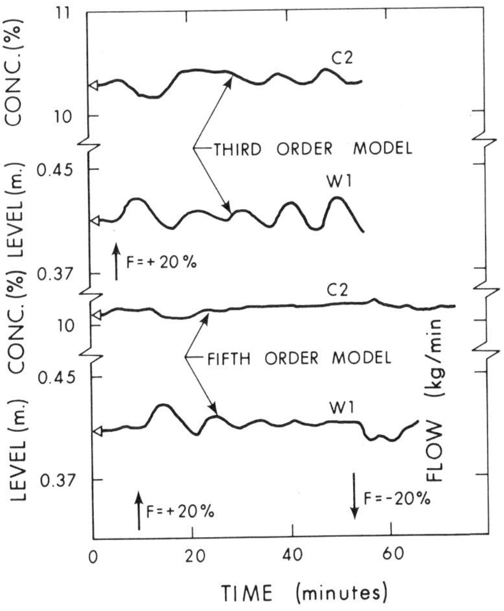


**Fig. 9. As per Figure 8 but using proportional plus integral feedback modes. Note that the offsets apparent in Figure 8 are eliminated, but the responses are more oscillatory.**

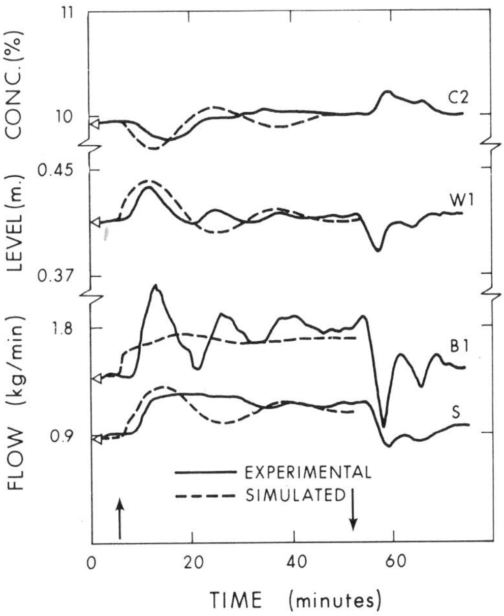


**Fig. 10. Comparison of simulated versus experimental evaporator responses to 20% step changes in feedflow with a 3RED10 incorporating FB+I modes.**

30PT controller had generally higher feedback gains which produced slightly more oscillatory responses and smaller offsets as shown by the curves in the top half of Figure 8.

Although they are included for other purposes, the curves in the top half of Figures 9 and 10 provide another comparison of the two approaches for developing low-order controllers. Figure 10 uses a 3RED10 proportional plus integral feedback controller developed by control law reduction. Figure 9 is the comparable 30PT controller. Both responses are to 20% disturbances in feedflow. Note that the responses in Figure 9 are more oscillatory and in fact the response of the liquid holdup W1 is unstable.

In summary, in the evaporator application the control law reduction approach tended to produce better low-order controllers since the controller gains were generally lower and the transient responses were less oscillatory with smoother changes in the manipulated variables.

#### Selection of Controller Order

The results plotted in the bottom half of Figures 8 and 9 show the response of the evaporator when using fifth-order optimal controllers and are directly comparable to previous work, for example, Newell et al., (1972a). Comparison of these results with those in the top half of Figures 8, 9, and 10 shows that, as expected, controllers based on a fifth-order model are better than the 3rd-order controllers designed using model reduction or control law reduction. However, there are compensating factors; for example, with the fifth-order controller, it is necessary to measure or estimate the first effect concentration C1 and the implementation of the controller is more difficult. The comparison of transient responses in Figure 10 for a 3RED10 controller shows that there are significant differences between simulated and experimental results. For example, the time delay evident in the experimental results is not adequately described by the process model.

Unfortunately there is no way of accurately predicting

the effect of model reduction or control law reduction on the actual closed-loop process responses. Guarantees of optimality, stability, integrity, etc., which may apply to the high-order model and controller are certainly lost when simplified controllers are applied to the actual process. Thus an iterative design approach followed by experimental evaluation and tuning may be required in complex applications. However, simulation is still an effective means for preliminary evaluations and comparison of alternatives.

## NOTATION

**A** = coefficient matrix in Equation (1)
**B** = coefficient matrix in Equation (1)
**C** = coefficient matrix in Equation (2)
**D** = coefficient matrix in Equation (1)
**d** = disturbance vector, $q \times 1$
**E** = coefficient matrix
**F** = coefficient matrix
**I** = identity matrix
$j$ = time interval counter
**K** = control matrix
$\mathbf{K}_1, \mathbf{K}_2$ = partitions of **K**
$l$ = dimension of $\mathbf{x}_1$
**M** = modal matrix
$\mathbf{M}_1, \mathbf{M}_2, \mathbf{M}_3, \mathbf{M}_4$ = partitions of **M**
$m$ = dimension of control vector
$N$ = number of data points
$n$ = dimension of state vector
$p$ = dimension of output vector
$q$ = dimension of disturbance vector
$t$ = time
**u** = control vector, $m \times 1$
**V** = $\mathbf{M}^{-1}$
$\mathbf{V}_1, \mathbf{V}_2, \mathbf{V}_3, \mathbf{V}_4$ = partitions of **V**
**x** = state vector, $n \times 1$
$\mathbf{x}_1, \mathbf{x}_2$ = partitions of **x**
**y** = output vector, $p \times 1$
**z** = canonical state vector, $n \times 1$
$\mathbf{z}_1, \mathbf{z}_2$ = partitions of **z**

### Greek Letters

$\boldsymbol{\alpha}$ = $\mathbf{M}^{-1} \boldsymbol{\varphi} \mathbf{M}$
$\boldsymbol{\alpha}_1, \boldsymbol{\alpha}_2$ = partitions of $\boldsymbol{\alpha}$
$\boldsymbol{\Delta}$ = coefficient matrix in Equation (3)
$\boldsymbol{\Delta}_1$ = partition of $\boldsymbol{\Delta}$
$\boldsymbol{\delta}$ = $\mathbf{M}^{-1} \boldsymbol{\Delta}$
$\boldsymbol{\delta}_1, \boldsymbol{\delta}_2$, = partitions of $\boldsymbol{\delta}$
$\boldsymbol{\eta}$ = $\mathbf{M}^{-1} \boldsymbol{\theta}$
$\boldsymbol{\eta}_1, \boldsymbol{\eta}_2$ = partitions of $\boldsymbol{\eta}$
$\boldsymbol{\theta}$ = coefficient matrix in Equation (3)
$\boldsymbol{\varphi}$ = coefficient matrix in Equation (3)
$\boldsymbol{\varphi}_1, \boldsymbol{\varphi}_2$ = partitions of $\boldsymbol{\varphi}$

### Subscript

$R$ = reduced-order

### Superscripts or Abbreviatioss

FB = proportional (state) feedback control
FF = proportional feedforward control
I = integral feedback control
SP = setpoint control
T = transpose
$\alpha$OPT = a control law of order $\alpha$ generated using the optimal-linear quadratic formulation and a model of order $\alpha$, for example, 3OPT
$\alpha$RED$\beta$ = a control law of order $\alpha$ generated by reducing an optimal control law of order $\beta$, for example, 3RED10
$-1$ = matrix inverse
$\cdot$ = time derivative

## LITERATURE CITED


Anderson, J. H., "Control of a Power Boiler Dynamic Model in the Presence of Measurement Noise and Random Input Disturbances," *Ind. Appl. of Dynamic Modelling,* IEE Conf. Publ. No. 57, Univ. of Durham, 42 (1969).

Chidambara, M. R., "Two Simple Techniques for Simplification of Large Dynamic Systems," 10th Joint Automatic Control Conf. Preprints, 669 (1969).

———., and E. J. Davison, "On 'A Method for Simplifying Linear Dynamic Systems'," *IEEE Trans. Auto. Control,* **AC-12,** 119 (1967a).

———., "Further Remarks on Simplifying Linear Dynamic Systems," *ibid.,* 213 (1967b).

———., "Further Comments on 'A Method for Simplifying Linear Dynamic Systems'," *ibid.,* 799 (1967c).

Davison, E. J., "A Method for Simplifying Linear Dynamic Systems," *IEEE Trans. Auto. Control,* **AC-11,** 93 (1966).

———., "The Simplification of Large Linear Systems," *Control,* **12,** 418 (1968a).

———., "A New Method for Simplifying Large Dynamic Systems," *IEEE Trans. Auto. Control,* **AC-13,** 214 (1968b).

———., "The Output Control of Linear Time-Invariant Multivariable Systems with Unmeasurable Arbitrary Disturbances," *ibid.,* **AC-17,** 621 (1972).

Fossard, A., "On a Method for Simplifying Linear Dynamic Systems," *ibid.,* **AC-15,** 261 (1970).

Friedly, J. C., *Dynamic Behavior of Processes,* Prentice-Hall, Englewood Cliffs, N. J. (1972).

Gantmacher, F. R., *Theory of Matrices,* Chelsea, New York (1959).

Gould, L. A., *Chemical Process Control: Theory and Applications,* Addison-Wesley, Reading, Mass. (1969).

Graham, E. U., "Simplification of Dynamic Models," Ph.D. thesis, Carnegie Inst. Technol., Pittsburgh (1968).

Lapidus, L., and R. Luus, *Optimal Control of Engineering Processes,* Blaisdell, Waltham, Mass. (1967).

Marshall, S. A., "An Approximate Method for Reducing the Order of a Linear System," *Control,* **10,** 642 (1966).

Newell, R. B., "Multivariable Computer Control of an Evaporator," Ph.D. thesis, University of Alberta, Edmonton (1971).

———., and D. G. Fisher, "Experimental Evaluation of Optimal Multivariable Regulatory Controllers with Model-Following Capabilities," *Automatica,* **8,** 247 (1972a).

———., "Model Development, Reduction and Experimental Evaluation for an Evaporator," *Ind. Eng. Chem. Design Develop.,* **11,** 213 (1972b).

——— and D. E. Seborg, "Computer Control Using Optimal Multivariable Feedforward-Feedback Algorithms," *AIChE J.,* **18,** 976 (1972c).

Nicholson, H., "Dynamic Optimization of a Boiler," *Proc. IEE,* **111,** 1479 (1964).

———., "Integrated Control of a Nonlinear Boiler Model," *Proc. IEE,* **114,** 1569 (1967).

Ogata, K., *State Space Analysis of Control Systems,* Prentice-Hall, Englewood Cliffs, N. J. (1967).

Rogers, R. O., and D. D. Sworder, "Suboptimal Control of Linear Systems Derived from Models of Lower Order," *AIAA J.,* **9,** 1461 (1971).

Sinha, N. K., and H. De Bruin, "Near-Optimal Control of High-Order Systems Using Low-Order Models," *Intern. J. Control,* **17,** 257 (1973).

Wilson, R. G., "Model Reduction and the Design of Reduced-Order Control Laws," Ph.D. thesis, University of Alberta, Edmonton (1974).

———., D. G. Fisher and D. E. Seborg, "Model Reduction for Discrete-Time Dynamic Systems," *Intern. J. Control,* **16,** 549 (1972).

Wilson, R. G., D. E. Seborg, and D. G. Fisher, "Modal Approach to Control Law-Reduction," 1973 Joint Automatic Control Conf. Preprints, 554 (1973).


## APPENDIX I

Many of the modal approaches for model reduction that have been reported in the literature as independent methods are in fact equivalent since they produce identical reduced-order models. In particular,

1. The formulations of Davison (1966) and Nicholson (1964) are equivalent.
2. Marshall's method (1966) and one of Chidambara's proposed methods (Chidambara and Davison, 1967b) are equivalent, as previously noted by Graham (1968).
3. The approaches of Fossard (1970) and Graham (1968) are equivalent to a revision of Davison's method derived by Chidambara and Davison (1967b).

Proofs that the resulting reduced-order models are equivalent were developed as part of this investigation and are available elsewhere (Wilson, 1974).

*Manuscript received October 24, 1973; revision received July 15 and accepted July 16, 1974.*

# Section 5: MULTIVARIABLE SERVO CONTROL

CONTENTS:



COMMENTS:

Although the most common control objective in continuous process plants is to maintain the process at the desired operating conditions in spite of external disturbances, there are many applications where "servo" control is desired. Servo control is defined as changing a process from one set of process operating conditions to another, normally steady-state, set of conditions. In some cases the desired trajectory is specified, in others only the final process conditions. The optimal multivariable regulatory control laws discussed in section 4 included provision for step changes in setpoints and/or model following. The model reference adaptive control systems described in the next section can also be used to implement desired changes in process conditions. The two papers in this section discuss optimal, open-loop changes in process conditions plus a means of combining the optimal open-loop policy with an existing feedback control system. In both cases the discrete optimal control policy $\{\underline{u}^*(kt) : k = 1,2,3 \ldots N\}$ is calculated *a priori* from the process model and a specified performance index. No on-line calculations are required and this approach proved to be simple to implement, practical and gave better responses than could be attained by other methods. Constraints on the state and/or control variables at each interval of time are easily implemented so the design can easily be restricted to a physically realistic operating range.

REFERENCES:

[1] Lapidus, L. and R. Luus, Optimal Control of Engineering Processes, Blaisdell, Waltham, Mass (1967).

[2] Athans, M. and P.L. Falb, Optimal Control, McGraw-Hill (1966).

# EXPERIMENTAL EVALUATION OF TIME-OPTIMAL, OPEN-LOOP CONTROL

By R. E. NIEMAN†‡ and D. G. FISHER†


## SYNOPSIS

This paper deals with the determination of an optimal means of driving a multivariable process from one state to another, subject to constraints on the control and state variables. The dynamic optimisation is based on a state-space process model plus a minimum time performance criterion, and is solved using standard linear programming techniques. Results from simulation and experimental studies on a pilot-plant evaporator operating under computer control are used to illustrate the improvement in performance and some of the factors that arise in practical applications.


## Introduction

During the last decade a great many theoretical and simulation studies have demonstrated the potential benefits of optimal multivariable control. Over the same time period a large number of industrial plants have installed real-time process control computers, thus having the hardware capability of implementing optimal control techniques. However, to date, there have been few industrial applications.

The purpose of this paper is to describe the application, at the University of Alberta, of optimal state-driving techniques to a pilot plant evaporator controlled by an IBM 1800 computer.

The following sections include a brief literature survey, a description of the problem formulation, and a discussion of the results of simulation and experimental studies.

An earlier version of this paper was presented at the Canadian National Conference on Automatic Control held at the University of Waterloo, Ontario, Canada, August, 1970.[1]

## Literature Review

Any optimal control procedure requires careful definition of the criteria. The time-optimal or minimum response time criterion is used in this work because it is particularly convenient for analysis, and in many instances closely approximates the actual physical objective.

Latour, Koppel, and Coughanowr,[2] present analytical expressions for the optimal switching times for any overdamped single input—single output system (SISO) that can be adequately represented by a second-order transfer function (with or without a pure time delay). They obtained significantly better performances using this method than with a well-tuned PID controller and the improvement increased with the magnitude of the change. However, for multivariable applications the two most popular approaches of determining the time optimal policy have been dynamic programming and Pontryagin's "Maximum Principle".[3] Zadeh and Whalen[4] showed that the time-optimal and fuel-optimal control problems could be reduced to linear programming problems for linear, discrete, time-invariant systems, and Bondarenko and Filimonov[5] present numerical results based on these two criteria.

Lesser and Lapidus[6] use the linear programming formulation to determine the time-optimal control of an absorber described by a sixth-order state difference equation. Inlet concentrations were used as the control variables in order to linearise the model and were constrained to keep the solution in a feasible region.

Lack and Enns[7] use a minimax objective function and compute the optimal control for a model of a nuclear reactor with 21 state variables. Sakawa and Hayashi[8] extend the discrete linear programming formulation to obtain approximate solutions for continuous systems. Torng[9] also considers the discrete regulator problem and points out that the linear programming solution might not be unique under all conditions. In a more recent paper Bashein[10] presents a modified simplex algorithm which requires only one application to solve the minimum time linear regulator problem, as opposed to the iterative solutions required with the standard algorithm. The computing time and storage requirements are also reduced, so the method might prove useful for implementation on small, real-time process computers.

## Process Equipment and Model

The optimal state driving control techniques described in this paper were applied to a pilot-plant evaporator at the University of Alberta. The unit operates with a nominal feed-rate of 300 lb/h of 3% aqueous triethylene glycol and gives a product of about 10%. A schematic diagram of the evaporator connected for operation in a double-effect, forward-feed mode of operation is shown in Fig. 1. The first effect is a calandria type unit with an 8 in diameter tube bundle. The second effect is a long tube vertical unit with three tubes and is operated with externally forced circulation. The second effect is operated under vacuum and utilises the vapour from the first effect as a heating medium.

The evaporator has conventional electronic instrumentation for about 14 control loops and the recording of over 30 temperatures. In-line refractometers are used for the continuous measurement of the feed and product concentrations. All transmitters and final control elements are connected to an IBM 1800 computer and were operated by a direct digital control (DDC) monitor program for the cases reported here. The computer also permits 24 h/day unattended operation of

† Department of Chemical and Petroleum Engineering, University of Alberta, Edmonton, Alberta T6G 2G6, Canada.

‡ *Present address:* c/o Department of Chemistry, Royal Military College, St. Jean, Quebec, Canada (direct correspondence to Professor Fisher).

Trans. Instn. Chem. Engrs., Vol. 51, No. 2, pp. 132-140 (1973).

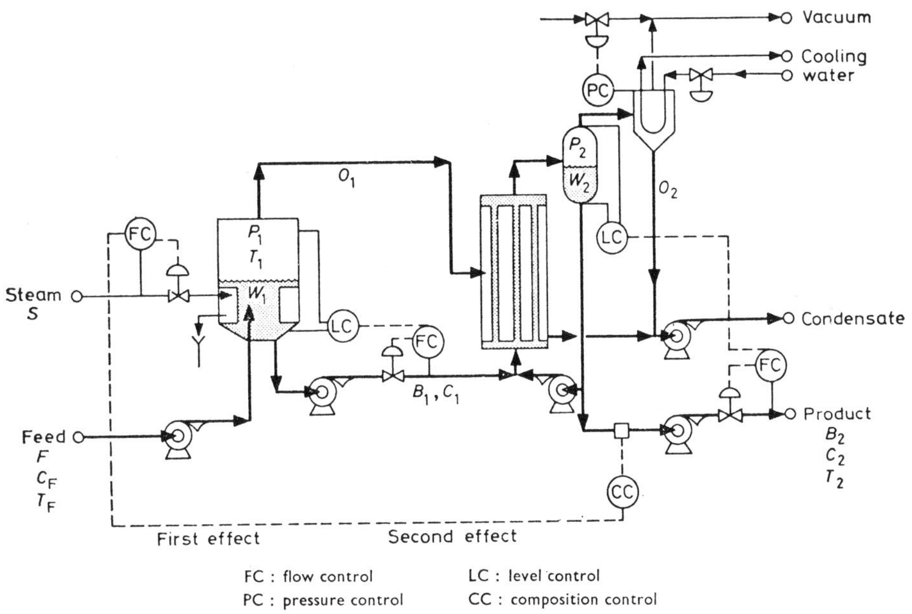


Fig. 1.—*Schematic flow diagram of the double-effect evaporator; all measurement transducers and final control elements are connected to a digital computer for data acquisition and control*

the evaporator and it handled the data acquisition, reduction, and plotting for the cases reported in this paper.

The evaporator can be adequately described by a set of five non-linear differential equations derived from material and energy balances around the principal components of the evaporator. These equations can then be linearised about the normal process operating conditions and put into a standard (fifth-order) state space form:

$$\left.\begin{aligned}\mathbf{x}(t) &= \mathbf{A}\mathbf{x}(t)+\mathbf{B}\mathbf{u}(t)\\ \mathbf{y}(t) &= \mathbf{C}\mathbf{x}(t)\end{aligned}\right\} \quad . \qquad (1)$$

The development and evaluation of this, and related models, has been published elsewhere.[11, 12] Single variable, transfer function models can also be developed and are used in later sections.

## Choice of Criteria

Normally the control criterion must be tailored to desired specifications and procedures; however, in this work there were no specific requirements. The variable of prime interest in the servo-mechanism control problem is the product concentration, $C_2$. However, the liquid levels must be maintained between certain limits so as to prevent liquid entrainment on one extreme, and pump cavitation on the other. The first-effect pressure must also be constrained. The minimum time criterion was chosen because it is computationally convenient, widely used in the literature, and is realistic for many plant operations, such as grade changes.

## Formulation of the Optimal Control Problem

For linear systems with constraints on the control variables the time-optimal control strategy is of the classic "bang-bang" nature in which the control variables are switched alternatively, between their upper and lower constraints. The control problem thus reduces to determining these optimal switching times.

For single variable systems, Latour *et al.*[2] have presented analytical expressions which define the time-optimal switching times for first- or second-order transfer function models with time delays. These relationships were used in obtaining cases I and II whose results are presented later.

For multivariable systems it is convenient to use a state space model as defined by equation (1), or by the following discrete form:

$$\mathbf{x}[(k+1)\,T] = \Phi(T)\,\mathbf{x}(kT)+\Delta(T)\,\mathbf{u}(kT) \quad . \qquad (2)$$

where $T$ is the sampling or control interval.

If the output vector, $\mathbf{y}$, is a linear function of the state (*e.g.*, $\mathbf{y} = \mathbf{C}\mathbf{x}$), then the problem may also be formulated in terms of $\mathbf{y}$. The optimal control problem is again to determine the trajectories (switching times) of the control variables, $\mathbf{u}$, needed to drive the system from its initial state, $\mathbf{x}_0$, to some desired final state, $\mathbf{x}_d$, while also minimising a particular objective function, such as process response time. This optimisation problem can be solved using standard linear programming (LP) algorithms if it is formulated as described below.

Repeated use of equation (2) permits the process state, $\mathbf{x}(NT)$, to be expressed in terms of the initial state and the control sequence as:

$$\mathbf{x}(NT) = \phi^{N}\,\mathbf{x}(0)+\phi^{N-1}\,\Delta\mathbf{u}(0)+\ldots+\Delta\mathbf{u}[(N-1)\,T] \quad . \qquad (3)$$

The maximum error between the value of the variables and their desired value is then defined as:

$$\lambda \geqslant |x_i(NT)-x_{di}| \qquad (i = 1, 2, \ldots, n) \quad . \qquad (4)$$

Alternatively, in a form more convenient for linear

programming:

$$\left.\begin{aligned} \mathbf{x}(NT)+\omega\lambda \geqslant \mathbf{x}_d \\ \mathbf{x}(NT)-\omega\lambda \leqslant \mathbf{x}_d \end{aligned}\right\} \quad (5)$$

where $\omega$ is the identity matrix or a matrix of weighting factors.

If $\mathbf{x}(NT)$ is replaced using equation (3), then equation (5) can be solved by a standard linear programming algorithm which determines the control vector, $[\mathbf{u}(kT), k = 1,2,3,\ldots,N]$, that will minimise $\lambda$. If the solution is successful then a smaller value of $N$ is assumed and the procedure repeated until the minimum time, $NT$, is obtained that will keep $\lambda$ within the desired limits. (Normally $\lambda \approx 0$ but can be larger if process specifications permit.)

Constant or time varying constraints on the state variables, at the sampling points, can be incorporated by expanding the problem. If the state variables are constrained at each sample time by upper $\mathbf{x}_u(k)$ and lower $\mathbf{x}_L(k)$ limits, then equation (3) is used to generate the state, at all time-intervals ($k = 1, 2, 3, \ldots, N$), and the formulation is as shown in equation (6). The number of LP variables remains as $(Nm+1)$, but the number of rows increases from 2n to $2n(1+N)$. (If the number of rows exceeds the number of variables then the dual LP problem is computationally more efficient.[15]) Obviously the use of state variable constraints rapidly increases the size of the LP problem. However,

$$\begin{bmatrix} \phi^{N-1}\Delta & \phi^{N-2}\Delta & \ldots & \Delta & \omega_N \\ \hdashline \phi^{N-1}\Delta & \phi^{N-2}\Delta & \ldots & \Delta & -\omega_N \\ \hdashline \Delta & 0 & \ldots & 0 & 0 \\ \phi\Delta & \Delta & \ldots & 0 & 0 \\ \vdots & & & & \\ \phi^{N-1}\Delta & \phi^{N-2}\Delta & \ldots & \Delta & 0 \\ \hdashline \Delta & 0 & \ldots & 0 & 0 \\ \phi\Delta & \Delta & \ldots & 0 & 0 \\ \vdots & & & & \\ \phi^{N-1}\Delta & \phi^{N-2}\Delta & \ldots & \Delta & 0 \end{bmatrix} \begin{bmatrix} \mathbf{u}_0 \\ \mathbf{u}_1 \\ \vdots \\ \vdots \\ \mathbf{u}_{N-1} \\ \lambda \end{bmatrix} \begin{matrix} \geqslant \\ \leqslant \\ \leqslant \\ \leqslant \\ \\ \\ \geqslant \\ \\ \\ \\ \end{matrix} \begin{bmatrix} \mathbf{x}_d-\phi^N\mathbf{x}_0 \\ \hdashline \mathbf{x}_d-\phi^N\mathbf{x}_0 \\ \hdashline \mathbf{x}_{u_1}-\phi\mathbf{x}_0 \\ \mathbf{x}_{u_2}-\phi^2\mathbf{x}_0 \\ \vdots \\ \mathbf{x}_{u_N}-\phi^N\mathbf{x}_0 \\ \hdashline \mathbf{x}_{L_1}-\phi\mathbf{x}_0 \\ \mathbf{x}_{L_2}-\phi^2\mathbf{x}_0 \\ \vdots \\ \mathbf{x}_{L_N}-\phi^N\mathbf{x}_0 \end{bmatrix} \quad (6)$$

control variable constraints can be incorporated without increasing the dimension of the problem.

## Implementation

The dynamic optimisation for the desired changeover was calculated off-line on an IBM 360/67 computer and hence computational time was not a critical factor. A typical running time with 5 state variables, 3 control variables, 40 sampling intervals and constraints on the state variables was about one minute. This optimisation yielded a set of discrete control vectors which were transferred to data files in the process computer. Implementation of this open-loop control action on a process computer, where execution time is critical, required no more than transferring values from the control vector, $\mathbf{u}(kT)$, to the appropriate DDC loops. The corresponding feedback control loops were of course inactive in this mode, and no on-line calculations were required. Therefore the choice of sampling intervals was not restricted by the calculational load of the control scheme. In this work the sampling time was left as the same value used in the normal regulatory control mode.

After the open-loop control policy was implemented a " bumpless " transfer was made to normal DDC regulatory control in all experimental runs and simulations. Any steady state error was then removed by conventional feedback control action.

## Discussion of Results

The results of the experimental and simulation studies are presented as a series of five cases ranging from the simplest single input–single output approach to multivariable applications based on a fifth-order evaporator model with state and control variable constraints. A summary of these cases is presented in Table I.

### *Case 1: Optimal control based on a first-order model*

The most important state variable in the evaporator model is the output concentration, $C_2$, which can be controlled by manipulating the input steam flow-rate, $S$. The multivariable evaporator model as described by equation (1) can be reduced[12] to yield a transfer function relationship of the following form:

$$\frac{C_2(s)}{S(s)} = \frac{K_p(\tau_3 s+1)}{(\tau_1 s+1)(\tau_2 s+1)} \approx \frac{K_p}{(\tau_4 s+1)} \quad (7)$$

Based on the first-order model, the time-optimal control policy is simply to step up the input variable, $S$, to its maximum value, and to hold it there until the controlled variable, $C_2$, reaches the desired value. The input variable is then changed to the value required to maintain the new (desired) steady state. It should be emphasised that the final steady-state value for the input variable, $S$, is *not* produced by the optimal control calculation, and must be found by experience, or from a model. In this work the steady-state relationship implied by equation (7) was not satisfactory, and a more accurate, non-linear steady-state model was developed for this purpose.[1]

The experimental response of the evaporator, when subjected to an " optimal " steam transient based on the assumption of first-order behaviour, is shown by the solid curves in Figs 2 and 3 for a step decrease and increase in the specified product concentration. In Fig. 3 the rate of change of $C_2$ is slower, and the duration of the transient is significantly longer because the upper constraint on the steam was 2·2 lb/min which permitted a much smaller change in driving force than was used for the results in Fig. 2.

The time, $t_1$, at which the steam should be changed to its final steady-state value can be calculated analytically from a first-order model of the process.[1,2] However, in this case it is simpler and more reliable to adopt a " real-time " approach and change the steam at the point when the concentration reaches the desired value: this is shown as time $t_1$ in Fig. 2. Reducing the steam to the correct steady-state value would maintain $C_2$ at the desired value if the evaporator behaved as

TABLE I.—*Summary of Simulated and Experimental Results*†

(1) Sim. = simulated = data taken using fifth-order non-linear model.
Exp. = experimental = data taken from experiments on the evaporator.

(2) $K_{NL}$ = non-linear gain.

| Case | Figure | Model Order | Model Origin | Data (1) | Constraints | Final Steady State (2) | Control of $W_1$ and $W_2$ |
|---|---|---|---|---|---|---|---|
| 1*a* | 2 | 1 | Theory | Exp. | $S \geqslant 1 \cdot 0$ | Exp. | Tight |
| 1*b* | 2 | | | | | Exp. | PI (average) |
| 1*c* | 2 | | | | | Exp. | PI (average) |
| 1*c* | 3 | 1 | Theory | Exp. | $S \leqslant 2 \cdot 2$ | Exp. | PI (average) |
| 1*d* | 3 | | | | | Exp. | PI (average) |
| 1*e* | 3 | | | | | Exp. | PI (average) |
| 2*a* | 4 | 2 | Theory | Sim. | $1 \cdot 0 \leqslant S \leqslant 2 \cdot 5$ | $K_{NL}$ | Tight |
| 2*b* | 4 | 2 | Fitted | Exp. | $1 \cdot 0 \leqslant S \leqslant 2 \cdot 5$ | Exp. | Tight |
| 2*c* | 4 | 2 | Fitted | Exp. | $1 \cdot 0 \leqslant S \leqslant 2 \cdot 5$ | Exp. | Tight |
| 3*a* | 5 | 2 | Theory | Exp. | $1 \cdot 0 \leqslant S \leqslant 2 \cdot 2$ | Exp. | Tight |
| — | 5 | | | | | Exp. | PI (average) |
| — | 5 | | | | | Exp. | PI (average) |
| 3*b* | 6 | 2 | Theory | Exp. | $1 \cdot 0 \leqslant S \leqslant 2 \cdot 2$ | Exp. | Tight |
| — | 6 | 2 | Theory | Exp. | | Exp. | PI (average) |
| — | 6 | 2 | Theory | Exp. | | Exp. | PI (average) |
| 4*a* | 7 | 5 | Theory | Sim. | $1 \cdot 0 \leqslant S \leqslant 3 \cdot 58$ <br> $P_1 \leqslant 10$ (lbf/in$^2$ gauge) | Model | Optimal |
| 4*b* | 7 | 5 | Theory | Exp. | Same | Exp. | Optimal |
| 5*a* | 8 | 5 | Theory | Sim. | $1 \cdot 0 \leqslant S \leqslant 3 \cdot 58$ <br> $P_1 \leqslant 10$ (lbf/in$^2$ gauge) <br> $25 \leqslant W_1 \leqslant 35$ <br> $30 \leqslant W_2 \leqslant 40$ | Model | Constrained |
| 5*b* | 8 | 5 | Theory | Exp. | Same | Exp. | Constrained |

† *Note:* Runs in Figs 2 to 6 are based on a feed-rate of 4·5 lb/min, those in Figs 7 and 8 on 5·0 lb/min.

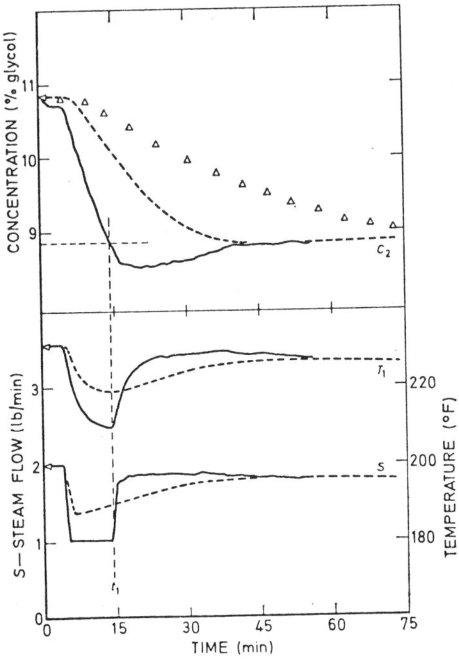


——— : optimal
— — — : closed-loop responses
△ △ △ : open-loop responses

Fig. 2.—*Comparison of experimental response of evaporator, using real-time switching of steam flow based on a "first-order policy", and open- and closed-loop responses*

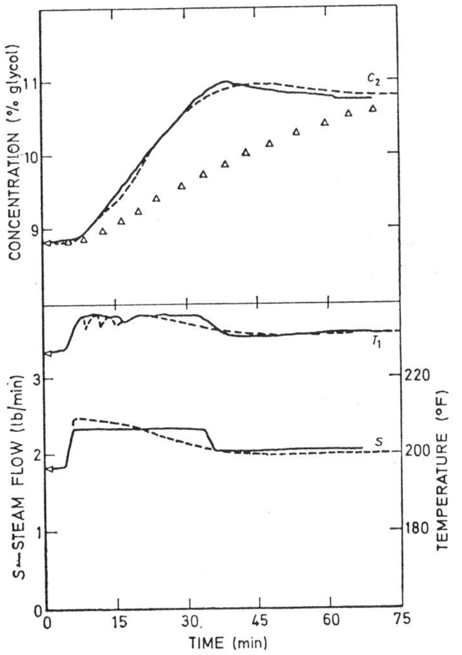


——— : optimal
— — — : closed-loop responses
△ △ △ : open-loop responses

Fig. 3.—*Comparison of experimental response of evaporator, using real-time switching of steam flow based on a "first-order policy", and open- and closed-loop responses*

a first-order process. However, the significant overshoot that occurs indicates that a higher-order model of the evaporator is required.

The true measure of the advantage of optimal control is obtained by comparing it with conventional methods. Conventional control on the evaporator is implemented by using an in-line refractometer to provide a continuous measurement of concentration which is fed back to a standard proportional-plus-integral controller which manipulates the inlet steam flow (see Fig. 1). This scheme has performed satisfactorily for years for regulatory control. The dashed curves in Figs 2 and 3 show the responses of the evaporator when a step-change is made in the setpoint of this controller. It would appear from Fig. 3 that the concentration control is almost as good as with the " optimal " techniques. However, the feedback controller did not incorporate constraints on the steam flow, and actually raised the steam flow to such a large value that the safety valve on the first-effect blew ($P_1 > 10$ lbf/in$^2$ gauge) and the extra steam resulted in extra water being boiled off in the first effect and vented (the opening and closing of the safety valve is reflected in the oscillations in the plot of the first-effect temperature, $T_1$). This action was considered undesirable, and was eliminated in the " optimal " runs by constraining the steam flow.

The triangular symbols in Figs 2 and 3 depict the open-loop responses of the evaporator when the steam was simply stepped from its initial steady-state value to the value required to produce the desired concentration. The controlled responses are obviously considerably faster.

Although not incorporated into the above analysis, the control policy adopted for the liquid levels in each effect of the evaporator has a significant effect on the concentration responses. In general " averaging " (low gain) control was better for normal regulator operations, and " tight " control was better when a change in the desired concentration was made. The multivariable cases discussed later consider both the concentration and liquid level control policies.

*Case II: Optimal control based on a second-order model*

In order to reduce the concentration overshoot which occurred in Case I, a second-order model was derived which was equivalent to the following transfer function:

$$\frac{C_2(s)}{S(s)} = \frac{K_p \exp(-\tau_d s)}{(\tau_1 s+1)(\tau_2 s+1)} \quad . \quad (8)$$

The model parameters listed in Table II were obtained by fitting the response of this second-order model to that of the non-linear, five-equation dynamic model, or to several open-loop, experimental evaporator responses.[1]

The simulated response of the five-equation, non-linear model when subjected to the switching times calculated analytically, based on model 1 in Table II, is shown by the solid line in Fig. 4.

The actual process response using precalculated switching times based on model 3 is shown by the points in Fig. 4. The

TABLE II.—*Parameters in Second-Order Fitted Models*

| Model | Basis | $\tau_1$ | $\tau_2$ | $\tau_d$† |
|---|---|---|---|---|
| 1 | Fifth-order non-linear model | 2 min | 25 min | 0 min |
| 2 | Several actual process responses | 12·1 min | 20·2 min | 0 min |
| 3 | Several actual process responses | 7·8 min | 22·3 min | 3 min |

† Specified by user.

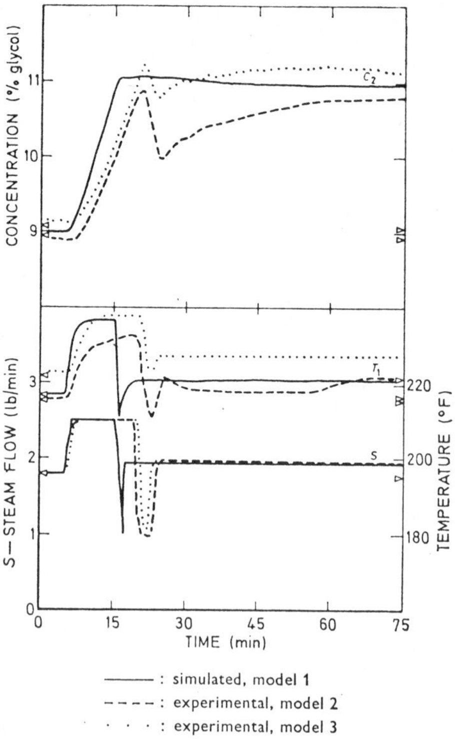


Fig. 4.—*Effect of model parameters on optimal control policies calculated from second-order models* (cf. Table II)

response of the outlet concentration, $C_2$, is slower than that of the model (an observation supported by other data), but the change in $C_2$ from the initial steady state to the first crossing of the desired value is smooth and fast. The experimental process responses using switching times based on model 2 are plotted in Fig. 4 using dashed lines.

The decrease in product concentration, $C_2$, shown by both the dotted and dashed curves in Fig. 4 is due to the steam remaining too long at the lower constraint before returning to the final steady-state value. This in turn can be traced directly to the values of $\tau_1 = 12{\cdot}1$ and $\tau_1 = 7{\cdot}8$ found in models 2 and 3 respectively *versus* the calculated value of $\tau_1 = 2$. The models all have approximately the same " goodness of fit " but the theoretical model and other experimental data suggest that the smaller value of $\tau_1$ in model 1 is more realistic. This emphasises the importance of " realistic " models rather than just a " good fit " to process data.

The liquid levels were under " tight " control and stayed relatively close to their setpoints.

*Case III: Optimal control using real-time switching*

" Bang-bang " control switching times calculated off-line are not always satisfactory due to modelling errors and/or real-time factors which affect the process response. Therefore, two methods were investigated which would modify the switching times based on the actual response(s) of the controlled variable(s):

(1). Analytical solution of the optimal switching curve.

(2). An intuitive or empirical method.

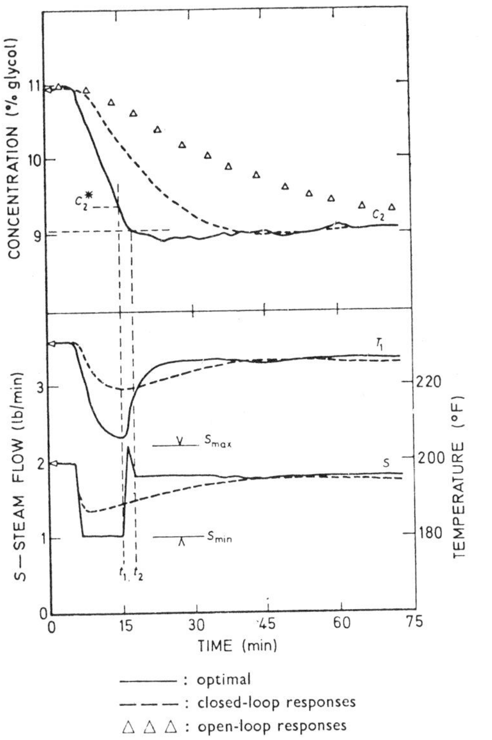


*Fig. 5.—Comparison of experimental response of evaporator, using real-time switching of steam flow based on a "second-order policy", and closed- and open-loop responses*

In the analytical approach $C_2$ and $dC_2/dt$ are estimated, on-line, from actual process measurements, and the steam is switched when the actual trajectory in the phase plane crosses the switching curve. Athans and Falb[13] present a thorough treatment of the basis for this approach. Implementation of this approach for control of the evaporator was not satisfactory because the on-line estimates of $C_2$ and $\dot{C}_2$ were not reliable enough. However, simulated results[1] obtained by using a second-order switching curve and the fifth-order, non-linear model of the evaporator showed an almost ideal response (similar to the $C_2$ response plotted as a solid line in Fig. 4), whereas the results based on switching times calculated *a priori* from the same second-order model showed that the steam is switched too late (thereby causing overshoot in $C_2$), and then remained too long at the lower constraint (thereby causing overcorrection in $C_2$).

The first method is ideal if an on-line computer is available to implement the switching and if satisfactory estimates of $C_2$ and $\dot{C}_2$ can be obtained.

However, it will be noted from Fig. 5 that the optimal policy for the steam can be characterised by $C_2^*$ and the interval $(t_2-t_1)$. Thus it is possible to implement the following procedure manually:

(1). $S \rightarrow S_{max}$ at $t = 0$.

(2). $S \rightarrow S_{min}$ when $C_2 = C_2^*$.

(3). $S \rightarrow S_2$ after time $(t_2-t_1)$.

This is an empirical procedure, and for a given process the parameter estimates can be improved with experience. Note that the allowable range of control, $S_{min} < S < S_{max}$, can initially be made very small and then increased to physical limits as confidence is gained. The experimental results obtained from the evaporator, plotted as solid lines in Figs 5 and 6, show that this "real-time", empirical approach resulted in the evaporator moving directly (at the maximum rate allowed by the constraints) to the desired value.

Daily cleaning of the refractometer prism, careful calibration of the refractometer, and accurate temperature compensation were required before on-line switching could be implemented.

The equivalent open-loop and closed-loop responses are shown in Figs 5 and 6 by the triangular points and the dashed-lines respectively. These are comparable to the analogous responses in Figs 2 and 3 and the same comments apply.

MULTIVARIABLE CONTROL

The single input–single output procedures used in Cases I, II and III are simple and convenient but do not properly account for:

(1). Multivariable processes.

(2). Interactions between process variables.

(3). Physical constraints on the state or output variables.

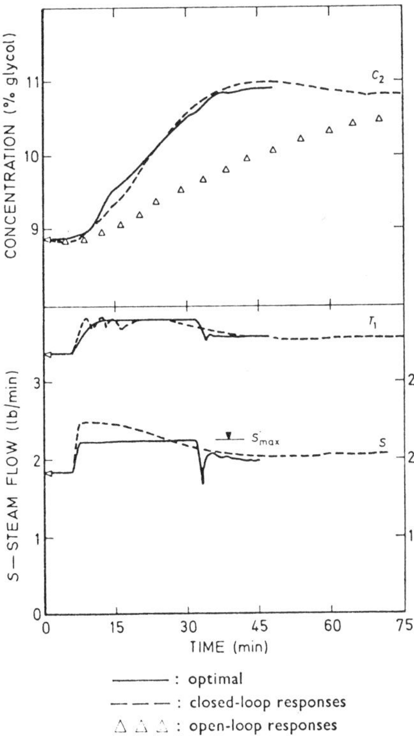


*Fig. 6.—Comparison of experimental response of evaporator, using real-time switching of steam flow based on a "second-order policy," and closed- and open-loop responses*

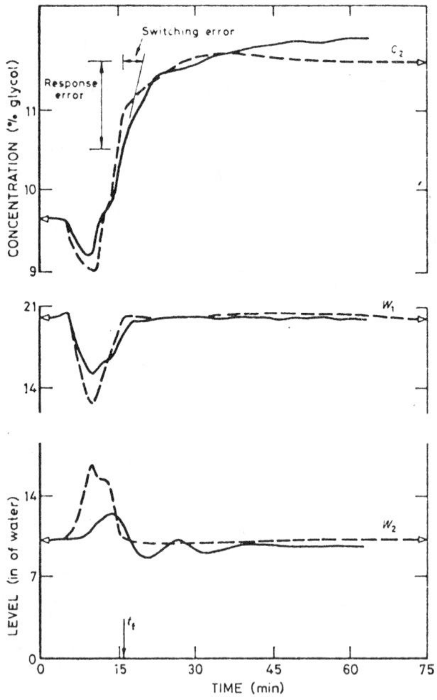


Fig. 7A.—*Responses of concentrations and liquid levels for " optimal " control of $C_2$, $W_1$, and $W_2$ calculated from a fifth-order model with a constraint on $P_1$*

For example, experience with the evaporator showed that:

(1). The stream from the first to the second effect ($B_1$) has almost as much and as fast an effect on $C_2$ as the steam flow-rate, $S$, and therefore should be considered when calculating optimal $C_2$ trajectories.

(2). Although they are not as critical, it is desirable to drive $W_1$ and $W_2$ as well as $C_2$ to their new steady-state values.

(3). The extended periods for which the steam was set to its upper constraint value, $S_{max}$, by the simpler optimal control formulations caused the first effect to over-pressurise and blow the safety valve. (This was handled in the previous cases by reducing the upper constraint, $S_{max}$, on the steam flow and by running at a feed-rate of 4·5 lb/min instead of the usual 5·0 lb/min. However, it is a trial-and-error search to find the correct value for $S_{max}$, and lowering it extends the process response time.) Therefore, the investigation was extended to include optimal control based on the linearised fifth-order process model and the linear programming formulation presented earlier.

*Case IV : Multivariable optimal control of $C_2$, $W_1$ and $W_2$*

Results using the optimal, multivariable, state-driving control policy calculated from a fifth-order linearised model are shown in Fig. 7. The dashed line is the response of the fifth-order non-linear model, the solid line is plotted from the response of the actual process.

The controlled variables are steam, $S$, the product from the first effect, $B_1$, and the product from the second effect, $B_2$. The steam transient includes a step to the upper steam constraint at time zero, a step down to a lower value so that the process will not violate the upper limit on first-effect pressure (*i.e.*, blow the safety valve), and finally a step down to the new steady-state value. Note that within the resolution of the one minute time interval used in the numerical optimisation calculations there is no step down below the final steady-state value. The outlet flows $B_1$ and $B_2$ are also switched between their upper and lower constraints to return the liquid levels to their setpoint values (only $B_1$ is plotted).

The $C_2$ concentration trajectory shown in Fig. 7 is distinctly different from Cases I, II, and III because of the " bang-bang " manipulation of $B_1$ and $B_2$. Note particularly that the initial $C_2$ response is in the wrong direction. However, in spite of these " anomalies " the overall transient in the experimental runs is better (shorter) than for the single variable cases based on *a priori* calculations of the switching times (*e.g.* Fig. 4). When comparing the actual experimental results for the single variable and multivariable cases it must be noted that the first- and second-order models were fitted to actual experimental responses, whereas the fifth-order model is theoretical. It has been observed, in other work, that the response of the fifth-order-model leads that of the actual process. Consequently the switching times calculated using this model are too " early " to fit the actual process response, and in view of the rapid rate of change of the concentration a small error in the switching time can have a large effect on the concentration. An intuitive way of looking at the discrepancy between the process and the model is to

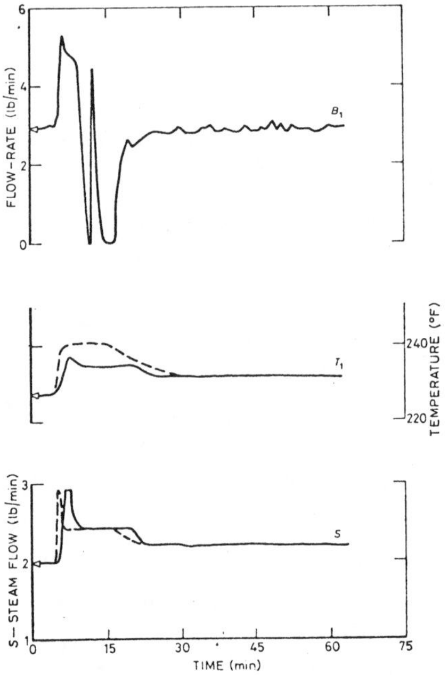


Fig. 7B.—*Responses of flow-rates and temperature for " optimal " control of $C_2$, $W_1$, and $W_2$ calculated from a fifth-order model with a constraint on $P_1$*

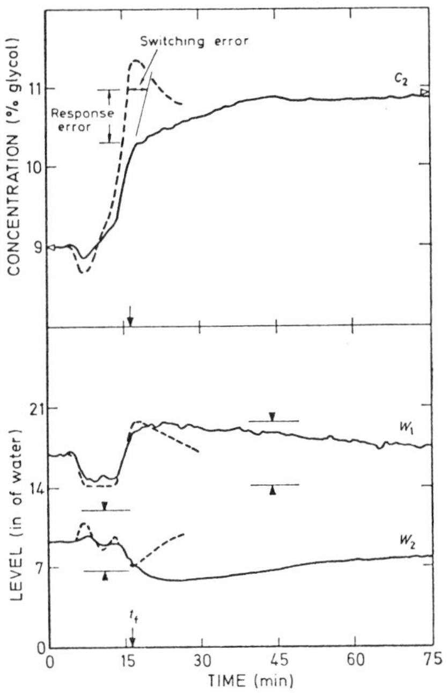


——— : experimental
— — — : simulated

Fig. 8A.—*Responses of the concentrations and liquid levels for "optimal" control of $C_2$ calculated from a fifth-order model, with constraints on $P_1$, $W_1$, and $W_2$*

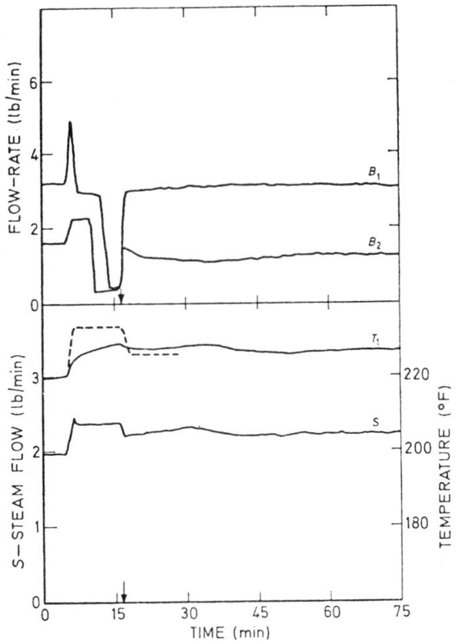


——— : experimental
— — — : simulated

Fig. 8B.—*Responses of the flow-rates and temperature for "optimal" control of $C_2$ calculated from a fifth-order model, with constraints on $P_1$, $W_1$, and $W_2$*

regard the final switch time, $t_f$ (arrow on time axis), to be about four minutes in error. With the rapid change of product concentration this results in a large response error.

*Case V: Multivariable optimal control of $C_2$*

The objective in Case IV was to drive $C_2$ and the two liquid levels to the new steady state. This objective is too severe, since in practice it is only necessary to keep the liquid levels within reasonable limits rather than return them exactly to their setpoints. The liquid level deviations are also too large in Fig. 7A. Therefore, in Case V the liquid level performance was given zero weight in the objective function, but upper and lower constraints were placed on the levels.

The experimental (solid) and simulated (dashed) results are shown in Fig. 8. It should be noted that the process operating conditions are different from those of Case IV. The steam trajectory is basically similar to Case IV but the manipulation of $B_1$ and $B_2$ is much less severe. The response of $C_2$ is better than in Case IV for $t < t_f$.

Again the experimental process response attained only a portion (65%) of the desired change in product concentration in the designated time, and the theoretical model led the experimental response.

The control policies, applied to the fifth-order non-linear model and to the pilot-plant evaporator, were derived from a fifth-order linear model obtained by linearising the non-linear model about nominal process operating conditions. Due to non-linearities, even simulated results show trajectories which do not close on the desired product concentration. These errors must be removed by feedback compensation when the control loops are reactivated ($t > t_f$). Similar phenomena are also evident with the experimental results, and in addition it is noted that the theoretical models (linear and non-linear) lead the process response. As demonstrated by the magnitude of the response errors in Figs 7 and 8, the accuracy of the process model is much more critical in the servo-mechanism control problem than for the regulatory control scheme where deviations about process operating conditions are small.[12] The use of a fitted process model to derive the optimal control, plus implementation of a convenient form of closed-loop compensation for modelling errors and for process disturbances, will be the subject of further investigation.[16]

## Conclusions

(1). Optimal, open-loop, state-driving methods produced significantly better responses on the pilot-plant evaporator than conventional control techniques.

(2). The linear programming formulation of the optimal control problem is flexible, practical, and permits direct introduction of constraints on the control and/or state variable. It also permits optimal multivariable problems to be solved by methods that are familiar and widely used in industry.

(3). The multivariable formulation of the evaporator problem accounted for the existence of process interactions and permitted the introduction of constraints, neither of which were handled by the simpler single variable techniques.

**Acknowledgments**

The authors wish to express their appreciation to the National Research Council of Canada for partial support of the project, the University of Alberta Computing Centre, and the Data Acquisition, Control and Simulation Centre in the Department of Chemical and Petroleum Engineering for their assistance and for the use of the computing facilities.

**Symbols Used**

*Process variables* (*Fig.* 1)

$B_1$ = first effect bottoms flow-rate (lb/min).
$B_2$ = second effect bottoms flow-rate (lb/min).
$C_1$ = glycol concentration in first effect (weight fraction).
$C_2$ = glycol concentration in second effect (weight fraction).
$C_F$ = glycol concentration in feed (weight fraction).
$F$ = feed flow-rate (lb/min).
$O_1$ = overheads first effect (lb/min).
$O_2$ = overheads second effect (lb/min).
$P_1$ = pressure in first effect vapour space (lbf/in$^2$ gauge).
$S$ = steam flow-rate (lb/min).
$T_1$ = first effect temperature (°F).
$T_F$ = temperature of the feed (°F).
$W_1$ = first effect liquid hold-up (lb).
$W_2$ = second effect liquid hold-up (lb).

*Variables in state-space models*

$\mathbf{A}$, $\mathbf{B}$, $\mathbf{C}$ = constant coefficient matrices in process model (equation (1)).
$\mathbf{x}$ = state variables vector ($n \times 1$).
$\mathbf{x}_d$ = desired state variables vector ($n \times 1$).
$\mathbf{x}_0$ = initial state variables vector ($n \times 1$).
$\mathbf{y}$ = output vector ($r \times 1$).
$\mathbf{u}$ = control variables vector ($m \times 1$).
$\Delta$ = control variables coefficient matrix in state difference equation $\left(= \int_0^T \exp[\mathbf{A}(T-t)]\, dt\, \mathbf{B}\right)$.
$\phi$ = fundamental matrix (= exp [$\mathbf{A}T$]).

*Other variables*

$K_p$ = process gain, product concentration to steam.
$T$ = sampling time.
$t_1$, $t_2$ = switching times.
$\tau$ = time constant in transfer function models.

**References**

1 Nieman, R. E. *Ph.D. Thesis*, 1971. University of Alberta.
2 Latour, P. R., Koppel, L. B., and Coughanowr, D. R. *Ind. & Eng. Chem., Process Des. & Dev.*, 1967, **6**, No. 4, p. 452.
3 Athans, M. *I.E.E.E. Trans. autom. Control*, 1967, **12**, No. 4, p. 580.
4 Zadeh, L. A., and Whalen, B. H. *I.R.E. Trans. autom. Control*, 1962, **7**, No. 4, p. 45.
5 Bondarenko, V., and Filimonov, I. *Prikl. Mat. Mekh.*, 1968, **32**, No. 1, p. 138.
6 Lesser, H. A., and Lapidus, L. *A.I.Ch.E. Jl*, 1966, **12**, No. 1, p. 143.
7 Lack, G., and Enns, M. *I.E.E.E. Trans. autom. Control*, 1967, **12**, No. 6, p. 749.
8 Sakawa, Y., and Hayashi, C., in: "*Proceedings of the International Federation of Automatic Control Symposium*", Tokyo, 1965, p. 53.
9 Torng, H. C. *J. Franklin Inst.*, 1964, **278**, No. 1, p. 28.
10 Bashein, G. *I.E.E.E. Trans. autom. Control*, 1971, **16**, No. 5, p. 479.
11 Andre, H., and Ritter, R. A. *Can. J. chem. Engng*, 1968, **46**, No. 4, p. 259.
12 Newell, R. B., and Fisher, D. G. *Ind. & Eng. Chem., Process Des. & Dev.*, 1972, **11**, No. 2, p. 213.
13 Athans, M., and Falb, P. L. "*Optimal Control*", 1966, chapter 7. (Toronto: McGraw-Hill).
14 Saucedo, R., and Schiring, E. E. "*Introduction to Continuous and Digital Control Systems*", 1968, chapter 12. (New York: MacMillan Co.).
15 Hadley, G. "*Linear Programming*", 1962. (Reading, Massachusetts: Addison-Wesley Publishing Co., Inc.).
16 Nieman, R. E., and Fisher, D. G. *Chem. Engng Commun.*, 1973, **1**, No. 2.

*The manuscript of this paper was received on 12 May, 1972*

# EXPERIMENTAL EVALUATION OF OPTIMAL MULTIVARIABLE SERVO CONTROL IN CONJUNCTION WITH CONVENTIONAL REGULATORY CONTROL†

RONALD E. NIEMAN and D. GRANT FISHER

*Department of Chemical and Petroleum Engineering, University of Alberta, Edmonton 7, Alberta, Canada*

*(Received May 22, 1972; in final form December 6, 1972)*

The problem considered is that of driving a multivariable process controlled by conventional feedback control algorithms, from some initial state, to a specified final state such that a linear performance index is minimized. The formulation is based on a linear, discrete, time-invariant model of the process plus a minimum-time or sum of the absolute errors criterion and is solved using a standard linear programming algorithm. Constraints on the state and/or control variables can be incorporated to guarantee practical, physically realizable control policies. Experimental and simulated results show that the technique is easy to implement, and that the use of different criteria produces significantly different process transients. Experimental data from a computer-controlled, pilot plant evaporator show that the method is practical and produces better results than conventional methods.

## INTRODUCTION

The purpose of this work was to develop, implement and evaluate optimal control techniques capable of driving an industrial process from some initial set of operating conditions to another set of desired conditions. Several techniques exist to calculate optimal control policies for a given process model, and numerical examples published in the literature show that these methods can produce significantly lower values of the performance index than can be obtained with conventional control. However, when the objective is implementation on a real process then additional factors that must be considered include: the effect of modelling errors; process noise and nonlinearities; the specification of a realistic performance index; constraints on the state and control variables; cost of implementation and methods of combining the optimal policies with existing regulatory algorithms. Some of these practical factors are illustrated in this paper by experimental data from a computer pilot plant evaporator. However, the previous work in this area and the formulation of the linear optimal control problem are presented first.

## LITERATURE REVIEW

Multivariable control techniques have been reviewed by Thoma (1971), Mendel and Gieseking (1971) and MacFarlane (1972) and are included in the recent textbooks such as those by Athans and Falb (1966), Lapidus and Luus (1967), Sage (1968), and Kuo (1970). The two most popular approaches for determining the optimal control policies of multivariable systems have been dynamic programming and Pontryagin's Maximum Principle. Numerical solution using search techniques, quasilinearization or invariant imbedding is usually required. These methods produce excellent results but often encounter numerical or dimensionality problems in real applications and are not easily adapted to systems with state and control constraints. Tabak (1970) reviews the application of mathematical programming techniques in optimal control theory. A linear programming formulation offers the advantage of being able to handle large multivariable systems with constraints on the control and/or state variables and is the only approach presented in this work. Zadeh and Wahlen (1962) first showed that the time, and fuel optimal control problems could be reduced to linear programming problems for linear discrete time invariant systems. Bondarenko and Filimonov (1968) present numerical results using these criteria. In the chemical engineering literature, Lesser and Lapidus (1966) simulated the time optimal control of an absorber. They indicated how state variable constraints could be introduced into the iterative linear programming approach but only control constraints were used in their numerical analysis. Jarvis and Wright (1971) calculated the switching policy required to obtain a specified changeover time on their multi-

† An earlier version of this paper was presented at the 21st Canadian Chemical Engineering Conference, Montreal, Canada, October (1971).

variable, simulated, application. Bashien (1971) presents a modified simplex algorithm which requires only one iteration to solve the minimum time problem.

Relatively little work has been reported on the use of linear programming techniques with criteria other than time and fuel optimal performance indices. Lack and Enns (1967) formulate a minimax criterion as a linear programming problem and compute the optimal control of a model of a nuclear reactor with twenty-one state variables. Propoi (1963) presents a numerical example involving calculation of the optimal control policy of an overdamped system with minimization of the transient response area as a criterion. This is equivalent to minimization of the sum of the absolute value of the error at all the sampling instants since there is no overshoot. Lorchirachoonkul and Pierre (1967) use a criterion whereby the absolute values of the errors are weighted in both time and space for multivariable distributed-parameter sampled-data systems.

## FORMULATION

Optimal control requires a process model, a performance index, and an algorithm for performing the optimization. This paper deals only with linear, discrete, time invariant, state space models, linear performance indices and relies on a standard linear programming technique for the optimization.

Performance criteria can be divided into those that are evaluated at a single point in time *vs.* those that are summed over a finite time interval. Time optimal control subject to constraints on state and/or control variables is an example of the first category and has been calculated using linear programming techniques by several workers, including Torng (1964) and Nieman and Fisher (1970). Nieman (1971) has shown that it is also possible to use a combination of single point criteria such as minimization of the rise time subject to limitations on the amount of over-shoot and a specified settle out time. These formulations are similar to the following development for linear, summed criteria and hence will not be discussed further.

Two popular sampled data performance indices are the sum of the absolute errors (SAE) and the sum of the absolute error times the time (STAE) which can both be written:

$$Z = \sum_{k=1}^{N} \sum_{i=1}^{n} e_i(k) \, |x_i(k) - x_{d_i}| \tag{1}$$

where $e_i(k)$ is the weighting factor assigned to $x_i(k)$, the $i$th state variable at the $k$th sampling instant. For equal weighting on all elements of $\mathbf{x}$, then $e_i(k) = 1$ for the SAE index and $e_i(k) = k$ for the STAE index and for time optimal control only $e_i(N)$ is nonzero.

Equation (1) can be easily transformed into a different form which is more compatible with linear programming by expressing the deviation between the actual and desired values of the state variables in terms of new variables $\epsilon_{1i}$ and $\epsilon_{2i}$. If only positive values are assigned to these variables then:
when $x_i(k) \geqslant x_{di}$;

$$\epsilon_{2i} = x_i(k) - x_{di} \qquad \text{and} \qquad \epsilon_{1i} = 0$$

and when $x_i(k) \leqslant x_{di}$;

$$\epsilon_{1i} = x_{di} - x_i(k) \qquad \text{and} \qquad \epsilon_{2i} = 0 \tag{2}$$

Thus Eq. (1) can be rewritten as

$$Z = \sum_{k=1}^{N} \sum_{i=1}^{n} e_i(k) \, \{\epsilon_{1i}(k) + \epsilon_{2i}(k)\} \tag{3}$$

The next step is to develop a mathematical model of the process that can be used to express the state variables in terms of the control variables and initial conditions. Most multivariable processes can be adequately represented, at least in a small region about the normal operating conditions, by the following linear, discrete, time invariant, controllable state space model with sampling time $T$:

$$\mathbf{x}([k+1]T) = \phi\mathbf{x}(kT) + \Delta\mathbf{u}(kT) \qquad \mathbf{x}(0) = \mathbf{x}_0 \tag{4}$$

where $\phi$ is the transition matrix and $\Delta$ is the control transfer matrix.

Repeated application of Eq. (4) makes it possible to define the state $\mathbf{x}(NT)$ in terms of the initial conditions $\mathbf{x}_0$ and the control sequence $\{\mathbf{u}(kT), k = 0, 1, 2, \ldots N-1\}$. Thus:

$$\mathbf{x}(NT) = \phi^N \mathbf{x}_0 + \phi^{N-1}\Delta\mathbf{u}(0) + \ldots + \Delta\mathbf{u}([N-1]T) \tag{5}$$

Equations (4) or (5) can be combined with a linear criterion such as those discussed above and rearranged into a standard linear programming format. The result expressed in vector/matrix form, based on Eqs. (2) and (5) is:

$$\left[\begin{array}{cccc|cccccccc} \Delta & 0\cdots 0 & & & \mathbf{e}_1 & -\mathbf{e}_1 & \mathbf{0} & \mathbf{0} & & \mathbf{0} & \mathbf{0} \\ \phi\Delta & \Delta\cdots 0 & & & \mathbf{0} & \mathbf{0} & \mathbf{e}_2 & -\mathbf{e}_2 & & \mathbf{0} & \mathbf{0} \\ \cdot & & & & & & & & & & \\ \cdot & & & & & & & & & & \\ \cdot & & & & & & & & & & \\ \phi^{N-1}\Delta & \phi^{N-2}\Delta\cdots\Delta & & & \mathbf{0} & \mathbf{0} & \mathbf{0} & \mathbf{0} & \cdots & \mathbf{e}_N & -\mathbf{e}_N \end{array}\right] \begin{bmatrix} \mathbf{u}_0 \\ \mathbf{u}_1 \\ \cdot \\ \cdot \\ \cdot \\ \mathbf{u}_{N-1} \\ --- \\ \epsilon_{11} \\ \epsilon_{21} \\ \epsilon_{12} \\ \epsilon_{22} \\ \cdot \\ \cdot \\ \cdot \\ \epsilon_{1N} \\ \epsilon_{2N} \end{bmatrix} = \begin{bmatrix} \mathbf{x}_d - \phi\ \mathbf{x}_0 \\ \mathbf{x}_d - \phi^2\ \mathbf{x}_0 \\ \cdot \\ \cdot \\ \cdot \\ \mathbf{x}_d - \phi^N x_0 \end{bmatrix} \tag{6}$$

where $\mathbf{e} = \mathbf{I}_{nxn}$ for equal weighting on the states and the coefficient matrix is $(Nn) \times (Nm + 2Nn)$. The linear programming problem is then to determine the control sequence $\{\mathbf{u}(k), k = 0, 1, 2, \ldots N-1\}$ which minimizes the criterion given by Eq. (3) subject to the linear constraints represented by Eqs. (6), (7) and (8),

$$\mathbf{0} \leqslant \mathbf{u}_L \leqslant \mathbf{u}(k) \leqslant \mathbf{u}_U \tag{7}$$

$$\begin{aligned} \boldsymbol{\epsilon}_1 &\geqslant \mathbf{0} \\ \boldsymbol{\epsilon}_2 &\geqslant \mathbf{0} \end{aligned} \tag{8}$$

Upper and lower control constraints, in the form of Eq. (7), appear in most practical problems and may be added directly to the linear program without increasing the size of the basis (see Hadley, 1962). State constraints may be easily added to Eq. (8) by simply putting upper limits on $\boldsymbol{\epsilon}_1(k)$ and $\boldsymbol{\epsilon}_2(k)$, in a manner analogous to the addition of control constraints, and there is no significant increase in execution time. Constraints on the derivatives of the state may also be added through a finite difference approximation. This is useful for removing inter-sample ripple and implementing constraints on maximum allowable rates of change of state variables.

## INTERFACING AN OPTIMAL CONTROL SYSTEM TO A CONVENTIONAL CLOSED-LOOP CONTROL SCHEME

If the control policy $\{\mathbf{u}^*(k), k = 1, 2, \ldots N\}$ is calculated as outlined in the previous section and applied to the actual, open-loop plant then one would expect a response that was "optimal" in the sense defined by the criterion, i.e. Eq. (1). However, factors such as modelling errors and unanticipated disturbances can produce offsets from the desired terminal conditions and/or undesirable response characteristics. These effects can be at least partially overcome by the conventional approach of introducing the desired step changes in process operating specifications, as setpoint changes to a feedback control system. In Figure 1 this would be equivalent to neglecting the top two "boxes" and letting $\mathbf{y}^*$ represent the desired step changes. If a model of the *closed-loop* system were available in the form given by Eq. (4) then the optimal input function $\{\mathbf{y}^*(k), k = 1, 2, 3, \ldots N\}$ could be calculated using the techniques outlined in the previous section. However, the closed-loop model, and hence the optimization problem, is more complicated than the analogous open-loop case. Locatelli and Rinaldi (1969) discuss criteria for choosing between open- and closed-loop implementations of optimal control and conclude that a rational *a priori* choice between the two approaches could only be made in special cases.

In the approach illustrated by Figure 1, the predicted optimal output trajectories are fed to the setpoints of the corresponding control loops at the same time that the optimal policy is added to the controller outputs. If the process response, $\mathbf{y}$, and the calculated response, $\mathbf{y}^*$, are identical the feedback control action, $\mathbf{u}_c$, will not change and the process response will

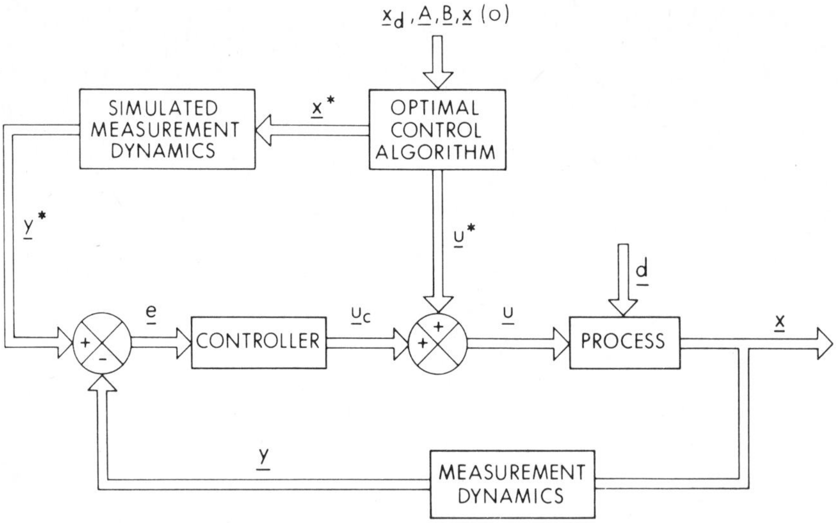


FIGURE 1 Addition of open-loop optimal control policy to an existing regulatory feedback control scheme.

proceed in a fashion identical to that obtained with optimal open-loop implementation. In the presence of disturbances or model inaccuracies the control is no longer optimal but deviations from the calculated trajectory are at least partially compensated for by feedback control action. In the case where $y = y^*$ then the design of the feedback controller is immaterial. However, in general, it should be designed to good regulatory (as opposed to servo) control. Since the optimal state trajectory approaches the desired state at final time, a "bumpless transfer" to the conventional regulatory control is made automatically.

## PROCESS DESCRIPTION AND MODELLING

The optimal control techniques developed in previous sections were evaluated by applying them to the pilot plant double effect evaporator described by Figure 2 and Table I. The three major control loops of the evaporator, as indicated in Fig. 2, are cascade control loops. The bottom flow rates from each effect are manipulated to control the respective liquid levels and the steam rate is manipulated to control the product concentration. The measurement transducers and final control elements are interfaced to an IBM

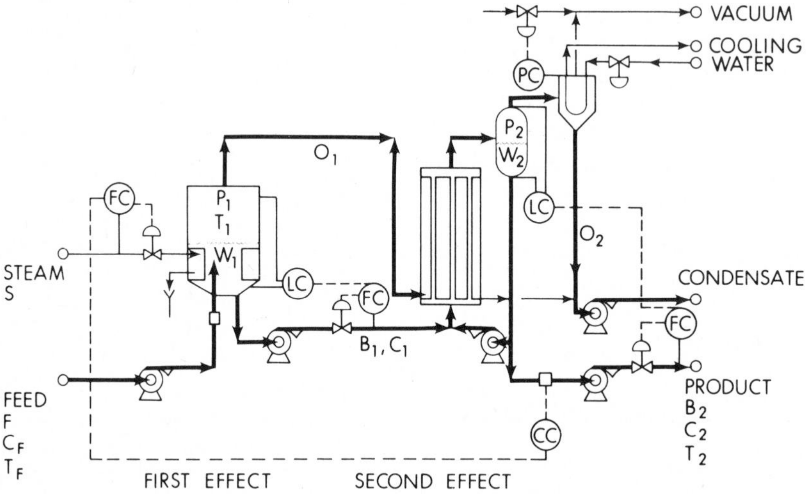


FIGURE 2 Schematic diagram of the double effect evaporator used for the experimental and simulation studies. Variables are defined in Table I.

1800 control computer which has an industrial type DDC monitor system. The process computer was used to facilitate data acquisition, data reduction and implementation of optimal control.

For modelling purposes the evaporator can be considered as a lumped parameter, time invariant system. Mass, heat and solute balances around the principal components of the evaporator can be reduced to five first-order, nonlinear differential equations, as described by Newell and Fisher (1972a). This theoretically derived fifth-order model exhibits a small but significant lead time when compared to the actual process. Since the accuracy of the model in open-loop state driving is critical, the five-equation nonlinear model was fitted to experimental data representing the process response to step changes in steam rate. Details of the quasilinearization algorithm employed and problems encountered have been documented by the authors, Nieman (1971) and Nieman and Fisher (1972). The fitted model exhibited considerably less lead than the theoretical model and produced a good estimate of the experimental open-loop response data. This nonlinear fitted model was then linearized and put in the form of Eq. (4) so that it would be compatible with the LP optimization algorithm described above. This 5-th order, Linearized, Fitted model is designated as 5LF. The point of linearization can have a significant effect on the optimal control of nonlinear processes, but in this work the normal steady state was used as the point of linearization because of convenience.

Since the variable of primary interest in the evaporator application is the product concentration, it is reasonable to consider "optimal" single variable control of C2 based on manipulation of the steam flow while the liquid levels are controlled by conventional single variable loops. Therefore, model 5LF was reduced and simplified to produce a second-order transfer function model, of the form defined by Eq. (9). This model gives a reasonable description of the open-loop product concentration responses to changes in inlet steam flow.

$$\frac{C2(s)}{S(s)} = \frac{K_p}{(44s+1)(1.85s+1)} \quad \text{Model 2LF} \qquad (9)$$

Several different sets of parameter values can be found which give essentially the same goodness-of-fit between the model response and experimental data. However, evaluations made as part of this study, but not reproduced here, show that each parameter set yielded a significantly different optimal control policy. In particular it was shown that $\tau_2$ should be close to the value of 1.85 min calculated by mathematical analysis of the unit, i.e. that it must be physically realistic as well as provide a good fit to open-loop response data. The process gain $K_p$ was calculated for each set of final steady state conditions using a nonlinear, algebraic function obtained by analytical reduction of the original nonlinear model.

TABLE I

Normal operating conditions

| | Process variables | | Normal steady state |
|---|---|---|---|
| x | Five-element state vector ($n$ = 5) | | |
| | W1 | Holdup in the first effect | 30 lb |
| | C1 | Concentration in the first effect | 4.85 wt % |
| | H1 | First effect solution enthalpy | 194 Btu/lb |
| | W2 | Holdup in the second effect | 35 lb |
| | C2 | Concentration in the second effect | 9.64 wt % |
| u | Six-element input vector ($m$ = 3 control variables) | | |
| | S | Inlet steam flow rate | 2.0 lb/min |
| | B1 | Bottoms flow rate from first effect | 3.3 lb/min |
| | B2 | Product flow rate from second effect | 1.66 lb/min |
| | F | Feed flow rate | 5.0 lb/min |
| | CF | Feed concentration | 3.0 wt % |
| | TF | Feed temperature | 190°F |

## DISCUSSION OF RESULTS

The main objective of the simulation and experimental studies was to investigate:

1) The cost and practicality of implementing these techniques on computer controlled process units.

2) The difference between simulated and experimental data, i.e. the observed effects of practical factors such as modelling errors and noise.

3) The effect of using different performance criteria and constraints.

4) The effectiveness of using the existing regulatory feedback control system to compensate for modelling errors and/or unmeasured disturbances.

The simulated and experimental responses of the evaporator are presented in Figures 3 through 7. The next section discusses several important factors related to the implementation and interpretation of the runs and is followed by a discussion of the significance and conclusions that can be drawn from this investigation.

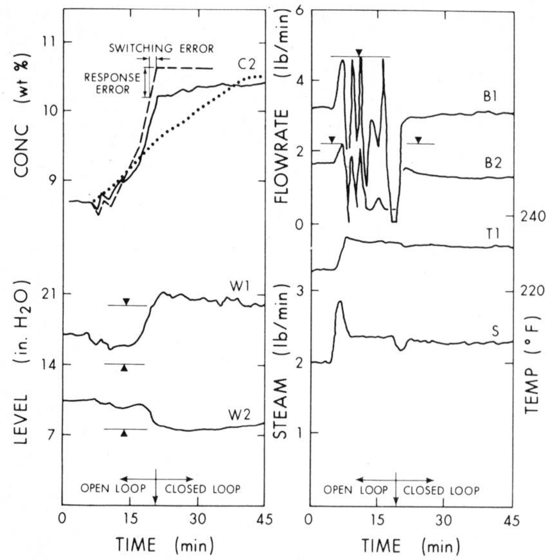


FIGURE 3 Comparison of simulated *vs.* experimental data and the use of a fifth- *vs.* second-order model. Only the concentration C2 is included in the criterion but W1 and W2 are constrained. Horizontal lines with arrow heads indicate constraints. --- SIM/TO/5LF/OL; — EXP/TO/5LF/OL; ··· EXP/TO/2LF/OL.

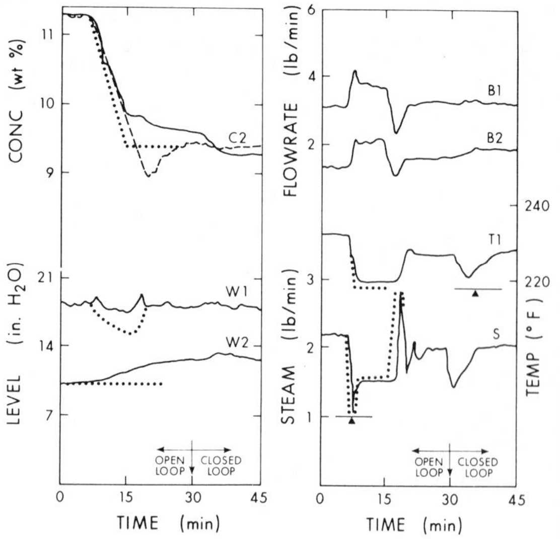


FIGURE 5 Comparison of open-loop *vs.* closed loop implementation of the optimal SAE control policy for a decrease in product concentration. --- EXP/SAE/5LF/OL; — EXP/SAE/5LF/OL; ··· SIM/SAE/5LF/OL ($y^*$).

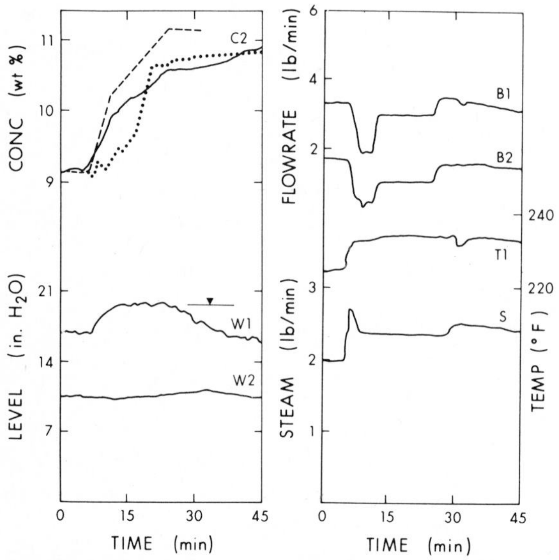


FIGURE 4. Comparison of simulated *vs.* experimental data plus the use of the sum-of-the-absolute errors (SAE) *vs.* the time-optimal (TO) criterion. The dotted curve is traced, slightly shifted, from the solid curve of Figure 3. --- SIM/SAE/5LF/OL; — EXP/SAE/5LF/OL; ··· EXP/TO/5LF/OL.

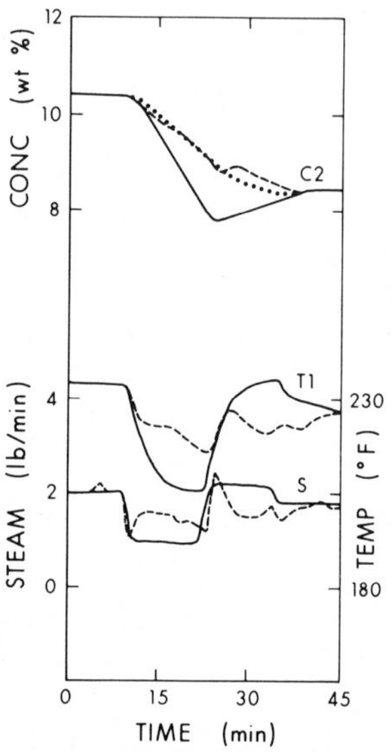


FIGURE 6 Comparison of open-loop *vs.* closed-loop implementation of the time optimal control policy for a decrease in product concentration. --- EXP/TO/2LF$_1$/CL; — EXP/TO/2LF$_1$/OL; ··· SIM/TO/2LF$_1$/OL ($y^*$).

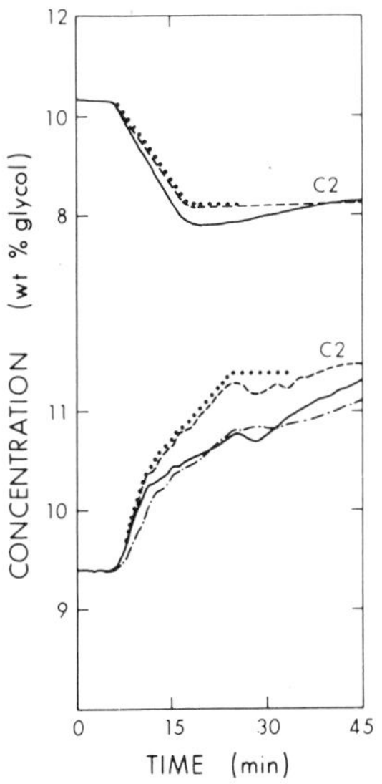


FIGURE 7 (Top) similar to Figure 6 except that the model used (equation 9) was more accurate. (Bottom) similar to Figure 5 except that the change in product concentration is positive. --- EXP/SAE/5LF/CL (model constraint on *S*); — EXP/SAE/5LF/CL (physical constraint on *S*); – · – EXP/SAE/5LF/OL; . . . SIM/SAE/5LF/OL.

## GENERAL COMMENTS ON IMPLEMENTATION AND DATA

The important conditions affecting the data in each figure are summarized by the code defined in Table II.

TABLE II

Code used to describe simulated and experimental results

| Data/Criterion/Model/Configuration | |
|---|---|
| Data: | SIM = simulated by applying **u*** to model 5NL |
| | EXP = experimental from actual evaporator runs |
| Criterion: | TO = time optimal |
| | SAE = sum of the absolute errors (equation 1) |
| Model: | 5LF = fifth order, linear fitted model |
| | 2LF = second order, linear fitted model |
| Configuration: | OL = open-loop |
| | CL = closed loop as per Figure 1 |

For example, –EXP/TO/5LF/CL indicates that the solid curve represents EXPerimental data; a control policy based on the Time Optimal criterion and the 5th order Linear Fitted model; the Closed-Loop implementation, defined by Figure 1.

Constraints on the state and/or control variables are indicated on the figures by horizontal lines with superimposed arrow heads. Most of the figures include a composite plot of several concentration responses and in order to simplify comparison some of the curves have been shifted slightly to eliminate small differences in initial concentrations. The reader is referred to the thesis by Nieman (1971) for details of the equipment, models, calculations and implementation plus additional data.

Most runs involve a change of approximately 2 wt % in the product concentration. The normal operating value is about 10% (Table I). If the liquid level deviations are not included directly in the performance criterion they are constrained to within approximately 15% of their steady state values. The first effect temperature is usually constrained to be less than 240°F to insure that the operating pressure does not exceed the equipment design limit of 10 psig.

Since all simulated and experimental data in this work were recorded at one point per minute and were plotted on a digital plotter, some peaks are diminished and step changes take on a ramp-like appearance. For example, the initial spike in the steam response is much reduced. Another contributing factor to this plot characteristic is the finite response time of the actual process variables.

If the optimal control calculations do not generate the control action required to hold the process at the final steady state then the action to take at the completion of the optimal transient must be determined separately. In the time optimal *open-loop* trials on the evaporator, the steam flow was simply set to a value (calculated from knowledge of the steady state process gain) which would maintain the concentration at the desired value and bumpless transfer was made to conventional proportional plus integral (P + I) feedback control. However, the levels W1 and W2, were frequently at their constraint limits rather than at the desired values and a bumpless transfer to the "averaging" P + I control which had proven best for normal operation resulted in very slow recovery of the levels. Faster recovery of the levels can be attained using tighter level control or including the levels in the optimization criterion.

From an intuitive point of view the optimal control policy for the problem of increasing the product concentration is to:

1) Maximize the steam flow so as to maximize the boil off of water;

2) minimize the bottoms flow from the first effect which reduces the dilution of C2 by B1;

3) lower the second effect level to reduce the holdup and permit faster response.

These actions are evident in Figure 3 but are complicated by the state constraints. The steam flow increased to its maximum value but is almost immediately reduced to about 2.4 pounds per minute to avoid violating the constraint on T1. The levels perform as expected except that the actual liquid level response, W1, exceeds the upper constraint specified in the optimization calculation. The B1 and B2 control policies are not easily interpreted and are not unique. It can be noted that just prior to the switching time denoted by the arrow on the time axis the steam flow is at its maximum (constrained) value and the bottoms flow B1 is at a minimum (and hence W1 is rising). Bottoms flow B2 is at the minimum value possible such that W2 (which is falling because the input stream, B1 has been reduced) will not violate the lower constraint before the switching time is reached. Thus the optimal policy seems to be to save up the available control action for one big "push" at the end, rather than to maintain an approximately constant rate of change in concentration, C2. Note particularly that the initial concentration response is in the "wrong" direction.

## COST OF IMPLEMENTATION

The incremental cost of adding "optimal" state driving control to an existing direct digital control (DDC) system is very small. The optimization problem is solved once, using an off-line computer. In the open-loop implementation the feedback regulatory controllers are disconnected and the calculated control policy, $\mathbf{u}^*(k)$, is sent out to the final control elements on the process inputs. At the completion of the transient the process is returned to DDC control. This technique involves less computer time than the original DDC algorithm. In the closed loop implementation illustrated by Figure 1 the precalculated values of $\mathbf{u}^*$ and $\mathbf{y}^*$ are simply incorporated into the standard DDC control algorithms. The only extra requirement is temporary storage space for $\mathbf{u}^*$ and $\mathbf{y}^*$.

If a control computer is not available then it may still be possible to implement the bang-bang, open-loop control policies manually. For example, "optimal" *single variable* control of the evaporator concentration, C2, by manipulation of the inlet steam, S, was easily implemented by manual manipulation using a conventional controller. However, manual implementation of the control policy for S, B1 and B2 as represented by the solid lines in Figure 3 is impractical.

## SIMULATED *vs.* EXPERIMENTAL RESULTS

If the process model is "perfect" then the experimental responses are identical to the simulated optimal responses. However, simplifications of the model structure, linearization, and errors in the parameter estimates all contribute to model inaccuracies. In addition, practical factors such as noise, unanticipated disturbances, the random aspects of many process characteristics, lead to differences between the model and process responses. The "degree of suboptimality" (DOS) produced by modelling errors, practical factors, etc. can be defined by relationships such as:

$$\mathrm{DOS} = 100\,\frac{(\mathrm{IP} - \mathrm{IP}^*)}{\mathrm{IP}^*} \qquad (10)$$

where IP is the index of performance obtained by applying the calculated control policy to the *actual process* and IP* is the index of performance obtained by applying the calculated control policy to the *model* used in the calculation of that policy. By definition, IP* is "optimal" or "best".

Relationships such as Eq. (10) provide a convenient, quantitative, means of comparing different strategies but are seldom satisfactory for multivariable experimental applications. "Engineering judgement" is usually the best overall basis for comparison. The DOS for the simulated (dotted) *vs.* experimental (solid) open-loop responses in Figure 5 is approximately 40% at the 25-min point. The DOS for analogous closed-loop data in Figure 5 is approximately 10%. A comparison of these values leads to the correct conclusion that the closed loop implementation is better than the open-loop implementation. However, a glance at the responses in Figure 5 leads to the same conclusion and shows that the high DOS (at the 25-min point) is primarily due to an offset in C2 (which is later removed by the closed-loop, feedback controller). The offset can be attributed to modelling error.

The effect of modelling errors shows up even more clearly in the C2 values reached by the process at the time of the final (precalculated) switch in steam flow. In Figure 3 the optimal simulated response (dashed curve) is very similar to the experimental response (solid line) but leads it by a minute or two. However, due to the rapid rate of change produced in C2 by this multivariable control policy the small *switching error* of $1\frac{1}{2}$ min leads to a relatively large *response error.* Thus the phase or time relationship between model and process responses is critical. For example, a theoretically derived fifth-order model of the evaporator was suitable as a basis for developing an optimal, multivariable, feedback, regulatory control system for the evaporator (Newell and Fisher, 1972a, b). However, even after this theoretically derived model was improved by fitting it to experimental response data (see previous discussion of model 5LF) it was still not entirely satisfactory for this work involving servo control.

The dotted curve in Figure 3 is the experimental process response produced from the time optimal control policy derived from a reduced model of the form represented by Eq. (9). Use of a second-order model meant that the time optimal control policy could be calculated analytically, e.g. see Latour *et al.* (1967) or Nieman (1971). However, when this model is used there is no direct means of putting a constraint on the first effect operating pressure and hence the user must select an upper constraint for the steam flow which, if maintained for a long period of time, will not increase the pressure beyond a safe limit. Because of this constraint the range of permissible steam flows is smaller than for the multivariable case and the rate of increase of C2 under the two different policies clearly shows the advantages of using the multivariable approach.

In general the bang-bang manipulation of the input variables that occurs in optimal servo control and the fact that relatively large deviations from the point of linearization are involved means that model accuracy is much more important than in regulatory control applications. For large changes in operating conditions or for very nonlinear processes it may be desirable to relinearize the process model at different points along the trajectory and do the optimization calculation in stages as was done by Choquette *et al.* (1970).

## THE EFFECT OF DIFFERENT PERFORMANCE CRITERIA AND CONSTRAINTS

A general conclusion based on the open-loop responses, the closed-loop responses, and other data not included here, is that the SAE criterion produces smoother transients than time-optimal. This is particularly obvious from a comparison of the manipulated variables B1 and B2 in Figures 3 and 4. Also, the time optimal C2 response (dotted curve Figure 4) increases almost exponentially, whereas the SAE response (Figure 4, solid curve) maintains a more uniform rate of change over the entire transient.

Constraints on the state and/or control variables are necessary in most applications to insure that the process does not violate equipment and/or operating specifications. For example, reducing the upper constraint on steam flow will slow down the rate of change in C2 when an increase in concentration is specified. Regardless of why they are used, constraints usually have a pronounced effect on the output responses and their effect must not be confused with other factors. In Figure 4 the sharp break in the SAE C2 response can be related to the increase in bottoms flow, B1, required to keep the level, W1, from exceeding its upper constraint. In Figure 5 the spike and the general increase in steam flow at approximately the 20-min mark is related to a lower constraint placed on the first effect temperature to prevent the operating pressure from going below atmospheric. (The corresponding effect on C2 is confused with the decrease in B1 that occurs about the same time.) It can be concluded that the choice of performance criterion and control and/or state variable constraints has a significant effect on the process transients and must be carefully selected by the control engineer.

## OPEN *vs.* CLOSED-LOOP IMPLEMENTATION

The approach illustrated in Figure 1 is a convenient means of combining an optimal open-loop state driving control policy with a conventional regulatory control system. A series of runs was performed to determine whether this approach would also be an effective means of compensating for modelling errors and/or measured disturbances.

In Figure 6 the simulated value of the outlet concentration, C2*, calculated from a second-order model (similar to, but not as accurate as model 2LF) is represented by the dotted line; the actual response under open-loop control by the solid line; and the response using the closed-loop implementation by the dashed curve. It is obvious that there is considerable modelling error (i.e. difference between solid and dotted curves) and that the closed loop implementation of Figure 1 is fairly successful in compensating for this error (i.e. the dashed curve, C2, follows the dotted curve C2*).

The concentration responses shown in the top half of Figure 7 are taken under conditions similar to those in Figure 6 except that model 2LF (Eq. 9) is used and an "unexpected" −10% step change in feed flow is introduced at the same time as the specified change in C2. The model response is more accurate than in Figure 6 and the closed-loop implementation is reasonably successful in compensating for both the modelling errors and the disturbance.

The data in Figure 5 and in the bottom half of Figure 7 represent the system response to the optimal control policy based on an SAE criterion and model 5LF. The dotted line in Figure 7 represents the "optimal" trajectory, C2*. The dashed line is the closed-loop implementation of this same policy on the evaporator. A constraint was placed on the steam flow, $S$, during the (off line) calculation of **u***  but was *not* implemented on the output of the feedback controller during the experimental run. Therefore, since the process response lagged the model resonse, feedback action increased the steam setpoint and this caused violations of the pressure constraint. The solid curve is an analogous closed-loop implementation except that the steam flow rate was limited to a value which maintained an acceptable pressure in the first effect vapor space. The improvement over the corresponding open-loop implementation (dash-dot curve) is small but significant.

The solid curve in Figure 5 is an open-loop experimental response for a setpoint change in the opposite direction. Notice that the closed-loop implementation (dashed line) resulted in overshoot of the product concentration. This resulted from the inter-action of the first effect bottoms with the product concentration. The first effect level is difficult to model and in this instance was higher than predicted causing feedback compensation which increased the first effect bottoms flow, and this in turn reduced the concentration C2. A multivariable feedback system, such as that reported by Newell and Fisher (1972), would perform better than the multiloop conventional control system of Figure 2 since interactions would be compensated for more rapidly. (No change would be required in the optimal control policy or method of implementation (i.e. Figure 1).) These runs indicate that the closed-loop approach can offer significant advantages over open-loop responses. The tuning of the feedback control loop would obviously have an effect on the closed-loop response but in most practical applications would be chosen to give good regulatory control during normal operation. At the completion of the optimal trajectory the concentration ($C2^*(t_f)$) approaches the desired setpoint and hence normal regulatory control can be easily resumed.

## COMPARISON WITH ALTERNATIVE APPROACHES

The techniques discussed in this paper can also be compared with the alternative of simply putting the desired changes into the setpoints of the feedback controllers. Nieman and Fisher (1970) have shown that for the evaporator application the time optimal control techniques are significantly better than putting setpoint changes into the conventional single variable controllers shown in Figure 2. However, Newell and Fisher (1972b) have shown that in the operating region where state and/or control constraints do not have to be considered, optimal multivariable feedback controllers will give process responses that are comparable to the ones obtained in this work. They do not require recalculation of the control for each new setpoint change are are not as sensitive to modelling errors. Thus for process applications this latter approach would probably be more convenient for *small* changes. However, when large setpoint changes are to be made, and where state and/or control constraints are essential, then the techniques discussed in this paper are recommended.

## CONCLUSIONS

The use of linear programming to solve the optimal control problem is flexible, convenient, and permits formulations which can handle different linear criteria and constraints on the state and control variables. The closed-loop, or modified model following approach presented in this work for combining state-driving and conventional feedback control is sub-optimal, but computationally simple and permits at least partial compensation for modelling errors and unanticipated load disturbances. Modelling errors, noise and unanticipated disturbances can produce significant deviations from the calculated optimal trajectories but the experimental data taken in this study show that implementation on real processes is practical and gives good control.

## REFERENCES

Athans, M. and Falb, P. L., *Optimal Control* (McGraw-Hill, New York, 1966).

Bashien, G., "A simplex algorithm for on-line computation of time optimal controls," *IEEE Trans. Auto. Cont.*, **AC-16**, 479 (1971).

Bondarenko, V. and Filimonov, I., "Applications of linear programming to extremal problems of control theory," *Appl. Math. Mech. J.*, **32**, No. 1, 138 (1968).

Choquette, P., Noton, A. R. M. and Watson, C. A. G., "Remote computer control of an industrial process," *Proc. IEEE*, **58**, No. 1, 10 (1970).

Hadley, G., *Linear Programming* (Addison-Wesley Publishing Co., Inc., Reading, Massachusetts, 1962), Chapter II.

Jarvis, R. C. F. and Wright, J. D., "Optimal state changeover control of a multivariable system using a minicomputer," *Can. J. Chem. Eng.* **50**, 520 (1972).

Kuo, B. C., *Discrete-Data Control Systems* (Prentice Hall, Englewood Cliffs, N.J. (1970).

Lack, G. and Enns, M., "Optimal control trajectories with minimax objective functions by linear programming," *IEEE Trans. Auto. Cont.*, **AC-12**, 749 (1967).

Lapidus, L. and Luus, R., *Optimal Control of Engineering Processes* (Blaisdell Publishing Co., Waltham, Mass., U.S.A., 1967).

Latour, P. R., Koppel, L. B. and Coughanowr, D. R., "Time-optimal control of chemical processes for set-point changes," *Ind. Eng. Chem., Proc. Des. Dev.*, **6**, No. 4, 452 (1967).

Lesser, H. A. and Lapidus, L., "The time-optimal control of discrete-time linear systems with bounded controls," *AIChE. J.*, **12**, No. 1, 143 (1966).

Locatelli, A. and Rinaldi, S., "Open- *vs.* closed-loop implementation of optimal control," *IEEE Trans. Auto. Cont.*, **AC-14**, 570 (1969).

Lorchirachoonkul, V. and Pierre, D., "Optimal control of multivariable distributed-parameter systems through linear programming," Proceedings of Joint Automatic Control Conference, Philadelphia, Penn., 702, June (1967).

Newell, R. B. and Fisher, D. G., "Model development, reduction and experimental evaluation for an evaporator," *Ind. Eng. Chem., Proc. Des. Dev.*, **11**, No. 2, 212 (1972a).

Newell, R. B. and Fisher, D. G., "Experimental evaluation of optimal multivariable regulatory controllers with model-following capabilities," *Automatica*, **8**, No. 3, 247 (1972b).

MacFarlane, A. G. J., "A survey of some recent results in linear multivariable feedback theory," *Automatica*, **8**, 455 (1972).

Mendel, J. M. and Gieseking, D. L., "Bibliography on the linear-quadratic-gaussian problem," *IEEE Trans. Auto. Cont.*, **AC-16**, 847 (1971).

Nieman, R. E., "Application of quasilinearization and linear programming to control and estimation problems," Ph.D. Thesis, Department of Chemical and Petroleum Engineering, University of Alberta, 1971.

Nieman, R. E. and Fisher, D. G., "Computer control using optimal state driving techniques," Proceedings of the National Conference on Automatic Control, University of Waterloo, August, 1970.

Nieman, R. E. and Fisher, D. G., "Parameter estimation using quasilinearization and linear programming," *Can. J. of Chem. Eng.*, Dec., **50**, 802 (1972).

Propoi, A. I., "Use of LP methods for synthesizing sampled-data automatic systems," *Automation and Remote Control*, **24**, 837 (1963).

Sage, A. P., *Optimum Systems Control* (Prentice Hall, Toronto 1968).

Tabak, D., "Application of mathematical programming techniques in optimal control," Proceedings of Joint Automatic Control Conference, Atlanta, Georgia, 648, June (1970).

Thoma, M., "Optimal multivariable control systems theory: a survey," 2nd IFAC Symposium on Multivariable Technical Control Systems, Dusseldorf, Oct. 11-13, 1971.

Torng, H. C., "Optimization of discrete control systems through linear programming," *J. Franklin Institute*, **278**, No. 1, 28 (1964).

Zadeh, L. A. and Wahlen, B. H., "On optimal control and linear programming," *I.R.E. Trans. Auto. Cont.*, **7**, No. 4, 45 (1962).

# Section 6: OTHER MULTIVARIABLE CONTROL TECHNIQUES

CONTENTS:



COMMENTS:

This section contains two papers which describe the application of Model Reference Adaptive Control (MRAC) systems [1] and two papers which extend the Smith predictor method of time delay compensation [2] to multivariable systems.

Adaptive control techniques are particularly attractive for industrial applications because instead of relying on *a priori* control strategies and parameters, the control system adapts itself on-line to reach the desired control objective. It is significant that in the experimental work the MRAC system was applied to the evaporator with the initial estimate of the multivariable feedback matrices defined as zero, i.e., $\underline{\underline{K}}^{FB} = \underline{\underline{0}}$ or an open-loop control policy. Although the evaporator is open-loop unstable, the MRAC system adapted fast enough to maintain good control of the evaporator and soon produced control matrices that gave performance equivalent to the optimal multivariable systems described earlier.

Many of the design methods based on modern control theory are not directly applicable to multivariable processes which contain time delays. In paper 6.3 the classical Smith predictor for single variable systems [2] is extended to multivariable systems which contain time delays in the control variables or the output variables. As in the classical method, time delays are eliminated from the characteristic equation of the closed-loop system. The evaporator application in paper 6.4 demonstrates that this approach is feasible for practical control problems and allows relatively large controller gains to be employed.

REFERENCES:

[1] Landau, I.D., "A Survey of Model Reference Adaptive Techniques - Theory and Applications", Automatica, 10, 353-379 (1974).

[2] Smith, O.J., "Closer Control of Loops with Dead Time", Chemical Engineering Progress, 53, 217-223 (1957).

# MODEL REFERENCE ADAPTIVE CONTROL BASED ON LIAPUNOV'S DIRECT METHOD

## Part I: Theory and Control System Design†

W. KENT OLIVER, DALE E. SEBORG‡ and D. GRANT FISHER

*Department of Chemical Engineering, University of Alberta, Edmonton, Alberta, Canada*

*(Received April 24, 1972; accepted for publication April 3, 1973)*

The Liapunov design approach for model reference adaptive control (MRAC) is extended to include adaptive integral and setpoint control modes. The feasibility of the Liapunov approach is illustrated by designing a multivariable MRAC system based on a fifth-order state space model of a pilot scale, double-effect evaporator. Simulation results are presented to demonstrate that the MRAC approach gives satisfactory control for a wide range of adaptive loop gains and despite very poor initial control policies.

## INTRODUCTION

Industrial applications of automatic control inevitably require on-line tuning of the controller constants to achieve a satisfactory degree of control. Furthermore, frequent readjustment of the controller constants may be required if process operating conditions change significantly. In an adaptive control system, the control parameters or control configuration are changed automatically during process operation to compensate for changing process conditions. In the model reference approach to adaptive control, the objective is to adjust the control parameters so that the process response approaches the response of a reference model when both are subjected to the same process disturbances or setpoint changes. The function of the reference model is to represent the desired closed-loop dynamic behavior of the process.

Model reference adaptive control (MRAC) offers several potential advantages for multivariable control problems. First, in contrast to some other adaptive control techniques, it does not require time-consuming, on-line calculations for process identification and optimization. Secondly, MRAC offers a feasible approach for tuning multivariable control systems. The availability of reliable tuning techniques is an important practical consideration in industrial applications of modern multivariable control techniques.

Historically, model reference adaptive control systems were developed in the late 1950's in the aircraft and aerospace industries for applications where adequate control could not be achieved using a fixed parameter control design. Eveleigh (1967) and Bristol (1970) have summarized practical applications of adaptive control techniques including MRAC. The early single variable MRAC systems had the same objective as later multivariable techniques: to minimize the error between the outputs of a reference model and of the physical process as shown in Figure 1. However, the early MRAC techniques used sensitivity functions and gradient-like search techniques to minimize a function of this error (Eveleigh, 1967). Gradient methods have been applied to single variable, reactor control problems by

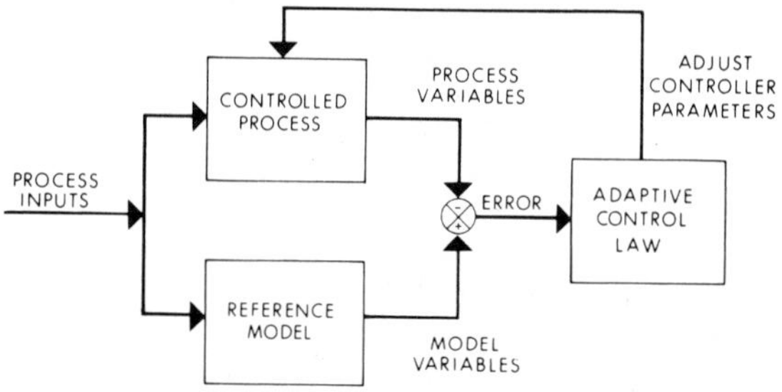


FIGURE 1 Block diagram of Model Reference Adaptive Control (MRAC).

† Paper presented at the New York AIChE Meeting, December 1972.

‡ To whom correspondence concerning this paper should be addressed.

Crandall and Stevens (1965), Casciano and Staffin (1967), and Ahlgren and Stevens (1971). Their simulation studies showed that MRAC gave improved control when reactor operating conditions varied greatly as during start-up or setpoint changes. However, as Parks (1966) has demonstrated, when gradient techniques are used the stability of the closed-loop system may be difficult to guarantee due to the nonlinear nature of the adaptive algorithm.

To overcome the stability problems associated with gradient methods, Butchart and Shackcloth (1965), Parks (1966), and Shackcloth (1967) have designed single variable adaptive control systems based on Liapunov stability considerations. An important advantage of this design method is that the closed-loop system is guaranteed to be stable. The Liapunov design approach has also been developed for linear, multivariable, time-invariant (Grayson, 1965; Winsor and Roy, 1968; and Porter and Tatnall, 1969) and time-varying (Porter and Tatnall, 1970a) processes. Recent simulation studies have investigated the effects of design parameters and reference model order in MRAC and in the closely related problem of model reference process identification (Gromyko and Sankovskii, 1969, Schooley and Pazdera, 1971; Pazdera and Pottinger, 1969; and Loan, 1971). Porter and Tatnall (1970b) have successfully applied their MRAC design method to a simple hydraulic servo-mechanism in the only experimental study of the Liapunov design approach reported to date.

In this investigation the basic feedforward-feedback adaptive control algorithms developed by Porter and Tatnall (1969) are extended to include integral and setpoint control modes. A MRAC system is then designed for a pilot-scale, double-effect evaporator using a fifth-order state space model. The feasibility of the proposed control system is demonstrated by typical simulation results. A related paper (Oliver *et al.*, 1972) presents additional simulated and experimental results plus a detailed evaluation of this adaptive control system.

## THEORETICAL DEVELOPMENT

The dynamic behavior of many systems can be adequately described by a state-space model of the form

$$\dot{\mathbf{x}}_p = \mathbf{A}\mathbf{x}_p + \mathbf{B}\mathbf{u} + \mathbf{D}\mathbf{d} \tag{1}$$

where $\mathbf{x}_p$ is the $n \times 1$ state vector, $\mathbf{d}$ is the $p \times 1$ disturbance vector, $\mathbf{u}$ is the $r \times 1$ control vector and $\mathbf{A}$, $\mathbf{B}$, and $\mathbf{D}$ are constant matrices of appropriate dimensions. Assuming a multivariable feedforward-feedback control law

$$\mathbf{u} = \mathbf{K}_{FB}\mathbf{x}_p + \mathbf{K}_{FF}\mathbf{d} \tag{2}$$

yields the closed-loop process model

$$\dot{\mathbf{x}}_p = \mathbf{A}_p\mathbf{x}_p + \mathbf{D}_p\mathbf{d} \tag{3}$$

where $\mathbf{A}_p \equiv \mathbf{A} + \mathbf{B}\mathbf{K}_{FB}$ and $\mathbf{D}_p \equiv \mathbf{D} + \mathbf{B}\mathbf{K}_{FF}$.

For future reference, define the control matrices $\mathbf{\Gamma}$ and $\mathbf{\Omega}$ as

$$\mathbf{\Gamma} \equiv \mathbf{B}\mathbf{K}_{FB} \tag{4a}$$

and

$$\mathbf{\Omega} \equiv \mathbf{B}\mathbf{K}_{FF} \tag{4b}$$

A linear, asymptotically stable, time-invariant model of the desired closed-loop plant behavior is postulated as

$$\dot{\mathbf{x}}_m = \mathbf{A}_m\mathbf{x}_m + \mathbf{D}_m\mathbf{d} \tag{5}$$

where $\mathbf{x}_m$ is the $n$-dimensional state vector for the reference model and $\mathbf{A}_m$ and $\mathbf{D}_m$ are specified to provide the desired transient and steady-state behavior of the closed-loop process. The error, $\mathbf{e}$, between the reference model and the process is defined as

$$\mathbf{e} \equiv \mathbf{x}_m - \mathbf{x}_p \tag{6}$$

The control objective in MRAC is to reduce the error to zero by appropriate adjustment of the elements of $\mathbf{K}_{FF}$ and $\mathbf{K}_{FB}$. The design method proposed by Porter and Tatnall (1969) can be summarized as follows:

Subtracting Eq. (3) from Eq. (5) yields

$$\dot{\mathbf{e}} = \mathbf{A}_m\mathbf{e} + (\mathbf{A}_m - \mathbf{A}_p)\mathbf{x}_p + (\mathbf{D}_m - \mathbf{D}_p)\mathbf{d} \tag{7}$$

Define matrices $\mathscr{A}$ and $\mathbf{\Lambda}$ by

$$\mathbf{A}_m - \mathbf{A}_p \equiv \mathscr{A} = [\alpha_{ij}] \tag{8}$$

and

$$\mathbf{D}_m - \mathbf{D}_p \equiv \mathbf{\Lambda} = [\lambda_{ij}] \tag{9}$$

Following Porter and Tatnall (1969), a Liapunov function is chosen to be the positive-definite quadratic form

$$V = \mathbf{e}^T\mathbf{P}\mathbf{e} + \sum_{i=1}^{n}\sum_{j=1}^{n}\frac{1}{\xi_{ij}}\alpha_{ij}^2 + \sum_{i=1}^{n}\sum_{j=1}^{p}\frac{1}{\nu_{ij}}\lambda_{ij}^2 \tag{10}$$

with $\xi_{ij} > 0$, $\nu_{ij} > 0$ and where $\mathbf{e}^T$ denotes the vector transpose of $\mathbf{e}$. The adaptive loop gains, $\xi_{ij}$ and $\nu_{ij}$, are design parameters which must be specified.

Matrix **P** is a positive-definite matrix which is the unique solution of the Liapunov matrix equation

$$\mathbf{A}_m{}^T\mathbf{P} + \mathbf{P}\mathbf{A}_m = -\mathbf{Q} \tag{11}$$

where **Q** is an arbitrary positive-definite $n \times n$ matrix. If the choice, $\mathbf{Q} = \mathbf{I}$, is made then the total time derivative of Eq. (10) is given by

$$\dot{V} = -\mathbf{e}^T\mathbf{e} + 2\sum_{i=1}^{n}\sum_{j=1}^{n}\left(\frac{1}{\xi_{ij}}\dot{\alpha}_{ij} + x_{pj}\mathbf{e}^T\mathbf{p}_i\right)\alpha_{ij}$$

$$+ 2\sum_{i=1}^{n}\sum_{j=1}^{p}\left(\frac{1}{\nu_{ij}}\dot{\lambda}_{ij} + d_j\mathbf{e}^T\mathbf{p}_i\right)\lambda_{ij} \tag{12}$$

where $\mathbf{p}_i$ denotes the $i$th column of matrix **P**. To ensure that the closed-loop system is stable, $\dot{V}$ must be at least negative-semidefinite, a requirement which is satisfied if the last two terms in Eq. (12) are identically equal to zero. These terms will equal zero if $\alpha_{ij} = 0$ and $\lambda_{ij} = 0$ (i.e. the model and the process matrices are identical) or if the bracketed terms in Eq. (12) are zero. This last condition and Eqs. (8) and (9) can be used to derive the following adaptive control algorithms

$$\dot{\gamma}_{ij} = x_{pj}\mathbf{e}^T\mathbf{p}_i\,\xi_{ij}\ (i,j = 1,\ldots,n) \tag{13}$$

$$\dot{\omega}_{ij} = d_j\,\mathbf{e}^T\mathbf{p}_i\nu_{ij}\ (i = 1,\ldots,n; j = 1,\ldots,p) \tag{14}$$

where $\gamma_{ij}$ and $\omega_{ij}$ denote the elements of $\mathbf{\Gamma}$ and $\mathbf{\Omega}$, respectively.

The MRAC design approach based on Liapunov's direct method can be considered to be nonlinear, since the algorithm in Eq. (13) is nonlinear with respect to $x_{pj}$. The method requires that measurements (or estimates) of all the process state variables, $\mathbf{x}_p$, and disturbances, **d**, be available and that all the elements of $\mathbf{\Gamma}$ and $\mathbf{\Omega}$ can be manipulated.

The multivariable adaptive control algorithm developed by Porter and Tatnall (1969) and summarized above considers only feedforward and proportional feedback control. For many processes, the addition of integral control action is essential in order to eliminate offsets resulting from sustained disturbances. Adaptive control during setpoint changes may also be desirable since process nonlinearities (not included in the linear process model) become important for setpoint changes. In the next two sections, methods for adding adaptive integral and adaptive setpoint control are proposed as an extension to the basic MRAC design method.

C

*Adaptive Integral Control*

Consider the open-loop state-space model of the process in Eq. (1). Suppose that it is desirable to eliminate offsets in $q$ of the state variables. If this $q$-dimensional subset of $\mathbf{x}_p$ is denoted by **y**, we can then write

$$\mathbf{y} = \mathbf{C}\mathbf{x}_p \tag{15}$$

where **C** is specified so that **y** is the desired subset of $\mathbf{x}_p$. Porter and Power (1970) and Newell (1971) have independently demonstrated that for zero offset in **y**, $q$ must be less than or equal to $r$, the dimension of the control vector, **u**. Next, define **z** by the equation

$$\mathbf{z} \equiv \int_0^t \mathbf{y}\,dt \tag{16}$$

Combining Eqs. (1), (15) and (16) gives the augmented system

$$\dot{\mathbf{x}}_p' = \mathbf{A}'\mathbf{x}_p' + \mathbf{B}'\mathbf{u} + \mathbf{D}'\mathbf{d} \tag{17}$$

where

$$\mathbf{x}_p' \equiv \begin{bmatrix}\mathbf{x}_p \\ \mathbf{z}\end{bmatrix}; \quad \mathbf{A}' \equiv \begin{bmatrix}\mathbf{A} & \mathbf{0} \\ \mathbf{C} & \mathbf{0}\end{bmatrix}; \quad \mathbf{B}' \equiv \begin{bmatrix}\mathbf{B} \\ \mathbf{0}\end{bmatrix};$$

$$\mathbf{D}' \equiv \begin{bmatrix}\mathbf{D} \\ \mathbf{0}\end{bmatrix} \tag{18}$$

A proportional feedback-feedforward control law for the augmented system can be written as

$$\mathbf{u} = \mathbf{K}'\mathbf{x}_p' + \mathbf{K}_{FF}\,\mathbf{d} \tag{19}$$

If $\mathbf{K}'$ is partitioned as $\mathbf{K}' = [\mathbf{K}_{FB}, \mathbf{K}_I]$ and Eq. (19) is expanded using Eq. (16), a proportional-integral-feedforward control law results

$$\mathbf{u} = \mathbf{K}_{FB}\mathbf{x}_p + \mathbf{K}_{FF}\mathbf{d} + \mathbf{K}_I\int_0^t \mathbf{y}\,dt \tag{20}$$

Substituting Eq. (19) into Eq. (17) yields

$$\dot{\mathbf{x}}_p' = [\mathbf{A}' + \mathbf{B}'\mathbf{K}']\mathbf{x}_p' + [\mathbf{D}' + \mathbf{B}'\mathbf{K}_{FF}]\mathbf{d} \tag{21}$$

Since Eq. (21) is in the same form as Eq. (3), the adaptive control algorithm of Eq. (13) can be used to adapt the elements of $\mathbf{K}'$ and hence, $\mathbf{K}_{FB}$ and $\mathbf{K}_I$. Thus Eqs. (20), (13) and (14) constitute an adaptive control algorithm in which all three modes: proportional, integral and feedforward can adapt with changing process conditions. The integral action will eliminate offsets in $q$ state variables, where $q \leqslant r$. However, the addition of integral control action increases the dimensions of the $\mathbf{\Gamma}$ matrix and hence increases the amount of on-line computation time required. This follows since the number of adaptive control equations in Eq. (13) is equal to the square of the augmented state vector dimension; thus $\mathbf{\Gamma}$ is an $n \times n$ matrix for proportional control but becomes an $(n + q) \times (n + q)$ matrix for proportional-integral control.

### *Adaptive Setpoint Control*

Linear state-space models are often derived by evaluating of linearized process model at a specified set of operating conditions. The design of a time-invariant multivariable control system can then proceed based on the linearized model. If the actual nonlinear process is subjected to large setpoint changes, it may be difficult to achieve adequate control using the time invariant control scheme. This follows since a control system designed for one set of operating conditions may not perform satisfactorily when the operating conditions are changed significantly. The addition of an adaptive setpoint mode to the previous feedforward–feedback control algorithm can help alleviate this problem.

Again consider a process whose open-loop dynamic behavior can be described by Eq. (1). Ideally, it would be desirable to be able to make arbitrary setpoint changes in all $n$ state variables. However, the number of degrees of freedom is $r$, the number of control variables. Thus the setpoints of only $q$ state variables, where $q \leq r$, can be changed in an arbitrary fashion and still have the system satisfy the steady-state version of Eq. (1).

A control law that accounts for setpoint changes is of the form

$$\mathbf{u} = \mathbf{K}_{FB}\mathbf{x}_p + \mathbf{K}_{FF}\mathbf{d} + \mathbf{K}_{sp}\mathbf{y}_{sp} \qquad (22)$$

where $\mathbf{y}_{sp}$ is the desired value of $\mathbf{y}$ and can be time-varying. Using the setpoint vector to augment the disturbance vector, Eq. (22) can be written as

$$\mathbf{u} = \mathbf{K}_{FB}\mathbf{x}_p + \mathbf{K}'_{FF}\mathbf{d}' \qquad (23)$$

where

$$\mathbf{K}'_{FF} \equiv [\mathbf{K}_{FF}, \mathbf{K}_{sp}] \qquad \text{and} \qquad \mathbf{d}' \equiv \begin{bmatrix} \mathbf{d} \\ \mathbf{y}_{sp} \end{bmatrix}$$

Combining Eqs. (1) and (23) and defining $\mathbf{D}' \equiv [\mathbf{D}, \mathbf{0}]$ yields

$$\dot{\mathbf{x}}_p = (\mathbf{A} + \mathbf{B}\mathbf{K}_{FB})\mathbf{x}_p + (\mathbf{D}' + \mathbf{B}\mathbf{K}'_{FF})\mathbf{d}' \qquad (24)$$

Since Eq. (24) is in the same form as Eq. (3), Eqs. (13) and (14) can be used to adapt the elements of $\mathbf{K}_{FB}$ and $\mathbf{K}'_{FF}$, respectively.

Both integral and setpoint control can be implemented simultaneously by proper partitioning of the augmented state and disturbance vectors. Newell and Fisher (1972a) have considered these extensions to the standard optimal regulator problem. The resulting control laws for both optimal control and MRAC are of the form

$$\mathbf{u} = \mathbf{K}_{FB}\mathbf{x}_p + \mathbf{K}_{FF}\mathbf{d} + \mathbf{K}_{I}\int_0^t \mathbf{y}\,dt + \mathbf{K}_{sp}\mathbf{y}_{sp} \qquad (25)$$

## DESIGN OF A MODEL REFERENCE ADAPTIVE CONTROL SYSTEM FOR A DOUBLE EFFECT EVAPORATOR

The theory described in the previous section was used to design a MRAC system for a pilot scale double-effect evaporator. Mathematical models of the evaporator have been developed (Newell and Fisher, 1972b; Andre and Ritter, 1968) based on material and energy balances for the two effects. The process model used in this investigation was derived by Newell (1971) and consists of a fifth-order state-space model with the state, disturbance and control vectors defined to be normalized, perturbation variables (see Appendix). Although the model was not empirically fitted, it gives reasonable agreement with experimental open-loop responses for disturbances which are 20% of the steady-state values (Newell and Fisher, 1972b).

The adaptive control law in Eq. (13) is expressed in terms of $\mathbf{\Gamma}$; however, in the evaporator application the variables which can be manipulated are not the 25 elements of $\mathbf{\Gamma}$ but rather the 15 elements of the $3 \times 5$ feedback control matrix, $\mathbf{K}_{FB}$. Thus, for purposes of implementation, an expression for $\mathbf{K}_{FB}$ in terms of $\mathbf{\Gamma}$ is desirable. By using the pseudo-inverse approach (Greville, 1959), the following approximate, least-squares solution to Eq. (4a) is obtained providing matrix $\mathbf{B}^T\mathbf{B}$ is nonsingular (or equivalently, providing that the rank of $\mathbf{B}$ is $r$)

$$\mathbf{K}_{FB} = (\mathbf{B}^T\mathbf{B})^{-1}\mathbf{B}^T\mathbf{\Gamma} \qquad (26)$$

Note that an exact solution of Eq. (4a) is not possible, since the 15 elements of $\mathbf{K}_{FB}$ would have to satisfy 25 scalar equations. This same restriction will arise in any application where the number of control variables is less than the number of state variables (i.e. $r < n$) and hence is the case most likely to occur in practical applications.

The physical limitations that $\mathbf{\Gamma}$ cannot be directly manipulated and that $r < n$ have a serious effect on the stability analysis, since $\dot{V}$ can no longer be made negative-semidefinite. For example, combining Eq. (4a) and Eq. (12) (with $n = 5$ and $p = 3$ for the evaporator application) gives an expression for $\dot{V}$ in terms of $\mathbf{K}_{FB} \equiv [k_{ij}]$ and $\mathbf{B} \equiv [b_{ij}]$.

$$\dot{V} = -\mathbf{e}^T\mathbf{e} + 2 \sum_{i=1}^{5} \sum_{j=1}^{5} - \frac{1}{\xi_{ij}} \sum_{l=1}^{3} (b_{il}\dot{k}_{lj} + x_{pj}\mathbf{e}^T p_i) \times \sum_{l=1}^{3} b_{il}k_{lj} \tag{27}$$

The double summation in Eq. (27) consists of 25 terms whose sum must be less than or equal to $\mathbf{e}^T\mathbf{e}$ in order to ensure that $\dot{V}$ is negative-semidefinite. However, only the 15 elements of $\mathbf{K}_{FB}$ are available for reducing the sum of 25 terms to less than or equal to $\mathbf{e}^T\mathbf{e}$. In general, $\dot{V}$ can no longer be made negative-semidefinite by forcing each of the 25 terms to be zero, as in the derivation of Eqs. (13) and (14).

In this application, an IBM 1800 digital computer that is interfaced to the evaporator was used to implement MRAC. Consequently, discrete-time versions of the reference model and the adaptive control algorithm were required. The reference model, which represented the desired closed-loop behavior, was chosen to be the discrete form of the open-loop process model plus an optimal control law which minimized a quadratic performance index for the discrete system (Oliver, 1972). The discrete adaptive control algorithm in Eq. (28) was derived using a simple backwards difference approximation to the derivative in Eq. (13).

$$\gamma_{ij}(k) = x_{pj}(k)\,\mathbf{e}^T(k)\,\mathbf{p}_i\xi_{ij}\,\Delta t + \gamma_{ij}(k-1) \qquad (i,j = 1, \ldots, n) \tag{28}$$

A review of the MRAC design just presented suggests the following procedure for the implementation of multivariable model reference adaptive control. At the $k$th sampling instant:

a) Sample the process states, $\mathbf{x}_p(k)$, and disturbances, $\mathbf{d}(k)$.

b) Calculate the model states, $\mathbf{x}_m(k)$, from the discrete reference model.

c) Form the error, $\mathbf{e}(k) = \mathbf{x}_m(k) - \mathbf{x}_p(k)$, and calculate the current control matrices $\mathbf{K}_{FB}(k)$, $\mathbf{K}_{FF}(k)$, etc.

d) Calculate the current values of the control variables from the discrete version of Eq. (25) and send these signals to the process.

This procedure was utilized in both the hybrid computer simulation and the experimental evaluation for the actual evaporator.

## SIMULATION RESULTS

A hybrid computer simulation of the evaporator plus MRAC system was carried out in order to evaluate the effects of the various design parameters. In the hybrid simulation, the open-loop process was simulated on an EAI 580 analog computer and the reference model response and adaptive control action were calculated on an EAI 640 digital computer. Details of the hybrid computer simulation are presented elsewhere (Oliver, 1972).

Figure 2 demonstrates that the process response

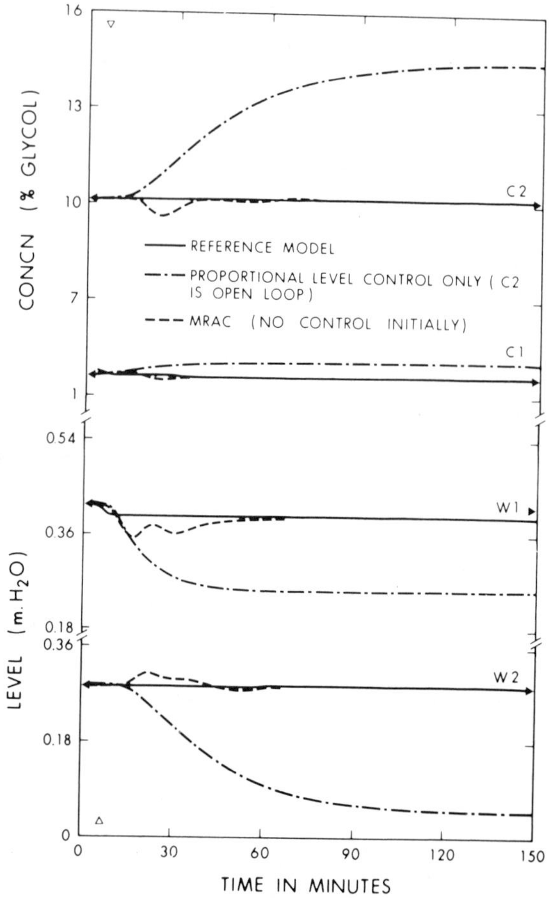


FIGURE 2 Comparison of the simulated evaporator response for MRAC with the simulated responses of the open-loop process and the reference model when a −20% step change in feed flow rate occurs (proportional feedback control, $\xi_{ij} = 10$).

for MRAC rapidly converges to the reference model response despite a very unfavorable initial control policy of no control (i.e. $\mathbf{K}_{FB}(0) = \mathbf{0}$). If this initial control policy were continued (and MRAC not used), the 20% feed flow disturbance would cause the first effect level to drain to zero in 45 min. Consequently, the process response would be even less satisfactory than the open-loop C2 response (with level control only) shown in Figure 2. Thus it is encouraging that MRAC can recover from even a disastrous initial control policy. (In Figure 2 and succeeding figures, the arrows along the vertical axis denote the initial steady state and the triangles along the horizontal axis denote the start of the step disturbance.)

The effect of the adaptive loop gains, $\xi_{ij}$, on the performance of the MRAC system is illustrated in Figure 3. In each of the three cases, an initial control policy of no control was used and the $n^2$ adaptive gains, $\xi_{ij}$, all had the same numerical value. All three responses are stable; however, when $\xi_{ij} = 1$, the responses appear sluggish and for $\xi_{ij} = 100$, the system responds in an oscillatory manner. For $\xi_{ij} = 10$, the process response adapts rapidly to the model response. Thus, as in conventional feedback control systems, there appears to be a range of gains which provides a satisfactory trade-off between speed of response and degree of oscillation. On the basis of the results shown in Figure 3, a gain of approximately ten would appear to be a reasonable choice, but no attempt was made to determine an "optimum" value of the adaptive loop gains.

A comparison of feedback and feedforward-feedback MRAC is presented in the top portion of Figure 4. The feedforward control matrix, $\mathbf{K}_{FF}$, that

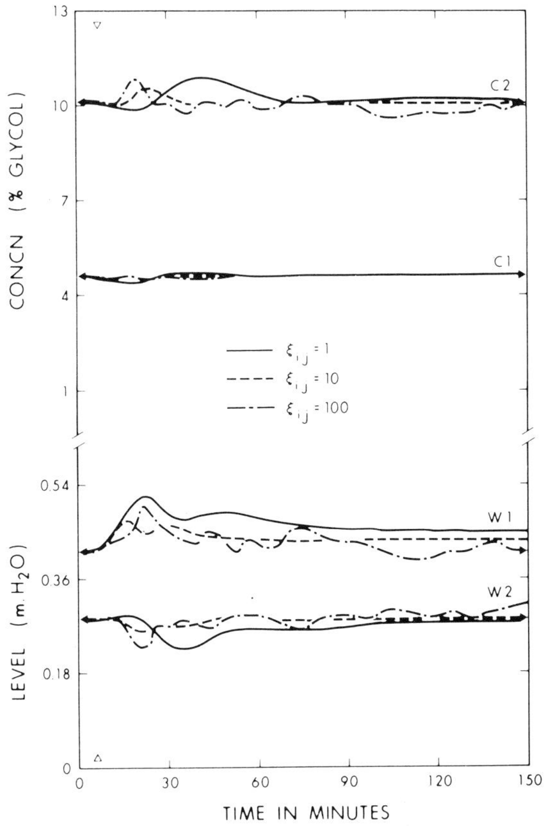


FIGURE 3 Effect of adaptive loop gains, $\xi_{ij}$, on simulated evaporator responses to a +20% step change in feed flow rate (proportional feedback control).

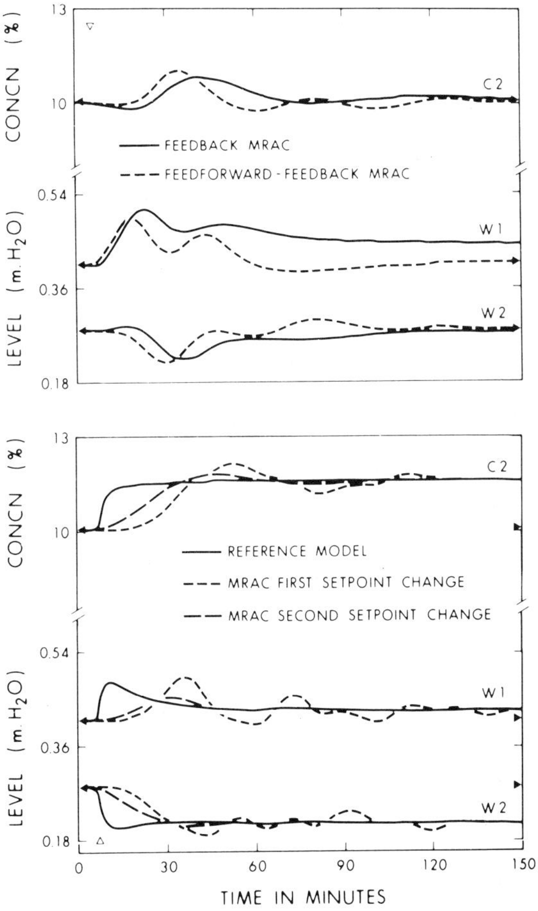


FIGURE 4 Comparison of simulated evaporator responses using MRAC. Top: proportional feedback control ($\xi_{ij} = 1$) *vs.* proportional feedback–feedforward control ($\xi_{ij} = 1$, $\nu_{ij} = 1$) for a +20% step change in feed flow rate. Bottom: proportional feedback-setpoint control for a +20% step change in the C2 setpoint ($\xi_{ij} = 1$, $\nu_{ij} = 10$).

was used in the reference model was designed to eliminate offsets in the three output variables (Newell *et al.*, 1972c). As indicated in Figure 4, the addition of adaptive feedforward control eliminated the offsets in the output variables and resulted in an improved W1 response. However, for this particular choice of adaptive loop gains, little improvement was noted in the responses of C2 and W2.

The ability of MRAC to achieve satisfactory control for setpoint changes is illustrated in the bottom portion of Figure 4 for two consecutive setpoint changes which occur 150 min apart. It is noteworthy that the second setpoint response is significantly improved over the first response due to a more favourable initial control policy, namely, the $\mathbf{K}_{FB}$ matrix resulting from the recovery to the first disturbance. (The initial control policy for the first setpoint change was no control, $\mathbf{K}_{FB}(0) = \mathbf{0}$).)

Figure 5 demonstrates the gradual improvement

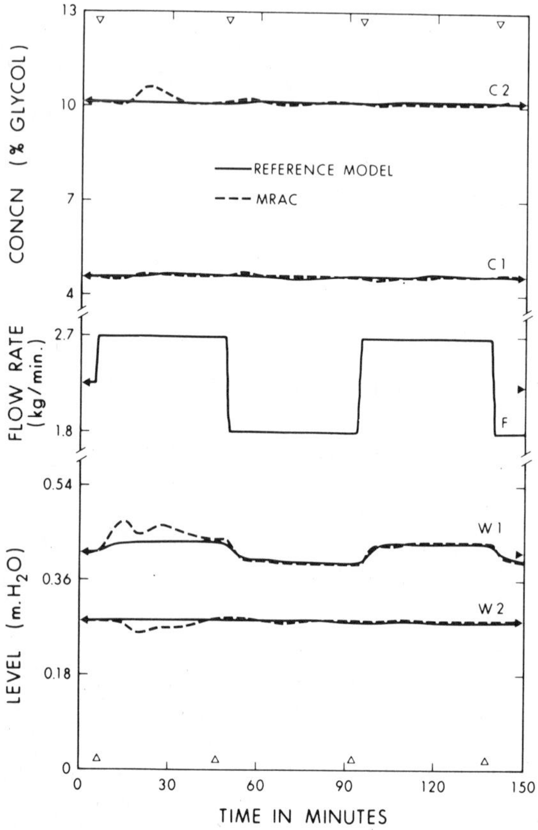


FIGURE 5 Comparison of the simulated evaporator response under MRAC with the reference model response when a series of 20% step changes in feed flow rate occurs (proportional feedback control, $\xi_{ij} = 10$).

in control that occurs when a disturbance repeatedly upsets the process. During the first step disturbance, MRAC causes the process to recover from the poor initial control policy of no control but the process response deviates significantly from the reference model response. After the second step disturbance in feed flow, the deviations become quite small and the model-following is excellent.

## SUMMARY AND CONCLUSIONS

The Liapunov design approach to multivariable, model reference adaptive control (MRAC) systems has been extended to include integral and setpoint control modes. A MRAC system was designed for a double effect evaporator using fifth order state-space models for both the process and the reference model. A hybrid computer simulation demonstrated that this MRAC technique is feasible and gives satisfactory control despite poor initial control policies and a wide range of adaptive gains. For a series of consecutive step disturbances, MRAC resulted in continually improved control with very good agreement obtained between process and reference model responses after the second disturbance.

## ACKNOWLEDGEMENT

Financial support from the National Research Council of Canada is gratefully acknowledged.

## REFERENCES

Ahlgren, T. D. and Stevens, W. F., "Adaptive control of a chemical process system," *AIChE J.*, **17**, 428 (1971).

Andre, H. and Ritter, R. A., "Dynamic response of a double effect evaporator," *Can. J. Chem. Eng.*, **46**, 259 (1968).

Bristol, E. H., "Adaptive control odyssey," Proceedings of the ISA Silver Jubilee Conference, Philadelphia, 1970, p. 561.

Butchart, R. L. and Shackcloth, B., "Synthesis of model reference adaptive system by Liapunov's second method," Proceedings of the IFAC Conference on the Theory of Self Adaptive Control System, Teddington, 1965.

Casciano, R. M. and Staffin, K. H., "Model-reference adaptive control system," *AIChE J.*, **13**, 485 (1967).

Crandall, E. D. and Stevens, W. F., "An application of adaptive control to a continuous stirred tank reactor," *AIChE J.*, **11**, 930 (1965).

Eveleigh, V. W., *Adaptive Control and Optimization Techniques* (McGraw-Hill, Toronto, 1967).

Fisher, D. G., Wilson, R. G. and Agostinis, W., "Description and application of a computer program for control systems design," *Automatica*, **8**, 737 (1972).

Grayson, L. P., "The status of synthesis using Liapunov's method," *Automatica*, **3**, 91 (1965).
Greville, T. N. E., "The pseudo-inverse of a rectangular or singular matrix and its application to the solution of systems of linear equations," *SIAM Review*, **1**, 38 (1959).
Gromyko, V. D. and Sankovskii, E. A., "Adaptive system with model and combined adjustment," *Automation and Remote Control*, **30**, 1959 (1969).
Loan, N. T., "Design and analysis of nonscanning adaptive systems," *Automation and Remote Control*, **32**, 912 (1971).
Newell, R. B., "Multivariable computer control of an evaporator," Ph. D. thesis, Department of Chemical and Petroleum Engineering, University of Alberta, 1971.
Newell, R. B. and Fisher, D. G., "Experimental evaluation of optimal multivariable regulatory controllers with model following capabilities," *Automatica*, **8**, 247 (1972a).
Newell, R. B. and Fisher, D. G., "Model development, reduction and experimental evaluation for an evaporator," *Ind. Eng. Chem. Proc. Design Develop.*, **11**, 213 (1972b).
Newell, R. B., Fisher, D. G. and Seborg, D. E., "Computer control using optimal, multivariable feedforward-feedback algorithms," *AIChE J.*, **18**, 976 (1972c).
Oliver, W. K., "Model reference adaptive control: hybrid computer simulation and experimental verification," M.Sc. thesis, Department of Chemical and Petroleum Engineering, University of Alberta, 1972.
Oliver, W. K., Seborg, D. E. and Fisher, D. G., "Model reference adaptive control based on Liapunov's direct method. Part II: hybrid computer simulation and experimental verification," *Chem. Eng. Commun.*, **1**, 133 (1973).
Parks, P. C., "Liapunov redesign of model reference adaptive control systems," *IEEE Trans. Auto. Control*, **AC-11**, 362 (1966).
Pazdera, J. S. and Pottinger, H. J., "Linear systems identification via Liapunov design techniques," Proceedings of the 9th Joint Automatic Control Conference, Boulder, 1969, p. 795.
Porter, B. and Power, H. M., "Controllability of multivariable systems incorporating integral feedback," *Electron. Lett.*, **6**, 689 (1970).
Porter, B. and Tatnall, M. L., "Performance characteristics of multivariable model reference adaptive systems synthesized by Liapunov's direct method," *Int. J. Control*, **10**, 241 (1969).
Porter, B. and Tatnall, M. L., "Stability analysis of a class of multivariable model-reference adaptive systems having time-varying process parameters," *Int. J. Control*, **11**, 325 (1970a).
Porter, B. and Tatnall, M. L., "Performance characteristics of an adaptive hydraulic servomechanism," *Int. J. Control*, **11**, 741 (1970b).
Schooley, D. J. and Pazdera, J. S., "The dynamic response of Liapunov designed adaptive model reference control systems," Proceedings of the 4th Hawaii Conference on Systems Science, 1971, p. 1060.
Shackcloth, B., "Design of model reference control systems using a Liapunov synthesis technique," *Proc. IEE*, **114**, 299 (1967).
Winsor, C. A. and Roy, R. J. "Design of model reference adaptive control systems by Liapunov's second method," *IEEE Trans. Auto. Control*, **AC-13**, 204 (1968).

# Appendix

## PROCESS VARIABLES

Each element of the **x**, **u**, **d** and **y** vectors is defined to be a normalized perturbation variable, e.g.

$$x_1 = \frac{W1 - W1_{ss}}{W1_{ss}}$$

where $W1_{ss}$ is the normal steady state value of W1.

| *State vector*, **x** | | *Normal steady-state values* |
|---|---|---|
| *W1* | First effect holdup | 20.8 kg |
| C1 | First effect concentration | 4.59% glycol |
| H1 | First effect enthalpy | 441 kJ/kg |
| W2 | Second effect holdup | 19.0 kg |
| C2 | Second effect concentration | 10.11% glycol |

| *Control vector*, **u** | | |
|---|---|---|
| S | Steam | 0.91 kg/min |
| B1 | First effect bottoms | 1.58 kg/min |
| B2 | Second effect bottoms | 0.72 kg/min |

| *Disturbance vector*, **d** | | |
|---|---|---|
| F | Feed flowrate | 2.26 kg/min |
| CF | Feed concentration | 3.2% glycol |
| hF | Feed enthalpy | 365 kJ/kg |

*Output vector*, **y**, *for integral and setpoint control*

$$\mathbf{y}^T = [W1, W2, C2]$$

# MODEL REFERENCE ADAPTIVE CONTROL BASED ON LIAPUNOV'S DIRECT METHOD

## Part II: Hybrid Computer Simulation and Experimental Application†

W. KENT OLIVER, DALE E. SEBORG‡ and D. GRANT FISHER

*Department of Chemical Engineering, University of Alberta, Edmonton, Alberta, Canada*

*(Received May 22, 1972; accepted for publication April 2, 1973)*

A Model Reference Adaptive Control (MRAC) system is evaluated via hybrid computer simulation and experimental application to a computer-controlled, pilot plant evaporator. In both the simulation and experimental studies, the MRAC system performed well and was insensitive to unmeasured process disturbances, to the choice of the initial control policy and to changes in plant operating conditions. In addition to providing an algorithm for adapting control systems to accommodate changing process parameters, MRAC also provides a systematic approach for tuning or developing multivariable control systems for time-invariant processes.

## INTRODUCTION

In this investigation a model reference adaptive control system designed for a double-effect evaporator (Oliver *et al.*, 1973) is evaluated using both simulated and experimental response data. The main objectives of the hybrid computer simulation were to investigate the effects of control system design parameters and to determine the sensitivity of the system to unmeasured disturbances, to the choice of the initial control policy and the adaptive loop gains, and to changes in the plant operating conditions. The experimental application utilized a digital computer that was interfaced to the pilot plant evaporator. The purpose of the experimental study was to confirm the results of the simulation study and to examine the practicality of MRAC for industrial applications.

## HYBRID COMPUTER SIMULATION

In the simulation study, the pilot plant evaporator was represented by the fifth-order, open-loop, state space model derived by Newell and Fisher (1972a) which is of the form

$$\dot{\mathbf{x}}_p = \mathbf{A}\mathbf{x}_p + \mathbf{B}\mathbf{u} + \mathbf{D}\mathbf{d} \tag{1}$$

The desired closed-loop behavior of the evaporator was defined by specifying a discrete-time reference model of the form

$$\mathbf{x}_m(k+1) = \phi_m \mathbf{x}_m(k) + \Delta_m \mathbf{d}(k) \tag{2}$$

The numerical values used for the model matrices in Eqs. (1) and (2) are available elsewhere (Oliver, 1972).

The continuous process model in Eq. (1) was simulated on the EAI 580 analog portion of the hybrid computer. The multivariable feedback control system used with the process model, the reference model, and the adaptive control calculations were all programmed for the digital portion of the hybrid system.

This particular approach to hybrid computer simulation offers several advantages for the evaluation of computer control techniques. First, it provides a realistic simulation of the actual process situation since the analog computer simulates the continuous plant and the digital computer closely imitates the operation of a process control computer. Secondly, certain aspects of the computer control application, including the timing considerations associated with sampling become, readily apparent from a hybrid simulation. Furthermore, the preparation of digital computer programs for the hybrid simulation tends to greatly reduce the amount of effort required to de-bug the corresponding programs for the process

† Paper presented at the New York AIChE Meeting, December 1972.

‡ To whom correspondence concerning this paper should be addressed.

control computer, particularly if the same programming language can be used in both the simulation and experimental studies.

## EXPERIMENTAL IMPLEMENTATION

Previous experimental studies by Newell *et al.* (1972b, c) and Nieman and Fisher (1973) have demonstrated that large improvements in control are possible with multivariable control techniques, and that multivariable control is easily implemented and requires only slightly more computer time than standard single-variable control. During routine operation, the evaporator is controlled using Direct Digital Control (DDC) algorithms for six single-variable control loops and four cascaded control loops. For multivariable control studies (including MRAC), the IBM 1800 DDC program is used to access the process variables (i.e. obtain measurements and store them in DDC tables) and to implement control through setpoint manipulation in the DDC loops.

The general multivariable control program developed by Newell (1971) executed the various steps shown in Figure 1. Since one of the states, C1, is not measured, the control program performed the necessary state estimation and provided filtering of the DDC measurements. The program also performed the units conversion necessary to change the measured state variables into normalized perturbation form prior to the control calculations, and to change the control signals from perturbation form to engineering units suitable for transmission to the process. The program was loaded into core from disk storage in response to a high priority timer interrupt and was executed at a control interval of 64 sec. This relatively large control interval could be used since the calculated values of $\mathbf{u}(k)$ were implemented as the setpoints of DDC flow control loops. (These DDC loops used sampling intervals of 1 sec.) The control calculations, state estimation, filtering and units conversion were programmed in FORTRAN with only simple assembler programs required for communication between the multivariable and DDC programs.

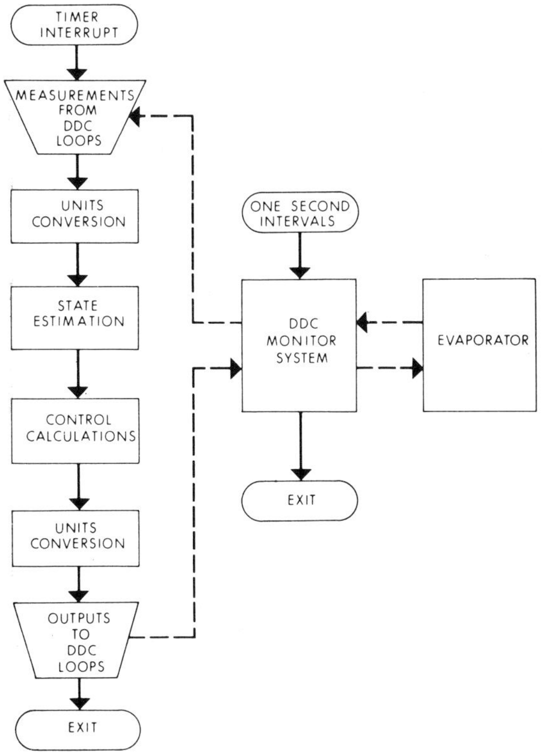


FIGURE 1 Block diagram for the implementation of multivariable control.

Further details of the experimental implementation of MRAC are available in the thesis of Oliver (1972). The important process variables and their normal steady-state values were presented in Part I (Oliver *et al.*, 1973). A schematic diagram of the pilot plant evaporator is available elsewhere (Nieman and Fisher, 1973; Newell *et al.*, 1972a, b, c).

## RESULTS

In the top half of Figure 2, experimental and simulated MRAC responses are compared with the reference model response for the same initial control policy, adaptive loop gains, and load disturbance. The experimental responses are in general agreement with the simulation results, although modelling errors are significant. For example, the experimental W1 response tends to lead the simulated response while the experimental C2 response is more sluggish than the simulated response. An extensive discussion of the modelling errors and simplifications associated with the theoretical process model of Eq. (1) has been reported by Newell and Fisher (1972a).

In Figures 2–8, the arrows along the horizontal axis designate the start of a step disturbance in feed flow and the arrows along the vertical axis denote the initial steady state.

### *Adaptive Loop Gains*

The adaptive loop gains $\xi_{ij}$, are important design parameters, since they determine the rate at which adaptation takes place. Simulation results in Part I

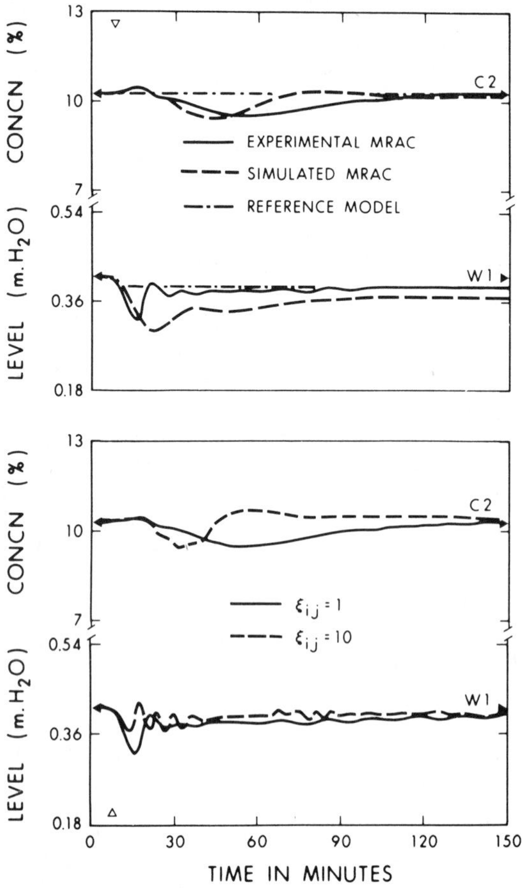


FIGURE 2 Top: comparison of experimental, simulated and reference model responses for a −20% step change in feed flow ($\xi_{ij} = 1$, no control initially). Bottom: experimental results demonstrating the effect of adaptive loop gain, $\xi_{ij}$, for a −20% step change in feed flow and no control initially.

indicated that small values of $\xi_{ij}$ produced sluggish responses while larger values gave more oscillatory results. The experimental responses shown in the bottom half of Figure 2 confirm this conclusion.

In applications of MRAC the "best" rate of adaptation (and hence "best" values for $\xi_{ij}$) may vary considerably depending on the control objectives in a particular situation. For example, a fast rate of adaptation is necessary if the adaptive control system is to provide satisfactory control during a single transient response (e.g., batch operation). In other situations, it may be preferable to employ a slower rate of adaptation to generate a $\mathbf{K}_{FB}$ matrix which reflects the recent history of the process. Very slow rates of adaptation should be used if it is desired to evolve a satisfactory multivariable control strategy for new process conditions and/or to generate control matrices which are "averaged" over a wide range of disturbances.

In many process control problems, the control of a few critical state variables is of considerable importance, while control requirements for the other state variables are less restrictive. Consequently, in designing multivariable control systems, a systematic approach for ensuring tight control of critical state variables is desirable. For example, in the optimal control design approach, control of the important variables can be emphasized by an appropriate choice of the weighting matrices used in the quadratic performance index (Newell and Fisher, 1972b). In MRAC, the desired closed-loop response is specified explicitly by the choice of the reference model. The response of the important state variables can also

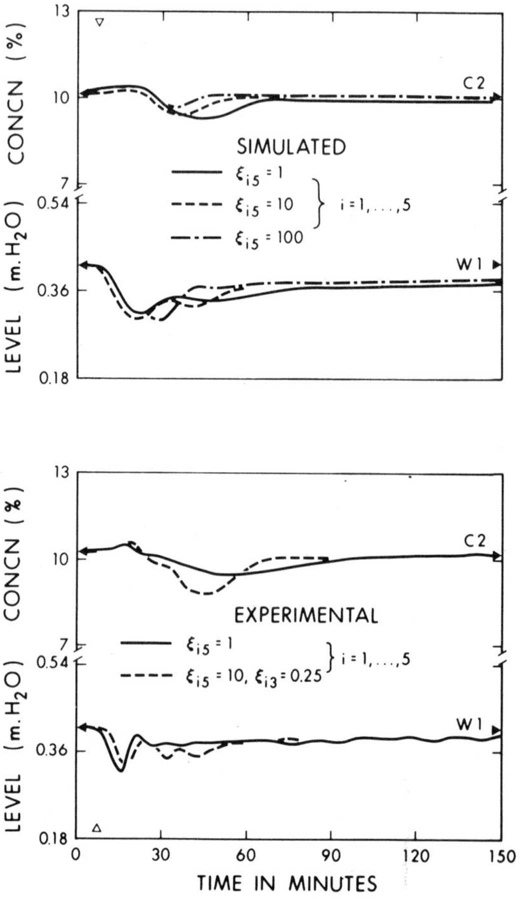


FIGURE 3 State weighting of the C2 response by using large adaptive loop gains in the fifth column of $\Xi$ (−20% step change in feed flow, $\xi_{ij} = 1$ for all other elements of $\Xi$, no control initially).

be tailored by the choice of values for the adaptive loop gains, as discussed below.

In the evaporator application product concentration, C2, was assumed to be the most important state variable, and consequently, the reference model was chosen to give excellent control of C2 (Oliver *et al.*, 1973) as illustrated by the reference model response in Figure 2. The possibility of ensuring tight control of C2 by an appropriate choice of the adaptive loop gains was also investigated.

The adaptive control algorithm in Eq. (3) was derived in Part I (Oliver *et al.*, 1973).

$$\gamma_{ij}(k) = x_{pj}(k)\,\mathbf{e}(k)^T \mathbf{p}_i \xi_{ij}\,\Delta t + \gamma_{ij}(k-1) \qquad (i,j = 1,\ldots,n) \tag{3}$$

and suggests that large values of $x_{pj}$ can be penalized by making the elements in the $j$th column of the adaptive loop gain matrix, $\Xi = [\xi_{ik}]$, relatively large. The simulation results in Figure 3 support this assertion, since C2 (i.e., $x_{p5}$) responds faster and with smaller deviations, when relatively large elements are used in the fifth column of $\Xi$. However, the experimental responses in Figure 3 are less conclusive, since the larger value of $\xi_{i5}$ = 10 results in a faster C2 response but produces a larger maximum deviation in C2.

### *Initial Control Policy*

Several simulation and experimental runs were made to determine the sensitivity of the adaptive control system to the choice of the initial control policy. In all cases, the MRAC system was able to recover rapidly from a disturbance even if a very poor initial control policy was used. As would be expected, the best process responses were obtained when the controller constants were initially set equal to values that gave satisfactory closed-loop control. However, even when the controller constants were initialized to zero, the MRAC system responded fast enough to keep the process response within reasonable limits and to generate a satisfactory set of controller constants during the transient response. The experimental results in Figure 4 compare the response starting with zero control (i.e., equivalent to an initial open-loop policy) to the response starting from a conventional multiloop control policy consisting of 3 proportional feedback loops. For the multiloop case, the initial value of the feedback control matrix was chosen to be

$$\mathbf{K}_{FB} = \begin{bmatrix} 0 & 0 & 0 & 0 & -4.89 \\ 3.52 & 0 & 0 & 0 & 0 \\ 0 & 0 & 0 & 15.8 & 0 \end{bmatrix}$$

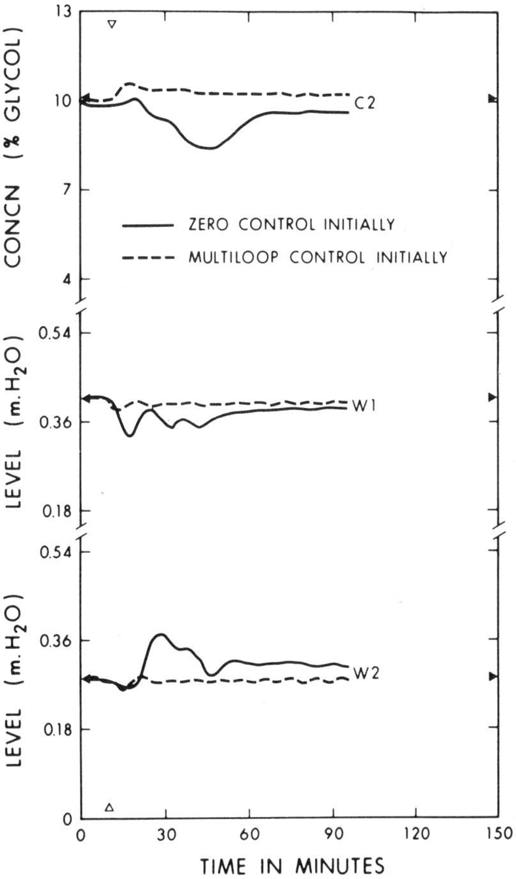


FIGURE 4 Experimental responses for different initial control policies and a −20% step change in feed flow ($\xi_{ij}$ = 1).

If this multiloop control system is used (without adaptive control), large offsets in the three output variables C2, W1 and W2 result and the responses are more oscillatory than the MRAC results shown in Figure 4 (Oliver, 1972).

These results suggest that MRAC provides a useful strategy for two practical problems that arise in applications of modern multivariable control techniques. First, MRAC is a promising method for tuning multivariable control schemes. Since multivariable schemes are usually designed using approximate process models, on-line tuning will generally be required. Without a systematic tuning technique like MRAC, on-line tuning of a multivariable control system may be difficult due to the large number of elements in the feedback control matrices. A second practical problem in many industrial situations is that suitable, dynamic, process models are not available. Consequently, a multivariable control system cannot be designed *a priori.* An alternative strategy is to use MRAC to evolve a satisfactory multivariable control system "on-line." The advantage of this approach is that the reference model is chosen simply to define the *desired* process behavior and does not have to be an accurate dynamic model of the *actual* process. The experimental results in Figure 4 demonstrate that MRAC provides a systematic approach for evolving a satisfactory multivariable control system using a conventional multiloop control system as the starting point.

*Repeated Disturbances*

Figure 5 illustrates that adaptive control tends to give improved control as disturbances continue to upset the process. The response to the first feed flow disturbance is poor because the initial control policy was open-loop, i.e. no control. The response to the second step change in feed flow is significantly improved due to the better initial control policy, namely, the $\mathbf{K}_{FB}$ matrix resulting from adaptive control during the recovery from the first disturbance. Figure 5 also shows the behavior of two of the control variables, B1 and B2, and the step disturbances in feed flow. The action of the control variables becomes smoother as the adaptation of the controller constants continues.

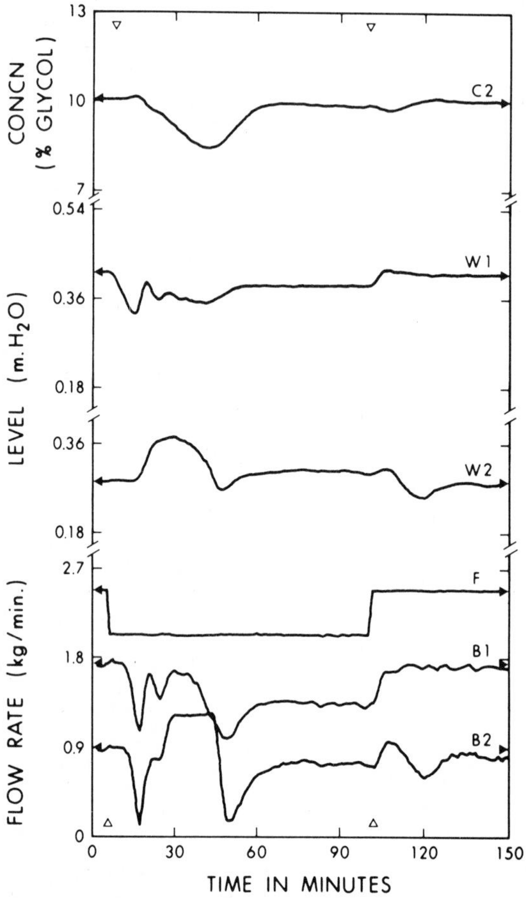


FIGURE 5 Experimental results demonstrating that control continues to improve as the process responds to a series of disturbances (two 20% step changes in feed flow; $\xi_{i3} = 0.25$, $\xi_{i5} = 10$, all other $\xi_{ij} = 1$; no control initially).

The behavior of one of the adapting elements, $\gamma_{51}$, is shown in Figure 6. Note that adaptation starts as soon as the feed disturbance is detected and proceeds very rapidly. By the time the process has recovered from the first feed disturbance, the value of $\gamma_{51}$ is close to the observed final value of 4.02. This behavior confirms that observed in the simulated runs.

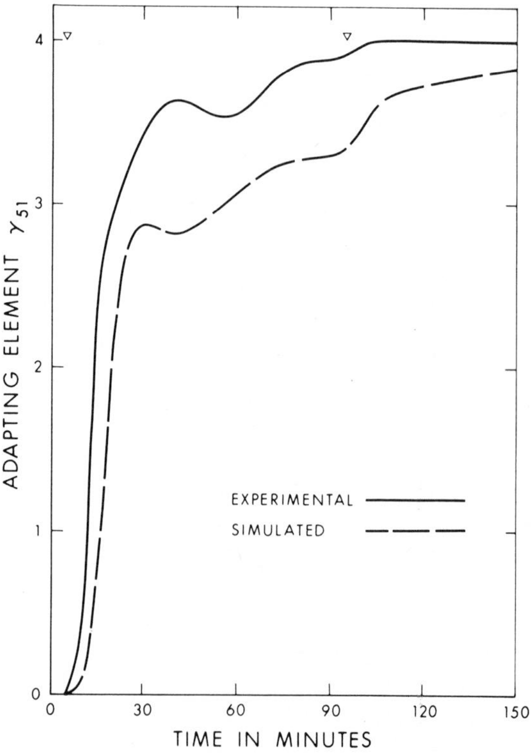


FIGURE 6 Comparison of experimental and simulated results showing the change in $\gamma_{51}$ during two 20% step changes in feed flow, down and then up ($\xi_{ij} = 0.25$, $\xi_{i5} = 10$, all other $\xi_{ij} = 1$, no control initially).

*Number of Adapting Elements of* $\Gamma$

In many computer control applications, the computational and storage requirements for the process control computer are a critical factor. For the present adaptive control technique, the computational load can be reduced by adapting only certain elements of $\Gamma$ rather than all $n^2$ elements as considered above.

Figure 7 presents simulation results for the case where only some elements of $\Gamma$ are allowed to adapt and the other elements are held constant (in this case, at a value of zero). Starting from an open-loop initial control policy, adequate control is still maintained if the adapting elements of $\Gamma$ are selected in a reasonable fashion. For example, in the evaporator application, at least one element in the first and fourth columns of $\Gamma$ must adapt (or at least be nonzero) in order to maintain control of the first and fourth state variables, the levels of both effects. This problem arises because the levels are not self-regulating and consequently, if all the elements in the first and fourth columns of $\Gamma$ are zero, these state variables become essentially uncontrolled. (The elements of $\Gamma$ that adapt in Figure 7 are summarized in Table I.)

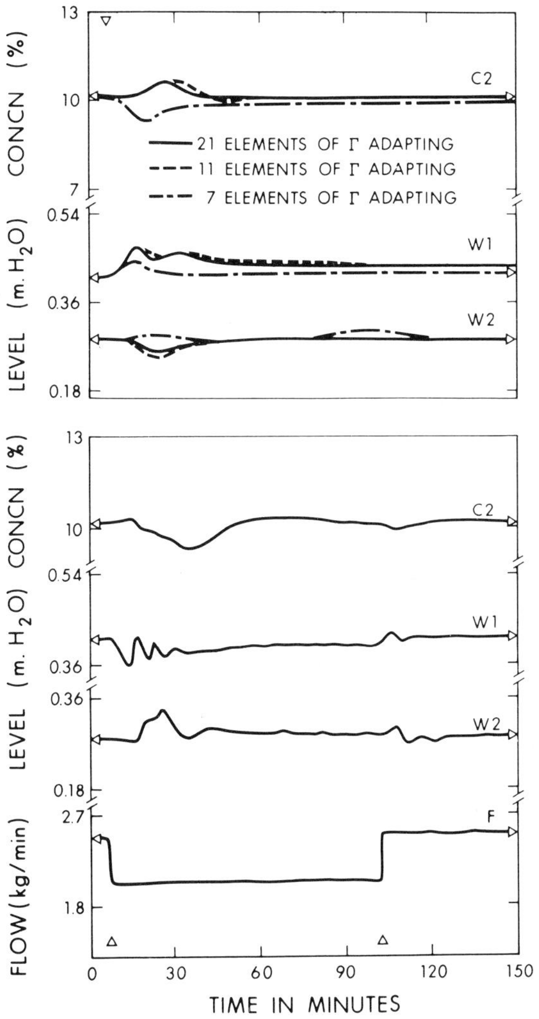


FIGURE 7 Top: simulated responses when only some elements of $\Gamma$ adapt (+20% step change in feed flow, no control initially). Bottom: experimental responses when only 11 elements of $\Gamma$ adapt (two 20% step changes in feed flow; $\xi_{ij}$ = 10, no control initially).

TABLE I

Adapting elements of matrix, $\Gamma$

| Number of elements adapting | Adapting elements of $\Gamma$ |
|---|---|
| 21 | All elements except $\gamma_{14}, \gamma_{24}, \gamma_{34}, \gamma_{54}$ |
| 11 | First and fifth columns of $\Gamma$ and $\gamma_{44}$ |
| 7 | $\gamma_{11}, \gamma_{44}, \gamma_{15}, \gamma_{25}, \gamma_{35}, \gamma_{45}, \gamma_{55}$ |

The simulated results in the top half of Figure 7 also illustrate that it becomes difficult to predict how the process will respond when the number of adapting elements of $\Gamma$ is reduced. For example, when only 7 elements of $\Gamma$ adapt, the C2 response is in the opposite direction to that produced by the other two responses. The experimental response in the bottom half of Figure 7 confirms that adequate control is maintained when only 11 elements of $\Gamma$ adapt and again demonstrates that control improves greatly for the second feed flow disturbance (cf. Figure 5).

*Effect of Unmeasured Disturbances*

In the derivation of the MRAC design equations in Part I (Oliver *et al.*, 1973), it was assumed that all elements of the disturbance vector, **d**, could be measured. However, in many practical problems it is not feasible to measure all of the disturbance variables and consequently, the sensitivity of the MRAC system to unmeasured process disturbances is of considerable interest. A comparison of the experimental response in the top half of Figure 8 with those in Figure 2 indicates that an undetected step disturbance in feed

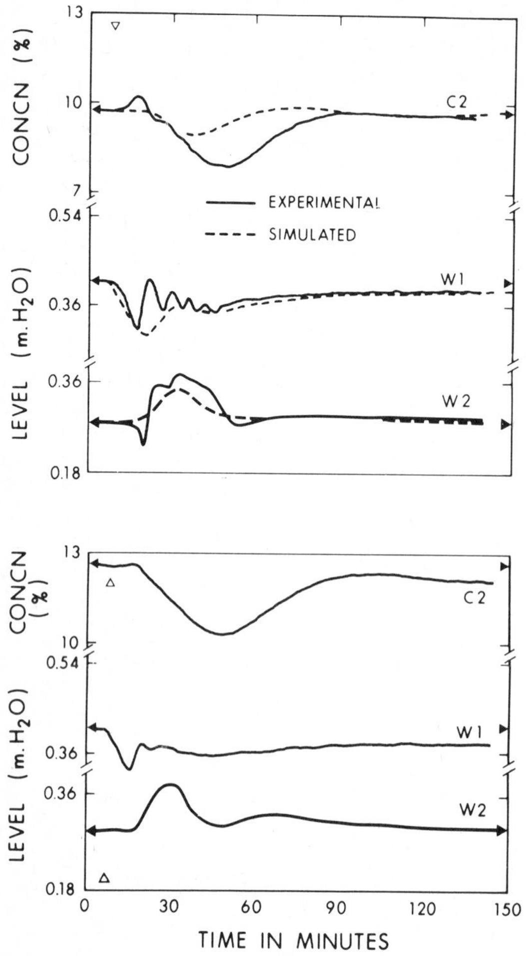


FIGURE 8 Top: Evaporator responses to a –20% feed flow disturbance and an unmeasured –10% feed composition disturbance ($\xi_{ij}$ = 1, no control initially, neither disturbance is "measured" in the simulation). Bottom: experimental response at operating conditions different from those used in the model linearization (–20% step change in feed flow; $\xi_{ij}$ = 1, no control initially).

composition results in a larger C2 deviation but the MRAC system is still able to bring the process back to the initial steady state in a reasonable period of time. Furthermore, the simulated results in Figure 8 show that adequate control is still maintained when neither the feed composition nor the feed flow disturbance is monitored.

### *Change in Process Operating Conditions*

As a further evaluation of the sensitivity of MRAC to modelling errors, the controlled process was operated at conditions significantly different from those about which the process model was linearized. (The linearized model and an optimal control policy were used to specify the closed-loop reference model, as described in Part I.) In the bottom half of Figure 8 a feed rate of 1.59 kg/min was used instead of the normal value of 2.27 kg/min and produced a steady-state product concentration of about 13% rather than the normal value of 10%. The experimental response in the bottom half of Figure 8 demonstrates that MRAC improves the poor initial control policy and generates a new feedback matrix to control the process at this "unusual" set of operating conditions.

### *Final Control Matrices*

For most transient responses, the elements of the feedback control matrix, $\mathbf{K}_{FB}$, became essentially constant after a few disturbances and the error between the process response and the model response was very small. However, in a few of the simulation runs, certain combinations of high adaptive loop gains, poor initial control policies and severe disturbances produced very large elements in the feedback matrix, $\mathbf{K}_{FB}$, and resulted in "bang-bang" control. These undesirable, large elements could easily be avoided if smaller adaptive loop gains were used or if the process was not subjected to large, sudden disturbances during the initial adaptation period. The use of better initial control policies also tended to prevent excessively large elements of $\mathbf{K}_{FB}$.

Several differences exist between the theory and application of MRAC which could explain why $\mathbf{K}_{FB}$ matrices with large elements sometimes occur. Since in this application there are fewer control variables than state variables, the following pseudo-inverse approach was used (Oliver *et al.*, 1973) to calculate the feedback matrix, $\mathbf{K}_{FB}$, from the adaptive control matrix, $\mathbf{\Gamma}$

$$\mathbf{K}_{FB}(k) = (\mathbf{B}^T\mathbf{B})^{-1}\mathbf{B}^T\mathbf{\Gamma}(k) \tag{4}$$

Unfortunately, a consequence of using this approximate, least-squares solution for $\mathbf{K}_{FB}$ is that the stability of the closed-loop system can no longer be guaranteed since the expression for $\dot{V}$ is no longer negative semi-definite (Oliver *et al.*, 1973).

A second difference between the MRAC theory and application is that the theoretical analysis on which the adaptive algorithm is based (Oliver, 1972) does not take into account constraints on the control variables; by contrast, both the hybrid simulation and the experimental application included such constraints. Gromyko and Sankovskii (1969) indicate that the stability of the closed loop system is guaranteed for only a certain region of the state space if there are constraints on the control variables. In the evaporator simulation, poor initial control policies or severe disturbances could cause control constraints to be reached and the system to become unstable.

Finally, the theory was developed for a continuous process and reference model, while the application of the theory used a discrete reference model and sampled values of the process variables. Consequently, the backwards difference approximation (Eq. 3) to the continuous adaptive algorithm could result in numerical instability problems and control matrices with large elements. It was observed during the simulation study that decreasing the sampling (or control) interval tends to prevent large elements in the final control matrices and the resulting bang-bang control (Oliver, 1972).

The simulation and experimental results suggest that the choice of small adaptive loop gains and a reasonable initial control policy will prevent the generation of control matrices with excessively large elements. Alternatively, the sampling interval can be reduced or a more accurate difference approximation used for the adapting algorithm, however, this will increase the computational load of the on-line computer.

## SUMMARY AND CONCLUSIONS

A multivariable model reference adaptive control system has been successfully applied to a pilot scale, double-effect evaporator. In both simulation and experimental studies, the MRAC system performed well and was able to maintain satisfactory process control despite even very poor initial control policies. An important result in this application was that the adaptive control system tended to improve control as successive disturbances upset the process. The MRAC system proved to be relatively insensitive to modelling errors including those due to linearizing the process model about a point not equal to the steady state process conditions. Similarly, unmeasured load disturbances did not affect the MRAC system significantly.

Satisfactory control was also achieved for a wide range of adaptive loop gains. The effect of the adaptive loop gains on the closed-loop response is similar to that of conventional controller gains: small values tend to slow down the rate of adaptation while large values increase the rate of adaptation and tend to yield more oscillatory responses. The control of important state variables can be emphasized by choosing some adaptive loop gains to be larger than others.

The experimental implementation of MRAC using a process control computer was not significantly more difficult than previous multivariable control studies (Newell, 1971) although execution time and storage requirements increased. If these requirements are prohibitive in industrial situations, adapting only certain key parameters or the application of adaptive control only when significant changes in the process occur, could be attempted.

In conclusion, model reference adaptive control is a practical and promising technique for industrial process control. In particular, MRAC provides a systematic method for tuning a multivariable control scheme or for evolving a satisfactory multivariable control system "on-line" using a conventional multi-loop control configuration as a starting point.

## ACKNOWLEDGEMENTS

The authors wish to thank the staff of the Data Acquisition, Control and Simulation (DACS) Centre in the Department of Chemical Engineering for their assistance in the use of the computer facilities. Financial support from the National Research Council of Canada is gratefully acknowledged.

## REFERENCES

Gromyko, V. D. and Sankovskii, E. A., "Adaptive system with model and combined adjustment," *Automation and Remote Control*, **30**, 1959 (1969).

Newell, R. B., "Multivariable computer control of an evaporator," Ph.D. thesis, Department of Chemical and Petroleum Engineering, University of Alberta, 1971.

Newell, R. B. and Fisher, D. G., "Model development, reduction and experimental evaluation for an evaporator," *Ind. Eng. Chem. Proc. Des. Develop.*, **11**, 213 (1972a).

Newell, R. B. and Fisher, D. G., "Experimental evaluation of optimal mutlivariable regulatory controllers with model-following capabilities," *Automatica*, **8**, 247 (1972b).

Newell, R. B., Fisher, D. G. and Seborg, D. E., "Computer control using optimal multivariable feedforward–feedback algorithms," *AIChE J.*, **18**, 976 (1972c).

Nieman, R. E. and Fisher, D. G., "Experimental evaluation of optimal multivariable servo control in conjunction with conventional regulatory control," *Chem. Eng. Commun.*, **1**, No. 2 (1973).

Oliver, W. K., "Model reference adaptive control: hybrid computer simulation and experimental verification," M.Sc. thesis, Department of Chemical and Petroleum Engineering, University of Alberta, 1972.

Oliver, W. K., Seborg, D. E. and Fisher, D. G., "Model reference adaptive control based on Liapunov's direct method. Part I: theory and control system design," *Chem. Eng. Commun.*, **1**, 125 (1973).

# An extension of the Smith Predictor method to multivariable linear systems containing time delays†

G. ALEVISAKIS and D. E. SEBORG

Department of Chemical and Petroleum Engineering,
University of Alberta,
Edmonton, Alberta, Canada

[Received 9 March 1972]

The classical Smith Predictor method for single variable systems is extended to a class of linear multivariable systems. Derivations of the multivariable Smith Predictor are presented for both continuous-time and discrete-time systems which contain time delays in the control variables and/or output variables. As in the classical method, use of the multivariable Smith Predictor eliminates the time delays from the characteristic equation of the closed-loop system.

## 1. Introduction

The detrimental effects of time delays on system stability and control are well known to both control system designers and personnel responsible for plant operation. From the classical viewpoint, the phase lag introduced by a time delay tends to reduce system stability and make satisfactory control more difficult to achieve. Furthermore, time delays greatly complicate control system design for multivariable systems since many design approaches are either not applicable to time delay systems or become very difficult to apply.

This study is concerned with the control of linear multivariable systems which contain time delays in the output variables and/or the control variables. A principal new result is the extension of the classical Smith Predictor method (Smith 1957, 1959) to a class of multivariable systems. By using the multivariable Smith Predictor, time delays can be eliminated from the characteristic equation of the system and thus design techniques developed for systems without time delays can then be applied.

In § 2 the classical Smith Predictor method for single variable systems is briefly reviewed. In §§ 3 and 4 a multivariable Smith Predictor is derived for continuous-time and discrete-time systems, respectively. Finally, a numerical example is presented in § 5 and the conclusions of this study are presented in § 6.

## 2. Smith Predictor for single variable control systems

In 1957, Smith (1957, 1959) proposed a control technique for single variable control systems which contain time delays. This technique, which became known as the Smith Predictor (or Smith Linear Predictor), is illustrated in figs. 1 and 2. The chief advantage of the Smith Predictor method is that time delays are eliminated from the characteristic equation of the closed-loop system. This is achieved by including a mathematical model of the process in the feedback loop around the controller.

† Communicated by the Authors.

International Journal of Control, Vol. 17, No. 3, pp. 541-551 (1973).

Fig. 1

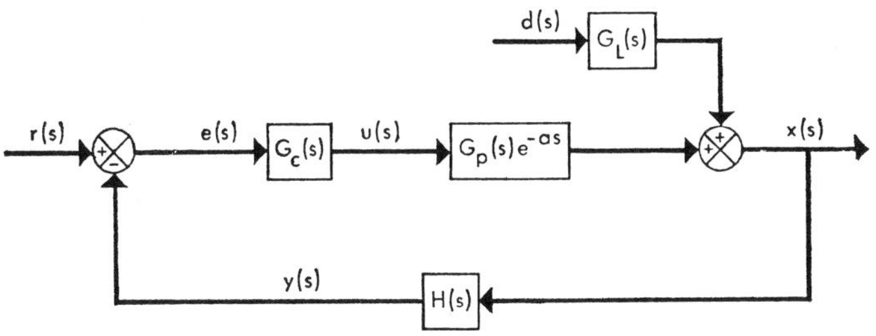


Single variable control system with a time delay in the process transfer function

Fig. 2

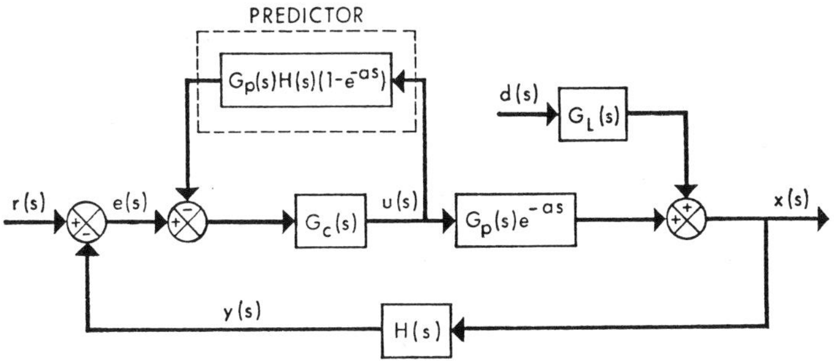


Smith predictor for the feedback control system of fig. 1.

The output of the predictor block in fig. 2 is the difference between two model responses : the response of the system without the time delay minus the response of the system with the time delay. If the process models were perfect, then the actual process response, $y(s)$, would be cancelled by the model response and the control action would be based on the response of the model without time delay. For the control system in fig. 2, the closed-loop transfer function for load changes is

$$\frac{x(s)}{d(s)} = G_L(s) - \frac{G_c(s)G_p(s)G_L(s)\exp(-as)}{1+G_c(s)G_p(s)H(s)}. \tag{1}$$

The characteristic equation for the closed-loop system in fig. 2 is given by

$$1+G_c(s)G_p(s)H(s)=0, \tag{2}$$

which is also the characteristic equation for the system in fig. 1 when the time delay is zero. Thus the Smith Predictor has successfully eliminated the time delay from the characteristic equation.

Later investigations considered applications of the Smith Predictor (Buckley 1960, Lupfer and Oglesby 1961, 1962, Nielsen 1969) and its use in designing sampled-data control systems (Gray and Hunt 1971). Buckley (1960) and Nielsen (1969) have made comparisons of the Smith Predictor method with conventional control techniques for systems which contain time delays. Apparently, extensions of the Smith Predictor method to multivariable control problems have not previously been reported.

## 3. Multivariable continuous-time systems

Consider the following linear, stationary, state-space model with time delays in the output variables and the control variables :

$$\dot{\mathbf{x}}(t) = \mathbf{A}\mathbf{x}(t) + \mathbf{B}\mathbf{u}(t-a) + \mathbf{D}\mathbf{d}(t), \tag{3}$$

$$\mathbf{y}(t) = \mathbf{C}_1\mathbf{x}(t) + \mathbf{C}_2\mathbf{x}(t-b), \tag{4}$$

where

$\mathbf{x}(t)$ = state vector of dimension $n$,
$\mathbf{u}(t)$ = control vector of dimension $m$,
$\mathbf{d}(t)$ = disturbance vector of dimension $p$,
$\mathbf{y}(t)$ = output vector of dimension $r$,
$a, b$ = constant time delays,

$\mathbf{A}$, $\mathbf{B}$, $\mathbf{C}_1$, $\mathbf{C}_2$ and $\mathbf{D}$ are constant real matrices of appropriate dimensions.

The time delays in this state-space model can be given the following physical interpretation. Time delay, $a$, is associated with the calculation and implementation of control, that is, a time delay in the manipulated variables. In many physical systems time delays are also associated with the measurement of certain state variables. A notable example in process control systems is chemical composition. This variable is, in general, difficult to measure and often requires a period of time to carry out the analysis, i.e. a time delay. The inclusion of time delay, $b$, in eqn. (4) provides a general model for systems in which some state variables can be measured instantaneously but time delays are involved in the measurement of the other state variables. Apparently, the output equation in eqn. (4) has not been widely used in previous investigations of time-delay systems despite its obvious practical importance.

### 3.1. *Time delay in output variables*

In this section we consider the special case where a time delay is present in the output variables but not in the control variables (i.e. $a=0$ in eqn. (3)). Here, the state-space model in eqns. (3) and (4) reduces to

$$\dot{\mathbf{x}}(t) = \mathbf{A}\mathbf{x}(t) + \mathbf{B}\mathbf{u}(t) + \mathbf{D}\mathbf{d}(t), \tag{5}$$

$$\mathbf{y}(t) = \mathbf{C}_1\mathbf{x}(t) + \mathbf{C}_2\mathbf{x}(t-b). \tag{6}$$

Assuming zero initial conditions and taking the Laplace transform of eqns. (5) and (6) gives, after rearrangement,

$$\mathbf{x}(s) = \mathbf{G}_{\mathrm{p}}(s)\mathbf{u}(s) + \mathbf{G}_{\mathrm{L}}(s)\mathbf{d}(s), \tag{7}$$

$$\mathbf{y}(s) = \mathbf{C}_1\mathbf{x}(s) + \mathbf{C}_2 \exp(-bs)\,\mathbf{x}(s), \tag{8}$$

where

$\mathbf{G}_{\mathrm{p}}(s) = (s\mathbf{I}-\mathbf{A})^{-1}\mathbf{B}$ = process transfer function matrix,
$\mathbf{G}_{\mathrm{L}}(s) = (s\mathbf{I}-\mathbf{A})^{-1}\mathbf{D}$ = load transfer function matrix.

Suppose a feedback control law of the form

$$\mathbf{u}(s) = -\mathbf{G}_c(s)\mathbf{y}(s) \tag{9}$$

is assumed, where $\mathbf{G}_c(s)$ is the matrix of feedback controller transfer functions. Then combining eqns. (7)–(9) gives the following characteristic equation :

$$|\mathbf{I} + \mathbf{G}_p(s)\mathbf{G}_c(s)[\mathbf{C}_1 + \mathbf{C}_2 \exp(-bs)]| = 0, \tag{10}$$

where $\mathbf{I}$ is an $n \times n$ identity matrix and $|\ |$ denotes the determinant.

Next, it will be demonstrated that the Smith Predictor method can be used to eliminate the time delay from the characteristic equation. Consider the block diagram shown in fig. 3. As for the single variable system in § 2, the

Fig. 3

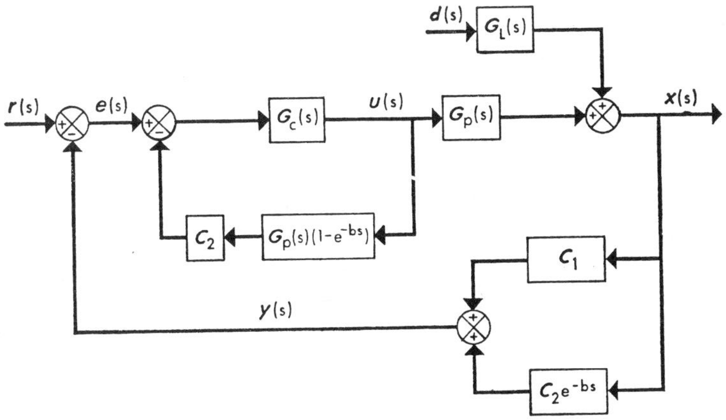


Smith Predictor for a multivariable system with a time delay in the output variables.

feedback loop around the controller contains a mathematical model of the process, both with and without time delays. From the block diagram in fig. 3, it follows that for $\mathbf{r}(s) = \mathbf{0}$,

$$\mathbf{u}(s) = -\mathbf{G}_c\mathbf{y}(s) - \mathbf{G}_c\mathbf{C}_2\mathbf{G}_p[1 - \exp(-bs)]\,\mathbf{u}(s), \tag{11}$$

where $\mathbf{G}_c$ and $\mathbf{G}_p$ denotes $\mathbf{G}_c(s)$ and $\mathbf{G}_p(s)$, respectively. Rearranging eqn. (11) gives the following control algorithm :

$$\mathbf{u}(s) = -[\mathbf{I} + \mathbf{G}_c\mathbf{C}_2\mathbf{G}_p]^{-1}\mathbf{G}_c[\mathbf{y}(s) - \mathbf{C}_2\mathbf{G}_p \exp(-bs)\,\mathbf{u}(s)]. \tag{12}$$

Combining eqns. (11), (7) and (8) gives

$$\mathbf{x}(s) = \mathbf{G}_L\mathbf{d}(s) - \mathbf{G}_p\mathbf{G}_c\mathbf{C}_1\mathbf{x}(s) - \mathbf{G}_p\mathbf{G}_c\mathbf{C}_2(\mathbf{x}(s) - \mathbf{G}_L\mathbf{d}(s)) - \mathbf{G}_p\mathbf{G}_c\mathbf{C}_2 \exp(-bs)\,\mathbf{G}_L\mathbf{d}(s) \tag{13}$$

or rearranging gives

$$\mathbf{x}(s) = [\mathbf{I} + \mathbf{G}_p\mathbf{G}_c(\mathbf{C}_1 + \mathbf{C}_2)]^{-1}[\mathbf{I} + \mathbf{G}_p\mathbf{G}_c\mathbf{C}_2(1 - \exp(-bs))]\mathbf{G}_L\mathbf{d}(s). \tag{14}$$

From eqn. (14) it is apparent that the characteristic equation for the closed-loop system is

$$|\mathbf{I}+\mathbf{G}_{\mathrm{p}}\mathbf{G}_{\mathrm{c}}(\mathbf{C}_1+\mathbf{C}_2)|=0, \tag{15}$$

which is also the characteristic equation for the system without time delays (i.e. $b=0$ in eqn. (6)). Thus the multivariable Smith Predictor has eliminated the time delay from the characteristic equation. The design of the controller transfer function matrix $\mathbf{G}_{\mathrm{c}}(s)$ can then proceed using design techniques developed for systems without time delays.

3.2. *Time delays in both control variables and output variables*

We now consider the general case where constant time delays occur in both the control variables and the measured outputs. The system of interest is given in eqns. (3) and (4). Assuming zero initial conditions and taking the Laplace transform of eqns. (3) and (4) gives

$$\mathbf{x}(s)=\mathbf{G}_{\mathrm{p}}\exp(-as)\,\mathbf{u}(s)+\mathbf{G}_{\mathrm{L}}\mathbf{d}(s), \tag{16}$$

$$\mathbf{y}(s)=\mathbf{C}_1\mathbf{x}(s)+\mathbf{C}_2\exp(-bs)\,\mathbf{x}(s). \tag{17}$$

Fig. 4

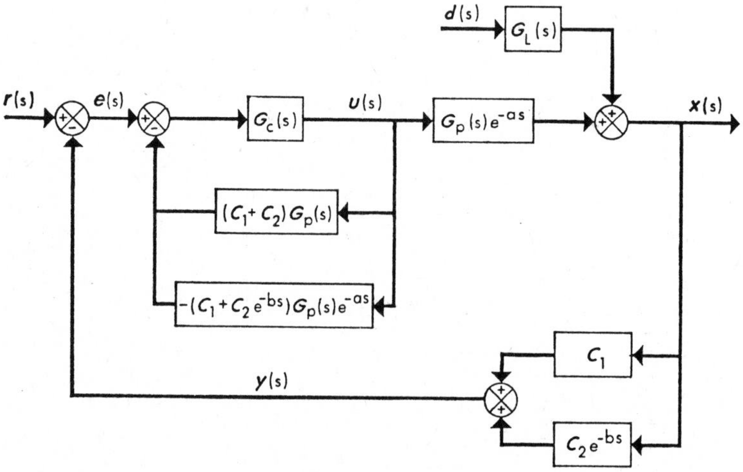


Smith Predictor for a multivariable system with time delays in both control and output variables.

The multivariable Smith Predictor for this system consists of two feedback loops around the controller as shown in fig. 4. From fig. 4 it follows that for $\mathbf{r}(s)=\mathbf{0}$,

$$\mathbf{u}(s)=-\mathbf{G}_{\mathrm{c}}\mathbf{y}(s)-\mathbf{G}_{\mathrm{c}}(\mathbf{C}_1+\mathbf{C}_2)\mathbf{G}_{\mathrm{p}}\mathbf{u}(s) \\ +\mathbf{G}_{\mathrm{c}}\mathbf{C}_1\mathbf{G}_{\mathrm{p}}\exp(-as)\,\mathbf{u}(s)+\mathbf{G}_{\mathrm{c}}\mathbf{C}_2\mathbf{G}_{\mathrm{p}}\exp[-(a+b)s]\,\mathbf{u}(s) \tag{18}$$

or rearranging gives the following control algorithm :

$$\mathbf{u}(s) = -[\mathbf{I} + \mathbf{G}_c(\mathbf{C}_1 + \mathbf{C}_2)\mathbf{G}_p]^{-1}\mathbf{G}_c[\mathbf{y}(s) - [\mathbf{C}_1 - \mathbf{C}_2 \exp(-bs)] \exp(-as)\, \mathbf{G}_p\mathbf{u}(s)]. \tag{19}$$

Combining eqns. (16)–(18) gives the result

$$\mathbf{x}(s) = [\mathbf{I} + \mathbf{G}_p\mathbf{G}_c(\mathbf{C}_1 + \mathbf{C}_2)]^{-1}[\mathbf{I} + \mathbf{G}_p\mathbf{G}_c\{[1 - \exp(-as)]\, \mathbf{C}_1 + \{1 - \exp[-(a+b)s]\}\, \mathbf{C}_2\}]\mathbf{G}_L\mathbf{d}(s). \tag{20}$$

From eqn. (20) it follows that the characteristic equation for the closed-loop system is

$$|\mathbf{I} + \mathbf{G}_p\mathbf{G}_c(\mathbf{C}_1 + \mathbf{C}_2)| = 0. \tag{21}$$

Thus the Smith Predictor is again successful in eliminating time delays from the characteristic equation. Furthermore, the resulting characteristic equation in eqn. (21) can easily be shown to be the characteristic equation for the system without time delays (i.e. the system in eqns. (3) and (4) with $a = b = 0$).

## 4. Multivariable discrete-time systems

A multivariable Smith Predictor can be developed in an analogous fashion for discrete-time systems which contain time delays. Consider the following stationary state-space model :

$$\mathbf{x}(n+1) = \boldsymbol{\phi}\mathbf{x}(n) + \boldsymbol{\Delta}\mathbf{u}(n-a) + \boldsymbol{\theta}\mathbf{d}(n), \tag{22}$$

$$\mathbf{y}(n) = \mathbf{C}_1\mathbf{x}(n) + \mathbf{C}_2\mathbf{x}(n-b), \tag{23}$$

where

$\mathbf{x}(n)$ = state vector of dimension $n$,
$\mathbf{u}(n)$ = manipulated vector of dimension $m$,
$\mathbf{d}(n)$ = disturbance vector of dimension $p$,
$\mathbf{y}(n)$ = output vector of dimension $r$,
$a, b$ = constant time delays which are integers (i.e. integer multiples of the sampling period, $T$),

$\boldsymbol{\phi}, \boldsymbol{\theta}, \boldsymbol{\Delta}, \mathbf{C}_1$ and $\mathbf{C}_2$ are constant matrices of appropriate dimensions.

As in the continuous-time case, a multivariable Smith Predictor algorithm can be developed by considering feedback loops around the controller. Suppose a proportional feedback control law is assumed of the form

$$\mathbf{u}(n) = -\mathbf{K}_c\mathbf{y}(n), \tag{24}$$

where $\mathbf{K}_c$ is a constant $m \times r$ matrix. Then a suitable algorithm for the Smith Predictor is

$$\mathbf{u}(n) = -\mathbf{K}_c\mathbf{y}(n) - \mathbf{K}_c\mathbf{p}(n), \tag{25}$$

$$\mathbf{p}(n) = \mathbf{C}_1\mathbf{p}_1(n) + \mathbf{C}_2\mathbf{p}_2(n), \tag{26}$$

where

$$\mathbf{p}_1(n)=\boldsymbol{\phi}\mathbf{p}_1(n-1)+\boldsymbol{\Delta}\mathbf{u}(n-1)-\boldsymbol{\Delta}\mathbf{u}(n-a-1) \quad \text{for} \quad n\geqslant 1, \tag{27}$$

$$\mathbf{p}_2(n)=\boldsymbol{\phi}\mathbf{p}_2(n-1)+\boldsymbol{\Delta}\mathbf{u}(n-1)-\boldsymbol{\Delta}\mathbf{u}(n-a-b-1) \quad \text{for} \quad n\geqslant 1 \tag{28}$$

and

$$\mathbf{p}_1(0)=\mathbf{p}_2(0)=\mathbf{0}.$$

Next it will be demonstrated that the Smith Predictor algorithm of eqns. (25)–(28) does eliminate the time delays from the characteristic equation. Taking the $z$ transform of eqns. (22), (23) and (25)–(28) gives

$$z\mathbf{X}(z)=\mathbf{M}\boldsymbol{\Delta}z^{-a}\mathbf{U}(z)+\mathbf{M}\boldsymbol{\theta}\mathbf{D}(z), \tag{29}$$

$$\mathbf{Y}(z)=\mathbf{C}_1\mathbf{X}(z)+\mathbf{C}_2z^{-b}\mathbf{X}(z), \tag{30}$$

$$\mathbf{U}(z)=-\mathbf{K}_c\mathbf{Y}(z)-\mathbf{K}_c\mathbf{P}(z), \tag{31}$$

$$\mathbf{P}(z)=\mathbf{C}_1\mathbf{P}_1(z)+\mathbf{C}_2\mathbf{P}_2(z), \tag{32}$$

$$\mathbf{P}_1(z)=z^{-1}[\boldsymbol{\phi}\mathbf{P}_1(z)+\boldsymbol{\Delta}\,\mathbf{U}(z)-\boldsymbol{\Delta}z^{-a}\,\mathbf{U}(z)], \tag{33}$$

$$\mathbf{P}_2(z)=z^{-1}[\boldsymbol{\phi}\mathbf{P}_2(z)+\boldsymbol{\Delta}\,\mathbf{U}(z)-z^{-(a+b)}\boldsymbol{\Delta}\,\mathbf{U}(z)], \tag{34}$$

where the notation $\mathbf{X}(z)\equiv z$ transform of $\mathbf{x}(n)$, is used. Equations (29)–(34) can be combined and rearranged to give the following expression :

$$\mathbf{X}(z)=[z\mathbf{I}+\mathbf{M}\boldsymbol{\Delta}\mathbf{K}_c(\mathbf{C}_1+\mathbf{C}_2)]^{-1}[\mathbf{I}+\mathbf{M}\boldsymbol{\Delta}\mathbf{K}_c\mathbf{C}_1(1-z^{-a})z^{-1} + \mathbf{M}\boldsymbol{\Delta}\mathbf{K}_c\mathbf{C}_2(1-z^{-(a+b)})z^{-1}]\mathbf{M}\boldsymbol{\theta}\mathbf{D}(z), \tag{35}$$

where $\mathbf{M}(z)$ is defined as

$$\mathbf{M}(z)\equiv(\mathbf{I}-z^{-1}\boldsymbol{\phi})^{-1} \tag{36}$$

Equation (35) implies that the characteristic equation for the system with the Smith Predictor is

$$|z\mathbf{I}+\mathbf{M}\boldsymbol{\Delta}\mathbf{K}_c(\mathbf{C}_1+\mathbf{C}_2)|=0. \tag{37}$$

It can easily be shown that eqn. (37) is also the characteristic equation for the original system in eqns. (22) and (23) when the time delays are zero (i.e. $a=b=0$). Thus the Smith Predictor algorithm in eqns. (25)–(28) can be used to eliminate time delays from the characteristic equation. The design of the feedback controller matrix $\mathbf{K}_c$ can then proceed using the wide variety of design techniques developed for systems without time delays.

## 5. Numerical example

To illustrate the operation of the multivariable Smith Predictor, consider the addition of time delays to the third-order system of Takahashi *et al.* (1968). Two cases will be considered :

(*a*) Time delay in the control variables.

(*b*) Time delays in both the control variables and the output variables.

The system equations for this example are (Takahashi *et al.* 1968)

$$\dot{\mathbf{x}}(t)=\mathbf{A}\mathbf{x}(t)+\mathbf{B}\mathbf{u}(t-a)+\mathbf{D}\mathbf{d}(t), \tag{38}$$

$$\mathbf{y}(t)=\begin{cases}\mathbf{C}\mathbf{x}(t) & \text{for case } (a),\\ \mathbf{C}_1\mathbf{x}(t)+\mathbf{C}_2\mathbf{x}(t-b) & \text{for case } (b).\end{cases} \tag{39}$$

where

$$\mathbf{A}=\begin{bmatrix}-3 & 1 & 0\\ 2 & -3 & 2\\ 0 & 1 & -3\end{bmatrix},\quad \mathbf{B}=\begin{bmatrix}1 & 0\\ 0 & 0\\ 0 & 1\end{bmatrix},\quad \mathbf{D}=\begin{bmatrix}1 & 0\\ 0 & 1\\ 0 & 0\end{bmatrix},$$

$$\mathbf{C}=\begin{bmatrix}0 & 1 & 0\\ 0 & 0 & 1\end{bmatrix},\quad \mathbf{C}_1=\begin{bmatrix}0 & 1 & 0\\ 0 & 0 & 0\end{bmatrix},\quad \mathbf{C}_2=\begin{bmatrix}0 & 0 & 0\\ 0 & 0 & 1\end{bmatrix}.$$

The corresponding discrete-time model for $T=0{\cdot}5$ can be obtained in the standard manner, using the analytical solution to eqn. (38) (Lapidus and Luus 1967) and is given in eqns. (40) and (41) :

$$\mathbf{x}(n+1)=\boldsymbol{\phi}\mathbf{x}(n)+\boldsymbol{\Delta}\mathbf{u}(n-a)+\boldsymbol{\theta}\mathbf{d}(n), \tag{40}$$

$$\mathbf{y}(n)=\begin{cases}\mathbf{C}\mathbf{x}(n) & \text{for case } (a),\\ \mathbf{C}_1\mathbf{x}(n)+\mathbf{C}_2\mathbf{x}(n-b) & \text{for case } (b),\end{cases} \tag{41}$$

where

$$\boldsymbol{\phi}=\begin{bmatrix}0{\cdot}284 & 0{\cdot}131 & 0{\cdot}0606\\ 0{\cdot}262 & 0{\cdot}344 & 0{\cdot}262\\ 0{\cdot}0606 & 0{\cdot}131 & 0{\cdot}284\end{bmatrix},\quad \boldsymbol{\Delta}=\begin{bmatrix}0{\cdot}276 & 0{\cdot}0148\\ 0{\cdot}103 & 0{\cdot}103\\ 0{\cdot}0148 & 0{\cdot}276\end{bmatrix},$$

$$\boldsymbol{\theta}=\begin{bmatrix}0{\cdot}276 & 0{\cdot}0516\\ 0{\cdot}103 & 0{\cdot}290\\ 0{\cdot}0148 & 0{\cdot}0516\end{bmatrix}.$$

A feedback control law of the form

$$\mathbf{u}(n)=-\mathbf{K}_c\mathbf{y}(n) \tag{42}$$

is desired. Since the Smith Predictor algorithm will be used, $\mathbf{K}_c$ can be designed using conventional design techniques for systems without time delays, and the state-space model in eqns. (40) and (41) with $a=b=0$. For example, a modification of the direct synthesis method of Porter and Crossley (1970) can be used to specify a desired closed-loop system matrix, $\mathbf{T}$, and a feedback control matrix, $\mathbf{K}_c$. This approach was used to design $\mathbf{K}_c$ as

$$\mathbf{K}_c=\begin{bmatrix}0{\cdot}0257 & 0{\cdot}0556\\ -0{\cdot}477 & -1{\cdot}03\end{bmatrix} \tag{43}$$

corresponding to the closed-loop system matrix, $\mathbf{T}=\boldsymbol{\phi}+\boldsymbol{\Delta}\mathbf{K}_c\mathbf{C}$, where

$$\mathbf{T}=\begin{bmatrix}0{\cdot}284 & 0{\cdot}131 & 0{\cdot}0606\\ 0{\cdot}262 & 0{\cdot}297 & 0{\cdot}200\\ 0{\cdot}0606 & 0 & 0\end{bmatrix}. \tag{44}$$

Fig. 5

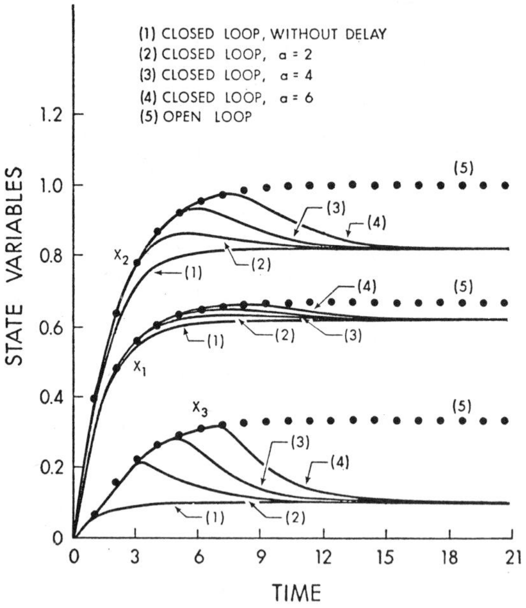


Load responses for multivariable system with time delay in the control variables.

Fig. 6

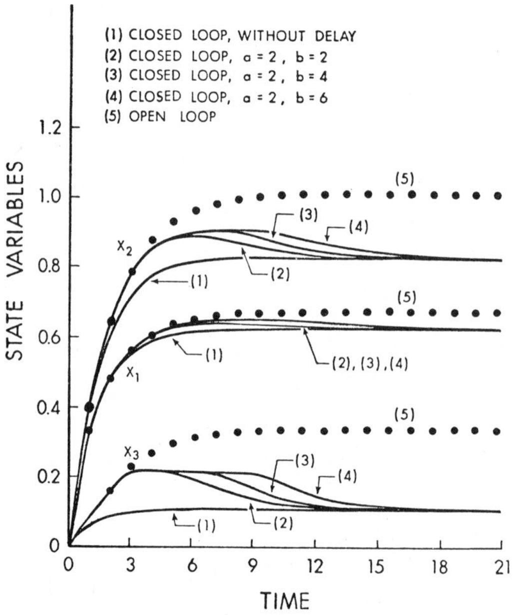


Load responses for multivariable system with time delays in both control and output variables.

Transient responses of the closed-loop system including the Smith Predictor are shown in figs. 5 and 6 for cases (*a*) and (*b*), respectively. All responses are for the feedback control matrix in eqn. (43), the Smith Predictor algorithm of eqns. (25)–(28) and a unit step change in the disturbance vector, **d**.

Figure 5 illustrates the effect of a delay in the control variables on the system response. The open and closed-loop responses for the system without time delays are given by curves 5 and 1, respectively. When the control variables are delayed as in curves 2–4, the system response initially follows the open-loop response, then changes direction and eventually approaches the closed-loop response for the undelayed system. Thus, by using the multivariable Smith Predictor and a feedback control matrix designed for the system without time delays, a satisfactory load response is obtained.

In fig. 6, time delays in both control variables and output variables are considered for the same disturbance and control scheme that were used in fig. 5. The responses are qualitatively similar to those in fig. 5 and give a satisfactory degree of control.

## 6. Conclusions

The classical Smith Predictor Method (Smith 1957, 1959) for single variable systems has been extended to a class of linear multivariable systems. Multivariable Smith Predictors have been derived for both continuous-time and discrete-time systems which contain time delays in either the control variables and/or the output variables. An important advantage of the multivariable Smith Predictor is that it eliminates time delays from the characteristic equation for the closed-loop system. This allows the control system designer to choose from the wide variety of synthesis techniques available for systems without delays as opposed to the much smaller number of techniques which are applicable to systems containing time delays. Another advantage of the multivariable Smith Predictor is that the discrete-time algorithm in eqns. (25)–(28) can be easily implemented on a real-time digital computer.

Several points require further investigation. Since the multivariable Smith Predictor contains a model of the system, its sensitivity to modelling errors is an important practical consideration. Consequently, simulation studies and experimental verification for actual physical systems would appear to be desirable. Furthermore, extensions of the multivariable Smith Predictor to a more general class of multivariable systems may prove to be useful. For example, systems in which only some of the control variables are delayed occur frequently in practical control problems. Such systems can be described by the following state-space model :

$$\dot{\mathbf{x}}(t) = \mathbf{A}\mathbf{x}(t) + \mathbf{B}_1\mathbf{u}(t) + \mathbf{B}_2\mathbf{u}(t-a), \tag{45}$$

$$\mathbf{y}(t) = \mathbf{C}_1\mathbf{x}(t) + \mathbf{C}_2\mathbf{x}(t-b). \tag{46}$$

This extension is the subject of a current investigation by the authors.

ACKNOWLEDGMENT

Financial support from IBM Canada Ltd. and the University of Alberta is gratefully acknowledged.

## References

Buckley, P. S., 1960, Proc. 1st International Federation of Automatic Control Congress, Moscow.

Gray, J. O., and Hunt, P. W. B., 1971, *Electronics Lett.*, **7,** 335.

Lapidus, L., and Luus, R., 1967, *Optimal Control of Engineering Processes* (Waltham, Mass. : Blaisdell).

Lupfer, D. E., and Oglesby, M. W., 1961, *I.S.A. Jl*, **8,** 11, 53 ; 1962, *I.S.A. Trans.*, **1,** 72.

Nielsen, G., 1969, *Proc. 4th International Federation of Automatic Control Congress*, Warsaw.

Porter, B., and Crossley, T. R., 1970, *Electronics Lett.*, **6,** 79.

Smith, O. J. M., 1957, *Chem. Engng Prog.*, **53,** 217 ; 1959, *I.S.A. Jl*, **6,** 2, 28.

Takahashi, Y., Thal-Larsen, H., Goldenberg, E., Loscutoff, W. V., and Ragetly, P. R., 1968, *J. bas. Engng*, **90,** 222.

# CONTROL OF MULTIVARIABLE SYSTEMS CONTAINING TIME DELAYS USING A MULTIVARIABLE SMITH PREDICTOR

G. ALEVISAKIS and D. E. SEBORG
Department of Chemical Engineering, University of Alberta, Edmonton, Alberta, Canada

(*Received* 26 *June* 1972; *accepted* 22 *May* 1973)

**Abstract**—A multivariable version of the classical Smith predictor is applied to a double effect evaporator containing time delays in the control variables and some of the output variables. As in the classical approach, the multivariable Smith predictor eliminates time delays from the characteristic equation of the closed-loop system. Simulation and experimental results for the pilot-scale evaporator are quite promising and demonstrate the feasibility of the multivariable Smith predictor for practical control problems.

## 1. INTRODUCTION

Time delays occur frequently in process control problems due to the distance-velocity lags associated with flowing fluids or the time required to complete a composition analysis. The detrimental effect of time delays on system stability and control are well-known to both control system designers and personnel responsible for plant operation. From the classical view point, time delays are undesirable since the control action is based on delayed information and the resulting phase lag tends to make the system less stable and more difficult to control. Furthermore, time delays in multivariable systems make control system design much more difficult since many design approaches are either not applicable to time delay systems or become unduly complicated.

In 1957 Smith[20, 21] proposed a "predictor" technique for single variable control systems which contain time delays. The basis for the "Smith predictor" method is illustrated in Figs. 1 and 2. The predictor consists of a feedback loop around the controller with the output of the "predictor" block representing the difference between two process model responses: the response of the system without time delays minus the response of the delayed system. If the process model were perfect, the model response (with the time delay) would cancel the actual process response, $y(s)$, and the control action would be based on the response of the undelayed system. Consequently, the time delay would then be eliminated from the characteristic equation of the system as demonstrated below.

For the feedback control system shown in Fig. 2, the closed-loop transfer function for load disturbances is

$$\frac{x(s)}{d(s)} = G_L(s) - \frac{G_c(s)G_p(s)G_L(s)e^{-as}}{1+G_c(s)G_p(s)H(s)}. \tag{1}$$

The characteristic equation for this system is

$$1+G_c(s)G_p(s)H(s)=0 \tag{2}$$

which is also the characteristic equation for the system in Fig. 1 when the time delay, $a$, is zero. Thus,

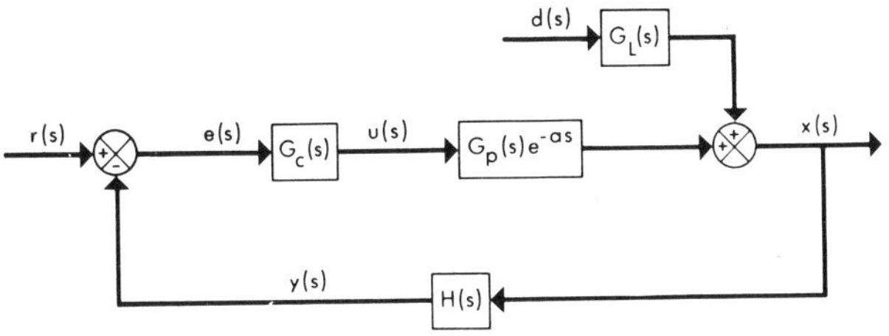


Fig. 1. Block diagram of a conventional feedback control system.

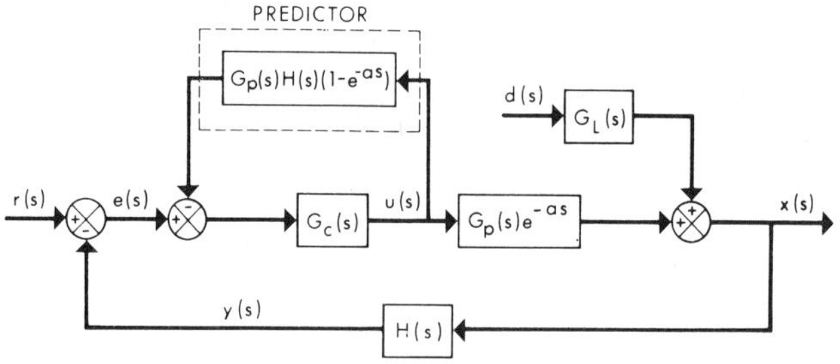


Fig. 2. Block diagram of the Smith predictor and the feedback control system of Fig. 1.

by using the Smith Predictor the time delay has been eliminated from the characteristic equation and consequently, larger controller gains can be used.

Later investigators[4, 10, 11, 15] have reported applications of the Smith Predictor and its use in designing sampled-data control systems[6]. Comparisons of the Smith predictor with conventional control techniques have been made by Buckley[4] and Nielsen[15]. Apparently, the only extension of the Smith predictor method to multivariable control problems is a previous paper by the present authors[2].

## 2. SMITH PREDICTOR FOR MULTIVARIABLE SYSTEMS

The design of control systems for multivariable processes has received considerable attention during the past 15 years. However, in most time-domain and frequency-domain design techniques it is assumed that time delays are not present. For example, in their present stage of development, modal analysis[5, 18], structural analysis[7], and the inverse Nyquist method[19] are not directly applicable to systems containing time delays. Although optimal control theory has been extended to include time delay systems[22, 23], both off-line and on-line computer requirements are significantly greater than for systems without time delays. Thus while the results of these theoretical analyses are of considerable value, the resulting optimal control laws are not yet attractive for actual applications.

In a recent paper[2], the authors have extended the Smith predictor method to multivariable processes which contain time delays in the measured output variables and/or in the control variables. The multivariable Smith predictor algorithm for a class of discrete-time systems is summarized below.

Consider the following stationary state-space model with time delays in all of the control variables and some of the output variables:

$$\mathbf{x}(i+1) = \vec{\boldsymbol{\phi}}\mathbf{x}(i) + \vec{\boldsymbol{\Delta}}\mathbf{u}(i-a) + \vec{\boldsymbol{\theta}}\mathbf{d}(i) \quad (3)$$

$$\mathbf{y}(i) = \vec{\mathbf{C}}_1\mathbf{x}(i) + \vec{\mathbf{C}}_2\mathbf{x}(i-b) \quad (4)$$

where $\mathbf{x}(i)$ = state vector of dimension $n$,
$\mathbf{u}(i)$ = control vector of dimension $m$,
$\mathbf{d}(i)$ = disturbance vector of dimension $p$,
$\mathbf{y}(i)$ = output vector of dimension $r$,
$a, b$ = constant time delays which are positive integers
$i$ = time interval counter ($i \geq 0$)
$\vec{\boldsymbol{\phi}}, \vec{\boldsymbol{\Delta}}, \vec{\boldsymbol{\theta}}, \vec{\mathbf{C}}_1$ and $\vec{\mathbf{C}}_2$ are constant matrices.

The time delays in the state-space model of Eqs. (3) and (4) can be given the following physical interpretation. Time delay, $a$, represents the delay associated with the calculation and implementation of changes in the control variables (e.g. situations where the control valves are located some distance from the process itself). In many physical systems, time delays are difficult to avoid in the measurement of certain state variables such as chemical composition. The inclusion of time delay, $b$, in Eq. (4) provides a useful model for physical systems where some state variables are measured instantaneously but others such as composition involve a measurement delay. Apparently, the output equation in Eq. (4) has not been widely used despite the many process control problems where time delays occur in the measurements of some (but not all) state variables. (Typically, in previous studies, either $\vec{\mathbf{C}}_1$ or $\vec{\mathbf{C}}_2$ is assumed to be zero.)

The multivariable Smith predictor consists of a feedback loop around the controller[2]. Assuming that proportional feedback control of the output variables is used, the control law for the predictor control scheme (and a setpoint of zero) is given by:

$$\mathbf{u}(i) = -\vec{\mathbf{K}}_c\mathbf{y}(i) - \vec{\mathbf{K}}_c\mathbf{p}(i) \quad (5)$$

$$\mathbf{p}(i) = \vec{\mathbf{C}}_1\mathbf{p}_1(i) + \vec{\mathbf{C}}_2\mathbf{p}_2(i) \quad (6)$$

where

$$\mathbf{p}_1(i) = \vec{\boldsymbol{\phi}}\mathbf{p}_1(i-1) + \vec{\mathbf{\Delta}}\mathbf{u}(i-1) - \vec{\mathbf{\Delta}}\mathbf{u}(i-a-1) \quad (7)$$

$$\mathbf{p}_2(i) = \vec{\boldsymbol{\phi}}\mathbf{p}_2(i-1) + \vec{\mathbf{\Delta}}\mathbf{u}(i-1) - \vec{\mathbf{\Delta}}\mathbf{u}(i-a-b-1) \quad (8)$$

$$\mathbf{p}_1(0) = \mathbf{p}_2(0) = \mathbf{0} \quad (9)$$

$\vec{\mathbf{K}}$ = proportional control matrix ($m \times r$).

An expression for the closed-loop response of the system for load changes can easily be derived after taking the $z$-transform of Eqs. (3–9). The resulting expression for the system response to a load change is given by:

$$\mathbf{X}(z) = (z\vec{\mathbf{I}}_n + \vec{\mathbf{M}}\vec{\mathbf{\Delta}}\vec{\mathbf{K}}_c(\vec{\mathbf{C}}_1 + \vec{\mathbf{C}}_2)]^{-1} \times [\vec{\mathbf{I}}_n + \vec{\mathbf{M}}\vec{\mathbf{\Delta}}\vec{\mathbf{K}}_c\vec{\mathbf{C}}_1(1 - z^{-a})z^{-1} + \vec{\mathbf{M}}\vec{\mathbf{\Delta}}\vec{\mathbf{K}}_c\vec{\mathbf{C}}_2(1 - z^{-(a+b)})z^{-1}]\vec{\mathbf{M}}\vec{\boldsymbol{\theta}}\mathbf{D}(z) \quad (10)$$

where $\mathbf{X}(z)$ denotes the $z$-transform of $\mathbf{x}(i)$, $\vec{\mathbf{I}}_n$ is the $n \times n$ identity matrix, and $\vec{\mathbf{M}}(z)$ is defined as

$$\vec{\mathbf{M}}(z) = (\vec{\mathbf{I}}_n - z^{-1}\vec{\boldsymbol{\phi}})^{-1}. \quad (11)$$

Equation (10) implies that the characteristic equation for the closed-loop system with the predictor is

$$|z\vec{\mathbf{I}}_n + \vec{\mathbf{M}}\vec{\mathbf{\Delta}}\vec{\mathbf{K}}_c(\vec{\mathbf{C}}_1 + \vec{\mathbf{C}}_2)| = 0 \quad (12)$$

which is also the characteristic equation for the system of Eqs. (3) and (4) when $a = b = 0$. Thus the Smith predictor algorithm has eliminated the time delays from the characteristic equation and the design of the control matrix, $\vec{\mathbf{K}}_c$, can proceed using design techniques developed for systems without time delays. The above analysis could also be extended to include other types of multivariable control techniques such as proportional plus integral control or to accommodate setpoint changes as well as load changes[14].

### 3. APPLICATION OF THE MULTIVARIABLE SMITH PREDICTOR TO A DOUBLE EFFECT EVAPORATOR

The theory described in the previous section was used to design a multivariable Smith predictor for a pilot-scale, double effect evaporator. The effectiveness of the predictor control algorithm was then evaluated via computer simulation and experimental studies.

The pilot plant evaporator is shown schematically in Fig. 3. The first effect is a calandria type unit with thirty-two tubes which are 0·0191 m dia and 0·457 m long. The feed to the first effect consists of 2·27 kg/min of 3 percent (by weight) triethylene glycol and is heated by 0·9 kg/min of saturated steam at 394°K. The second effect has externally forced circulation through three 0·0254× 1·83 m tubes fed by the first effect product and heated by condensing vapour from the first effect. The second effect vapour is totally condensed and vacuum control is used to maintain the necessary pressure differential between the two effects. The primary control objective is to maintain the product concentration, $C_2$, at the nominal steady state value of 10 per cent glycol despite disturbances in the feed conditions.

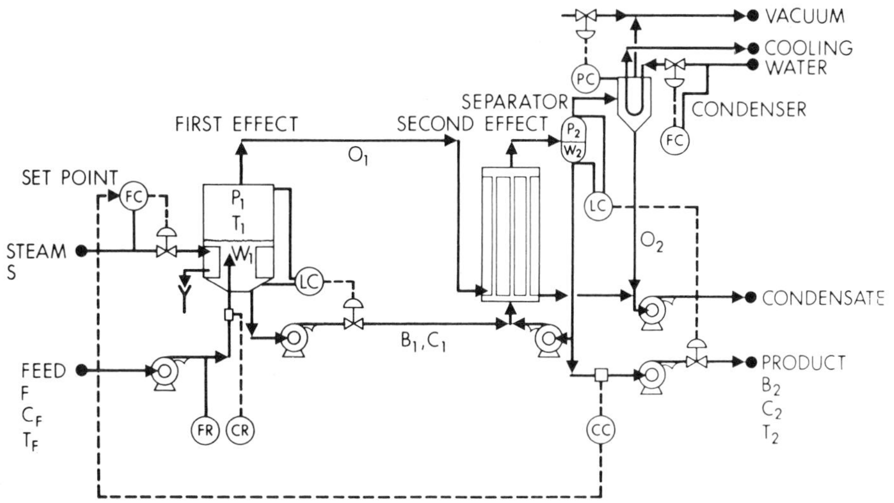


Fig. 3. Schematic diagram of the double effect evaporator and a multiloop control scheme.

Mathematical models of this evaporator have been developed[3, 13] based on linearized material and energy balances for the two effects. The model used in this investigation is due to Newell and Fisher[13] and consists of a fifth order, state-space model expressed in terms of normalized, perturbation variables. The state, control and disturbance vectors and their normal steady-state values are presented in the Appendix. Of the five state variables, only the first effect concentration, $C_1$, is not measured.

For computer control applications, it is convenient to use discrete-time, rather than continuous-time, state space models due to the discrete nature of a digital computer. Consequently, the discrete-time model in Eqs. (3) and (4) was used to design the Smith predictor. The numerical values of the coefficient matrices for a sampling interval of $T = 64$ sec and the normal operating conditions are given in the Appendix. This discrete state-space evaporator model has also been used in previous investigations of multiloop and inferential control [9], optimal feedback control[12–14], optimal setpoint control[16], model-reference adaptive control[17], and Kalman filtering[8].

The experimental evaluation of the multivariable Smith predictor control scheme was conducted with an IBM 1800 computer that is interfaced to the pilot plant evaporator. A standard multivariable control program written by Newell[12] was modified and used to estimate state variable, $C_1$, and perform unit conversions from engineering units to perturbation form and vice versa, for use in the control calculations. The program was loaded into core from disk storage in response to a high priority timer interrupt at a control interval of 64 sec. Calculated and measured variables were also stored on disk at each control interval. Due to slow disk operations, the average execution time of the multivariable control program was 5 sec.

## 4. SIMULATION AND EXPERIMENTAL RESULTS

A simulation study was undertaken to evaluate the performance of the multivariable Smith predictor for a variety of predictor algorithms, control systems and operating conditions. The predictor was used in conjunction with two types of feedback control systems. The first, a multiloop feedback control system, is shown schematically in Fig. 3 and consists of three single variable proportional feedback controllers in which the control variables, $S$, $B_1$ and $B_2$, are paired with the controlled variables, $C_2$, $W_1$ and $W_2$, respectively. This multiloop control law can be written as:

$$\mathbf{u}(i) = \vec{\mathbf{K}}_c \mathbf{x}(i). \tag{13}$$

A satisfactory feedback matrix for multiloop control of the evaporator when no time delays are present is given by[17]:

$$\vec{\mathbf{K}}_c = \begin{bmatrix} 0 & 0 & 0 & 0 & -4{\cdot}89 \\ 3{\cdot}52 & 0 & 0 & 0 & 0 \\ 0 & 0 & 0 & 15{\cdot}8 & 0 \end{bmatrix}. \tag{14}$$

A second control scheme, optimal multivariable feedback control, was considered using the feedback control matrix designed by Newell and Fisher[14] based on a quadratic performance index:

$$\vec{\mathbf{K}}_c = \begin{bmatrix} 5{\cdot}09 & -1{\cdot}48 & -2{\cdot}68 & 0 & -14{\cdot}6 \\ 3{\cdot}95 & 0{\cdot}36 & 0{\cdot}21 & 0 & 7{\cdot}39 \\ 5{\cdot}31 & 1{\cdot}19 & -0{\cdot}11 & 15{\cdot}8 & 18{\cdot}8 \end{bmatrix}. \tag{15}$$

The actual pilot plant evaporator does not normally contain any significant time delays since transport lags are small and product composition can be measured continuously with an on-line refractometer. In this investigation, a time delay in the output variable, $C_2$, was introduced by basing the control calculations on past measurements retrieved from disk storage. While this method of introducing a time delay is somewhat artificial, it does correspond to industrial situations where a major time delay in many control loops is the "analysis time" associated with analytical instruments such as gas chromatographs.

In the simulation and experimental runs reported below, the evaporator responses were obtained for a + 20 per cent step disturbance in feed flow which was introduced at the second sampling interval (i.e. at $t = 128$ sec). For convenience, it was assumed that the time delay was an integer multiple of the 64 sec sampling interval.

Experimental and simulated responses of the evaporator (without time delays) under multiloop control are shown in the top half of Fig. 4. The multiloop control scheme provides a satisfactory degree of control with small offsets resulting since proportional control is used. Reasonable agreement between simulated and experimental responses was obtained for $C_2$ and $W_1$, but not for $W_2$, indicating that modelling errors exist due to process nonlinearities[13]. In Fig. 4 and succeeding figures, the arrows along the vertical axis denote the initial steady state.

The results in the bottom half of Fig. 4 and in Fig. 5 demonstrate that the multivariable Smith predictor provides satisfactory control when measure-

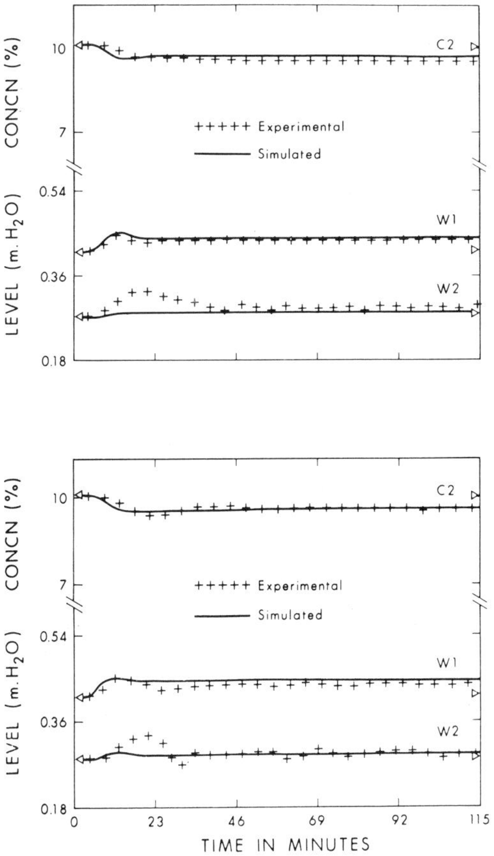


Fig. 4. Top: comparison of experimental and simulated evaporator responses for multiloop control and no time delays. Bottom: comparison of experimental and simulated evaporator responses when the multivariable predictor is used to compensate for a measurement delay of 256 sec in $C_2$.

ment delays of 256 and 512 sec, respectively, occur in product concentration, $C_2$. Furthermore, the response of the primary controlled variable, $C_2$, is only slightly more oscillatory than the response of the undelayed system (cf. top half of Fig. 4). By contrast, when the predictor technique was not used, these large measurement delays and the multiloop control scheme produced very poor responses [1].

The situation where time delays exist in both the control variables and the output variables was also considered in the simulation study. The simulated responses in Fig. 6 illustrate the excellent control that results when the Smith predictor is used with optimal multivariable control to compensate for even very large time delays in $C_2$ and the control variables. In each case, the response of the unde-

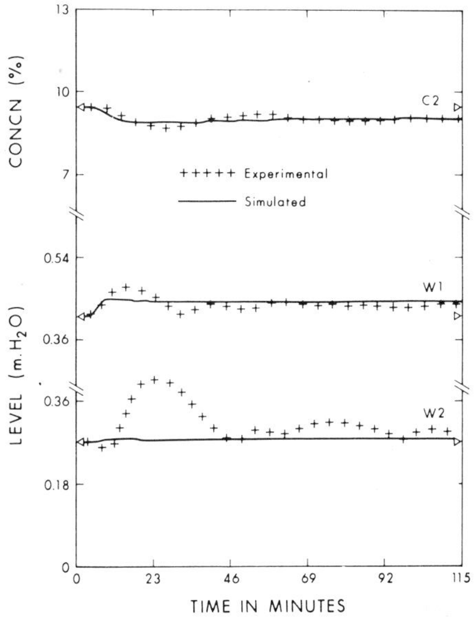


Fig. 5. Comparison of experimental and simulated evaporator responses when the multivariable predictor is used to compensate for a measurement delay of 516 sec in $C_2$.

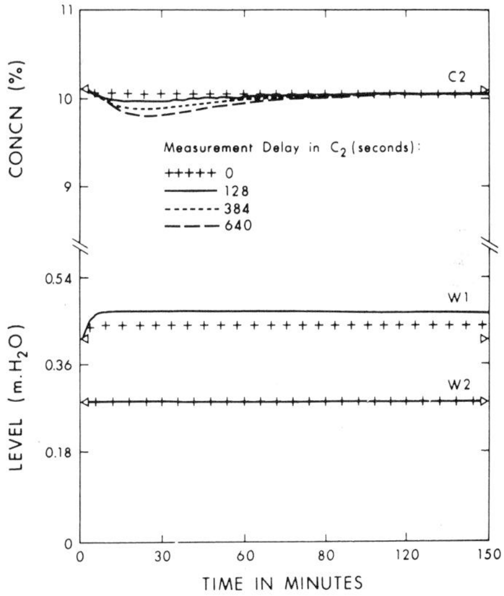


Fig. 6. Simulated evaporator responses when the multivariable predictor and an optimal feedback controller are used to compensate for a time delay of 128 sec in the control variables and various measurement delays in $C_2$.

layed system is followed quite closely although the maximum deviation in $C_2$ becomes larger as the time delays increase. As predicted theoretically, the predictor responses eventually approach the response of the undelayed system, although for $W_1$, this convergence is very slow. The simulation results also indicated that if the predictor is not used, the system is unstable for the conditions considered in Fig. 6[1].

An advantage of the Smith predictor is that it allows higher controller gains to be used, as demonstrated by the simulation results in the top half of Fig. 7. When a measurement delay of 384 sec in $C_2$ exists and the predictor is not used, the nominal controller gain of $k_{15} = -4{\cdot}89$ results in an excessively oscillatory $C_2$ response. Reducing this controller gain to values of $-2{\cdot}0$ and $-3{\cdot}0$ reduces the oscillations but results in more sluggish responses with larger offsets. However, application of the Smith Predictor and the nominal controller gain of $-4{\cdot}89$ results in an improved transient response.

A series of simulation runs was made to determine the sensitivity of the multivariable Smith predictor to inaccurate estimates of the actual time delays. Typical results are shown in the bottom half of Fig. 7 for the case where the actual time delay in the control variables is 256 sec but the assumed time delay used in the predictor algorithm was 192, 256 or 320 sec, respectively. The simulated responses are still satisfactory for inaccurate estimates of the time delays although the controller gain, $k_{34}$, had to be reduced to a value of $5{\cdot}0$ since the nominal value of $k_{34} = 15{\cdot}8$ resulted in unstable responses. However, if the predictor is not used, the closed-loop system is unstable for even the relatively low gain of $k_{34} = 5{\cdot}0$.

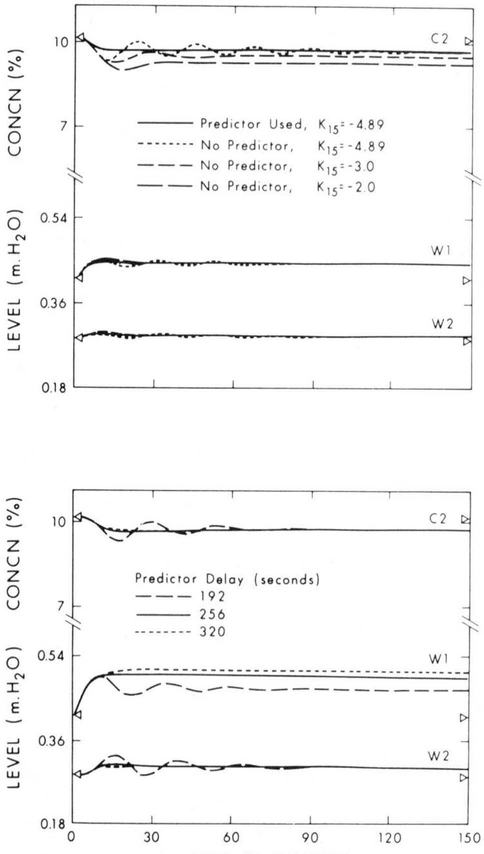


Fig. 7. Top: simulated responses for multiloop control with and without the predictor for several controller gains and a measurement delay of 384 sec in $C_2$. Bottom: simulated responses for multiloop control when a time delay of 256 sec exists in the control variables and various time delays are used in the predictor ($k_{34} = 5{\cdot}0$).

These results and similar results for multivariable feedback control[1] indicate that although the Smith predictor is quite sensitive to inaccurate estimates of the time delays when large controller gains are used, this sensitivity can be substantially reduced by using smaller controller gains. Furthermore, the smaller controller gains are often larger than the values that are available when the predictor is not used. Alternatively, one can attempt to estimate uncertain or time-varying delays on-line. A simple estimation algorithm based on the deterministic model has been considered by Alevisakis[1].

## 5. CONCLUSIONS

A multivariable extension[2] of the classical Smith predictor method for single variable control systems[20, 21] has been applied to a pilot scale, double effect evaporator. An important advantage of the multivariable Smith predictor is that it eliminates time delays from the characteristic equation of the closed loop system and thus allows larger controller gains to be used. It also permits the control system designer to choose from the wide variety of synthesis techniques that are available for systems without time delays as opposed to the much smaller number of techniques which are applicable to systems containing time delays. Another advantage of the multivariable Smith predictor is that the discrete-time algorithm in Eqs. (5)–(8) can be easily implemented on a process control computer.

In the simulation and experimental studies, the multivariable Smith predictor resulted in satisfactory control of the evaporator for large time delays in the output variables and/or the control variables, and for both multiloop and multivariable feedback controllers. The multivariable Smith predictor allowed higher controller gains to be employed than did conventional feedback control systems, even when inaccurate estimates of the time delays were used in the predictor algorithm.

*Acknowledgement*—Financial support from IBM Canada Ltd. and the National Research Council of Canada is gratefully acknowledged.

## NOTATION

- $a$ time delay in control variables
- $b$ time delay in output variables
- $B_1$ first effect bottoms flow
- $B_2$ second effect bottoms flow
- $\vec{\mathbf{C}}_1, \vec{\mathbf{C}}_2$ output coefficient matrices
- $C_1$ first effect concentration
- $C_2$ second effect concentration
- $C_F$ feed concentration
- $d$ load variable
- $\mathbf{d}$ load vector
- $\mathbf{D}$ $z$-transform of load vector
- $e$ error
- $F$ feed flow
- $G$ transfer function
- $h_1$ first effect enthalpy
- $h_F$ feed enthalpy
- $H$ output transfer function
- $i$ time interval counter
- $\vec{\mathbf{I}}_n$ identity matrix
- $k_{ij}$ controller gain (element of control matrix, $\vec{\mathbf{K}}_c$)
- $\vec{\mathbf{K}}_c$ control matrix
- $m$ dimension of control vector
- $\vec{\mathbf{M}}$ matrix defined in Eq. (11)
- $n$ dimension of state vector
- $p$ dimension of load vector
- $\mathbf{p}$ predictor output vector
- $\mathbf{P}$ $z$-transform of predictor output vector
- $r$ dimension of output vector
- $s$ Laplace transform variable
- $S$ steam flow rate
- $t$ time
- $T$ sampling (or control) interval
- $u$ control variable
- $\mathbf{u}$ control vector
- $\mathbf{U}$ $z$-transform of control vector
- $W_1$ first effect holdup
- $W_2$ second effect holdup
- $x$ state variable
- $\mathbf{x}$ state vector
- $\mathbf{X}$ $z$-transform of state vector
- $y$ output variable
- $\mathbf{y}$ output vector
- $\mathbf{Y}$ $z$-transform of output vector
- $z$ $z$-transform variable

*Greek symbols*

- $\vec{\mathbf{\Delta}}$ control coefficient matrix
- $\vec{\boldsymbol{\theta}}$ disturbance coefficient matrix
- $\vec{\boldsymbol{\phi}}$ state coefficient matrix

*Subscripts*

- $c$ controller
- $L$ load
- $p$ process
- $\rightarrow$ matrix

## REFERENCES


[1] ALEVISAKIS G., M.Sc. thesis, University of Alberta, Edmonton, Canada 1972.

[2] ALEVISAKIS G. and SEBORG D. E., *Int. J. Control* 1973 **17** 541.

[3] ANDRE H. and RITTER R. A., *Can. J. Chem. Engng* 1968 **46** 259.

[4] BUCKLEY P. S., *Proc.* 1*st IFAC Congress*, Moscow 1960. Butterworths, London 1963.

[5] GOULD L., *Chemical Process Control.* Addison-Wesley, Reading, Mass. 1969.

[6] GRAY J. O. and HUNT P. N. B., *Electron. Lett.* 1971 **7** 131.

[7] GREENFIELD G. G. and WARD T. J., *Ind. Engng Chem. Fundls.* 1967 **6** 571.

[8] HAMILTON J. C., SEBORG D. E. and FISHER D. G., *A.I.Ch.E. Jl.* (in press).

[9] JACOBSON B. A., M.Sc. thesis, University of Alberta, Edmonton, Canada 1970.

[10] LUPFER D. E. and OGLESBY M. W., *ISA Trans.* 1962 **1** 72.

[11] LUPFER D. E. and OGLESBY M. W., *ISA Jl.* 1963 **8** 73.

[12] NEWELL R. B., Ph.D. thesis, University of Alberta, Edmonton, Canada 1971.

[13] NEWELL R. B. and FISHER D. G., *Ind. Engng Chem. Proc. Des. Dev.* 1972 **11** 213.

[14] NEWELL R. B. and FISHER D. G., *Automatica* 1972 **8** 247.

[15] NIELSEN G., 4*th IFAC Congress*, Warsaw 1969.

[16] NIEMAN R. E. and FISHER D. G., *Trans. Instn. Chem. Engrs.* 1973 **51** 132.

[17] OLIVER W. K., SEBORG D. E. and FISHER D. G., *Chem. Engng Commun.* (in press).

[18] PORTER B. and CROSSLEY R., *Modal Control: Theory and Applications.* Taylor and Francis, London 1972.

[19] ROSENBROCK H. H., *Proc. IEE* 1969 **116** 1929.

[20] SMITH O. J. M., *ISA Jl.* 1959 **6** 28.

[21] SMITH O. J. M., *Chem. Engng Progr.* 1957 **53** 217.

[22] SOLIMAN M. A. and RAY W. H., *Automatica* 1971 **7** 681.

[23] SOLIMAN M. A. and RAY W. H., *Int. J. Control* 1972 **15** 609.


## APPENDIX

The state, control and disturbance vectors and their normal steady-state values are:

| State vector, **x** | | Normal steady-state value |
|---|---|---|
| $W_1$ | first effect holdup | 20·8 kg |
| $C_1$ | first effect concentration | 4·59% glycol |
| $h_1$ | first effect enthalpy | 441 kJ/kg |
| $W_2$ | second effect holdup | 19·0 kg |
| $C_2$ | second effect concentration | 10·11% glycol |
| Control vector, **u** | | |
| $S$ | steam | 0·91 kg/min |
| $B_1$ | first effect bottoms | 1·58 kg/min |
| $B_2$ | second effect bottoms | 0·72 kg/min |
| Disturbance vector, **d** | | |
| $F$ | feed flowrate | 2·26 kg/min |
| $C_F$ | feed concentration | 3·2% glycol |
| $h_F$ | feed enthalpy | 365 kJ/kg |

These nominal operating conditions and a sampling time of $T = 64$ sec result in the following discrete-time evaporator model when the state variables are written in normalized, perturbation form, e.g.

$$x_1 = \frac{W_1 - \bar{W}_1}{\bar{W}_1}$$

where $\bar{W}_1$ = steady-state value of $W_1$.

*State-space model* (cf. Eqs. (3) and (4))

$$\vec{\boldsymbol{\phi}} = \begin{bmatrix} 1 & -0{\cdot}0008 & -0{\cdot}0912 & 0 & 0 \\ 0 & 0{\cdot}9223 & 0{\cdot}0871 & 0 & 0 \\ 0 & -0{\cdot}0042 & 0{\cdot}4376 & 0 & 0 \\ 0 & -0{\cdot}0009 & -0{\cdot}1052 & 1 & 0{\cdot}0001 \\ 0 & 0{\cdot}0391 & 0{\cdot}1048 & 0 & 0{\cdot}9603 \end{bmatrix}$$

$$\vec{\boldsymbol{\Delta}} = \begin{bmatrix} -0{\cdot}0119 & -0{\cdot}0817 & 0 \\ 0{\cdot}0116 & 0 & 0 \\ 0{\cdot}1569 & 0 & 0 \\ -0{\cdot}0138 & 0{\cdot}0848 & -0{\cdot}0406 \\ 0{\cdot}0137 & -0{\cdot}0432 & 0 \end{bmatrix}$$

$$\vec{\boldsymbol{\theta}} = \begin{bmatrix} 0{\cdot}1182 & 0 & -0{\cdot}0050 \\ -0{\cdot}0351 & 0{\cdot}0785 & 0{\cdot}0049 \\ -0{\cdot}0136 & -0{\cdot}0002 & 0{\cdot}0662 \\ 0{\cdot}0012 & 0 & -0{\cdot}0058 \\ -0{\cdot}0019 & 0{\cdot}0016 & 0{\cdot}0058 \end{bmatrix}$$

# Section 7: ON-LINE ESTIMATION AND FILTERING

CONTENTS:



COMMENTS:

Many control techniques such as the "optimal-quadratic" approach considered earlier require that all of the state variables be available for use in the control laws. In cases where all the state variables cannot be measured directly it is necessary to use a state estimator such as the Kalman filter or the Luenberger observer presented in this section.

Although the observer is intended for deterministic systems and the Kalman filter for stochastic systems, they can both be used to generate estimates of unmeasured states. In the evaporator application, both approaches performed well in the simulation study but the Kalman filter proved to be more flexible, robust and practical for use on real systems.

REFERENCES:

[1] Special Issue on the Linear-Quadratic-Gaussian Problem, IEEE Transactions on Automatic Control, AC-16, December, 1971.

[2] Sage, A.P. and J.L. Melsa, Estimation Theory with Applications to Communications and Control, McGraw-Hill (1971).

# An Experimental Evaluation of Kalman Filtering

The effectiveness of the stationary form of the discrete Kalman filter for state estimation in noisy process systems was demonstrated by simulated and experimental tests on a pilot plant evaporator. The filter was incorporated into a multivariable, computer control system and resulted in good control despite process and/or measurement noise levels of 10%. The results were significantly better than those obtained when the Kalman filter was omitted or replaced by conventional exponential filters. In this application the standard Kalman filter was reasonably insensitive to incorrect estimates of initial conditions or noise statistics and to errors in model parameters. The filter estimates were sensitive to unmeasured process disturbances. However this sensitivity could be reduced by treating the noise covariance matrices R and Q as design parameters rather than noise statistics and selecting values which result in increased weighting of the process measurements relative to the calculated model states.

**JAMES C. HAMILTON**
**DALE E. SEBORG**
**and**
**D. GRANT FISHER**

Department of Chemical Engineering
University of Alberta
Edmonton, Alberta T6G2G6, Canada

## SCOPE

Many of the design techniques based on modern control theory assume that values are available for all the state variables in the system of interest. However, in most practical situations, it is not feasible to measure all state variables and, furthermore, the measurements that are available often contain significant amounts of random noise and/or systematic errors. In these situations, on-line estimation techniques can be used to estimate the unmeasured state variables and to reduce the effects of noise. Sequential estimation techniques, or filters as they are commonly called, produce estimates of the true process values from noisy process measurements and values calculated from a suitable process model.

The Kalman filter has probably received more attention in the recent literature than any other state estimation technique. It has been applied successfully in the aerospace industry and more recently there have been a number of theoretical investigations and simulation studies of its use in process control applications. However, there have been very few reported applications to actual industrial processes.

In this investigation the effectiveness of the stationary form of the discrete Kalman filter is evaluated through simulation studies and experimental application to a pilot scale, double effect evaporator. The Kalman filter is derived using a fifth-order state space model of the evaporator and used to provide state estimates for an optimal multivariable feedback controller. The effects of design parameters, errors in the model parameters, and incorrect process statistics on the performance of the filter are examined.

## CONCLUSIONS AND SIGNIFICANCE

The Kalman filter proved to be a practical and relatively easy to implement addition to the multivariable computer control system for a pilot scale, double effect evaporator. Simulation and experimental studies demonstrated that the stationary form of the discrete Kalman filter provided satisfactory state estimates even in the presence of significant noise levels, uncertain noise statistics, and significant modeling errors. However, the filter was sensitive to unmeasured step disturbances to the process.

The performance of the filter was found to depend on the accuracy of the assumed process model and the choice of the weighting matrices used in the design of the filter. The process model should, subject to reasonable complexity and model order, be as accurate as possible. Fortunately in this application the filter proved to be relatively insensitive to changes of ±25% in key model parameters. Theoretically, the weighting matrices should be made equal to the noise covariance matrices, but in practice noise statistics are not known exactly and estimates must be used. In the evaporator application the Kalman filter performed well when the matrix elements were equal to the actual noise statistics. However, it was also found that the elements of the covariance matrices could be treated as design parameters and chosen to improve the filter performance. For example, if significant modeling errors or unmeasured process disturbances are anticipated, it is desirable to make the elements of the process noise covariance matrix larger than the actual values that define the process noise statistics. This strategy leads to a filter design which places greater weighting on the process measurements and less weighting on the process model. Consequently, the effects of modeling errors and unmeasured process disturbances on filter performance are reduced.

In both the experimental and simulation studies, the use of a Kalman filter in the multivariable control scheme resulted in significantly improved control over the cases where no filter or a conventional exponential filter was used. Although not universally applicable, the results of this experimental investigation should provide useful guidelines for future applications of Kalman filtering.

---

Correspondence concerning this paper should be addressed to D. E. Seborg or D. G. Fisher. J. C. Hamilton is with Imperial Oil Enterprises Ltd., Calgary, Alberta.

AIChE Journal, Vol. 19, No. 5, pp. 901-909 (1973).

## PREVIOUS WORK

The Kalman filter has received extensive study since publication of the classic papers by Kalman (1960) and Kalman and Bucy (1961). Despite a voluminous literature including numerous applications in the aerospace industry, it is only recently that the applicability of the Kalman filter and related nonlinear filters to chemical engineering problems has been demonstrated. However, the chemical engineering applications to date have typically consisted of simulation studies involving relatively simple process models.

The literature includes evaluations of several nonlinear filters such as the extended Kalman filter for state and parameter estimation. Simulation studies include applications to continuous stirred tank reactors by Seinfeld et al. (1969), Seinfeld (1970), and Wells (1971); to tubular and packed bed reactors by Gavalas and Seinfeld (1969), Joffe and Sargent (1972), McGreavy and Vago (1972), and Vakil et al. (1972); to heat exchangers by Coggan and Noton (1970), and Coggan and Wilson (1971b); and to a basic oxygen furnace by Wells (1970). In several of these studies, such as those by Seinfeld (1970) and Wells and Larson (1970) the filters were implemented as part of a feedback control scheme and resulted in significantly better control. The feasibility of implementing Kalman filters on small process control computers has been demonstrated by Coggan and Wilson (1971a).

Goldmann and Sargent (1971) have recently reported a detailed study of the factors affecting the performance of the Kalman filter for two simulated chemical processes. The authors considered only measurement noise in their simulations but investigated the sensitivity of the technique to errors in the design matrices, plant modeling errors, and autocorrelated measurement noise.

Only a few industrial applications of Kalman filtering have been reported in the field of process control. Åström (1970), Sastry and Vetter (1969), and Sastry et al. (1969) were concerned with applications to papermaking while Noton et al. (1968, 1970) reported the use of an extended Kalman filter in parameter and state estimation for an industrial multireactor system. Wells and Wismer (1971) have also used an extended Kalman filter to estimate the carbon content in a basic oxygen furnace.

The purpose of this investigation is to evaluate the effectiveness of a Kalman filter in a multivariable computer control scheme for a pilot scale evaporator. Of particular interest are several factors which affect the performance of the Kalman filter including poor estimates of the noise covariance matrices, unmeasured process disturbances, poor initial state estimates, and errors in the model parameters. Both simulated and experimental results are presented.

## THEORY

The mathematical formulation of both the optimal control problem and the optimal estimation problem is well known and has been presented in texts and publications such as those by Bryson and Ho (1969) and Athans (1971). General reviews of filtering theory have been presented by Bucy (1970) and Rhodes (1971); of least squares theory by Swerling (1971); and of the Linear-Quadratic-Gaussian Problem by Mendel and Gieseking (1971). The following is a summary of the derivation of the discrete, stationary, standard Kalman filter used in this work.

Consider the discrete, linear, deterministic time-invariant process model in Equations (1) and (2):

$$\mathbf{x}[(k+1)T] = \mathbf{\Phi}(T)\ \mathbf{x}(kT) + \mathbf{\Delta}(T)\ \mathbf{u}(kT) + \mathbf{\Theta}(T)\ \mathbf{d}(kT) \quad (1)$$

$$\mathbf{y}(kT) = \mathbf{H}\ \mathbf{x}(kT) \quad (2)$$

In the optimal control problem, the design objective is to determine the control policy $\mathbf{u}(kT)$, $k = 0, 1, 2 \ldots N$, which minimizes a performance index such as the widely used quadratic performance index:

$$J = \mathbf{x}^T(N)\ \mathbf{S}\ \mathbf{x}(N) + \sum_{k=1}^{N-1} [\mathbf{x}^T(k)\ \mathbf{Q}_1\ \mathbf{x}(k) + \mathbf{u}^T(k-1)\ \mathbf{R}_1\ \mathbf{u}(k-1)] \quad (3)$$

where $N$, $\mathbf{Q}_1$, $\mathbf{R}_1$ and $\mathbf{S}$ are design parameters which must be specified a priori and $\mathbf{x}(k)$ is used to denote $\mathbf{x}(kT)$ etc. If the optimal control policy which minimizes this performance index is denoted by $\mathbf{u}^\circ(k)$, the optimal control law is given by

$$\mathbf{u}^\circ(k) = \mathbf{K}_{FB}(k)\ \mathbf{x}(k) \quad (4)$$

where the time-varying feedback control matrix $\mathbf{K}_{FB}(k)$ can be obtained from the solution of a matrix Ricatti equation (Bryson and Ho, 1969). However, an important simplification occurs if control over an infinite period of time is assumed [that is, $N \to \infty$ in Equation (3)]. In this special case, the controller matrix $\mathbf{K}_{FB}(k)$ in Equation (4) becomes a constant matrix and the optimal control law is

$$\mathbf{u}^\circ(k) = \mathbf{K}_{FB}\ \mathbf{x}(k) \quad (5)$$

The control law in Equation (5) is attractive for on-line computer control calculations since only a single, time invariant matrix, $\mathbf{K}_{FB}$ must be stored rather than a sequence of $N$ time-varying matrices as in Equation (4).

In order to implement the optimal control laws defined by Equations (4) and (5), current values of all $n$ state variables are required. If it is not feasible to measure all state variables (the usual case), then some type of state estimation technique is necessary.

In the optimal estimation problem, the objective is to calculate state estimates $\hat{\mathbf{x}}(k)$ from noisy measurements of the input and output variables such that the estimates minimize a specified performance index. It can be formulated as follows. As the stochastic process of interest, consider the deterministic process model of Equations (1) and (2) with the inclusion of random process noise $\mathbf{w}(k)$ and measurement noise $\mathbf{v}(k)$, such that

$$\mathbf{x}(k+1) = \mathbf{\Phi}\ \mathbf{x}(k) + \mathbf{\Delta}\ \mathbf{u}(k) + \mathbf{\Theta}\ \mathbf{d}(k) + \mathbf{\Gamma}\ \mathbf{w}(k) \quad (6)$$

$$\mathbf{y}(k) = \mathbf{H}\ \mathbf{x}(k) + \mathbf{v}(k) \quad (7)$$

If it is assumed that $\mathbf{v}(k)$ and $\mathbf{w}(k)$ are zero-mean, uncorrelated white-noise sequences, then the covariance matrices satisfy

$$\mathrm{cov}\ [\mathbf{w}(k), \mathbf{w}(j)] = E[\mathbf{w}(k)\ \mathbf{w}^T(j)] = \mathbf{Q}(k)\ \delta_{kj} \quad (8)$$

$$\mathrm{cov}\ [\mathbf{v}(k), \mathbf{v}(j)] = E[\mathbf{v}(k)\ \mathbf{v}^T(j)] = \mathbf{R}(k)\ \delta_{kj} \quad (9)$$

$$\mathrm{cov}\ [\mathbf{w}(k), \mathbf{v}(j)] = \mathrm{cov}\ [\mathbf{v}(k), \mathbf{w}(j)] = \mathbf{0} \quad (10)$$

where $\delta_{kj}$ is the Kronecker delta and $\mathbf{Q}(k)$ and $\mathbf{R}(k)$ are the covariance matrices for the process noise and measurement noise, respectively.

The performance specifications for a suitable state estimation algorithm (filter) include:

1. That it produce a sequential estimate of the state $\mathbf{x}(k)$ which is linear in the measurement $\mathbf{y}(k)$ and is up-

dated as each new set of measurements is obtained, and

2. The filter output $\hat{\mathbf{x}}(k)$ be a minimal variance estimate in the sense it minimizes the following performance index:

$$J = E\{[\mathbf{x}(k) - \hat{\mathbf{x}}(k)]^T [\mathbf{x}(k) - \hat{\mathbf{x}}(k)]\} \quad (11)$$

The solution to this optimal estimation problem is the Kalman filter:

$$\hat{\mathbf{x}}(k) = \bar{\mathbf{x}}(k) + \mathbf{K}(k)[\mathbf{y}(k) - \mathbf{H}\bar{\mathbf{x}}(k)] \quad (12)$$

where $\bar{\mathbf{x}}(k)$ is calculated from the deterministic model using the state estimate plus the measured process inputs from the previous time interval, that is,

$$\bar{\mathbf{x}}(k) = \mathbf{\Phi}\,\hat{\mathbf{x}}(k-1) + \mathbf{\Delta}\,\mathbf{u}(k-1) + \mathbf{\Theta}\,\mathbf{d}(k-1) \quad (13)$$

Although different in notation and form, Equations (12) and (13) are equivalent to those presented by Sage and Melsa (1971). Assuming that the process is statistically stationary (that is, $\mathbf{Q}$ and $\mathbf{R}$ are constant matrices) the gain matrix $\mathbf{K}(k)$ can be calculated from the following recursive relations:

$$\mathbf{K}(k) = \mathbf{P}(k)\,\mathbf{H}^T\mathbf{R}^{-1} = \mathbf{M}(k)\,\mathbf{H}^T\,(\mathbf{H}\,\mathbf{M}(k)\,\mathbf{H}^T + \mathbf{R})^{-1} \quad (14)$$

$$\mathbf{M}(k+1) = \mathbf{\Phi}\,\mathbf{P}(k)\,\mathbf{\Phi}^T + \mathbf{\Gamma}\,\mathbf{Q}\,\mathbf{\Gamma}^T \quad (15)$$

$$\mathbf{P}(k) = (\mathbf{I} - \mathbf{K}(k)\,\mathbf{H})\,\mathbf{M}(k) \quad (16)$$

and the assumed initial conditions:

$$\mathbf{M}(0) = \mathbf{P}(0) = E\{[\mathbf{x}(0) - \hat{\mathbf{x}}(0)]\,[\mathbf{x}(0) - \hat{\mathbf{x}}(0)]^T\} \quad (17)$$

Starting with the initial conditions in Equation (17), Equations (14) to (16) are evaluated in a sequential manner until $\mathbf{K}(k)$ converges to a constant value, $\mathbf{K}$. If it is assumed that the observation time $NT$ is long compared to the dominant time constants of the process, then the following stationary form of Equation (12) may be used:

$$\hat{\mathbf{x}}(k) = \bar{\mathbf{x}}(k) + \mathbf{K}\,[\mathbf{y}(k) - \mathbf{H}\bar{\mathbf{x}}(k)] \quad (18)$$

where $\mathbf{K}$ is the limiting solution of Equations (14) to (16) as $k \to \infty$.

There is a significant practical advantage in using the stationary version of the Kalman filter since only one matrix $\mathbf{K}$ need be stored instead of a large set of matrices $\{\mathbf{K}(k)\}$. A further consequence of using the stationary form of the filter is that the covariance of the error in the initial state estimate $\mathbf{P}(0)$ has no effect on the gain matrix $\mathbf{K}$.

Finally, it should be noted that according to the Separation Theorem (Bryson and Ho, 1969), the optimal control policy for the stochastic system in Equations (6) and (7) consists of the optimal control law for the deterministic system [Equation (4) or (5)] with $\mathbf{x}(k)$ replaced by the optimal estimate $\hat{\mathbf{x}}(k)$ from the Kalman filter. That is,

$$\mathbf{u}^*(k) = \mathbf{K}_{FB}\,\hat{\mathbf{x}}(k) \quad (19)$$

The Separation Theorem is valid if both the process and measurement noise are Gaussian, an assumption which has not been necessary up to this point.

## DESCRIPTION OF PROCESS AND PROCESS MODEL*

The pilot plant evaporator used as the subject of all the simulated and experimental investigations reported in this paper is a double-effect unit and is normally operated with a feed rate of approximately 2.27 kg/min of 3% aqueous triethylene glycol. The primary control objective is to maintain the product concentration $C2$ at a constant value of approximately 10% in spite of disturbances in the feed flow rate $F$, the feed concentration $CF$, and/or the feed temperature $TF$. The liquid holdups $W1$ and $W2$ must also be maintained within acceptable operating limits. The primary control, or manipulated variables, are the inlet steam flow S, the bottoms flow from the first effect $B1$, and the product flow from the second effect $B2$. The unit is heavily instrumented and can be controlled either by conventional electronic instruments or by an IBM1800 digital computer. When single variable controllers are used the normal configuration is to control $C2$ by manipulating S, $W1$ by $B1$ and $W2$ by $B2$. In state space terminology the control problem is to maintain the state vector $\mathbf{x}$ (or alternatively the output vector $\mathbf{y}$) equal to the desired value in spite of disturbances $\mathbf{d}$ by manipulating the control vector $\mathbf{u}$.

A fifth-order state space model of the evaporator was derived by Newell and Fisher (1972a) based on linearized material and energy balances. The discrete model equations are of the form given in Equations (1) and (2) and are expressed in terms of normalized perturbation variables (see Appendix). The definitions of the process variables and their normal steady state values are in the Notation. Of the five state variables in this model, only the first effect concentration $C1$ is not measured.

The process noise vector $\mathbf{w}$ in Equation (6) was assumed to consist of six elements corresponding to the three disturbances $\mathbf{d}$ and the three manipulated variables $\mathbf{u}$. For the simulation studies described in the next section, each element of $\mathbf{w}$ and $\mathbf{v}$ in Equations (6) and (7) was assumed to be a Gaussian noise sequence with zero mean and a standard deviation of 0.1. Thus relatively high noise levels were considered in this investigation (compare Figure 1).

## SIMULATION STUDY

The effect of the following factors on the performance of the Kalman filter was determined by digital simulation:

1. Different noise covariance matrices, $\mathbf{Q}$ and $\mathbf{R}$.
2. Unmeasured process disturbances.
3. Incorrect estimates of the initial state.
4. Errors in model parameters.

In both the simulation and experimental studies, diagonal $\mathbf{Q}$ and $\mathbf{R}$ matrices were assumed with each matrix having equal diagonal elements. That is,

$$\mathbf{Q} = q\,\mathbf{I}_s \quad \text{and} \quad \mathbf{R} = r\,\mathbf{I}_l \quad (20)$$

The physical interpretation of the $\mathbf{Q}$ and $\mathbf{R}$ matrices in Equation (20) is that the noise levels of the individual signals are identical and statistically independent. Once the ratio $r/q$ is specified, the Kalman filter gain matrix $\mathbf{K}$ is uniquely determined by the asymptotic solution of Equations (14) to (16). Hence in the following discussion, this ratio rather than the absolute values of $q$ and $r$ will be cited. Gaussian noise sequences with standard deviations of 0.1 were used in all simulation runs, and hence the theoretically correct value of $r/q$ was 1.0.

The effectiveness of the Kalman filter in providing estimates of the entire state vector from noisy measurements

* A schematic diagram of the evaporator is available in Newell et al. (1972 a, b, c).

is illustrated by Figure 1. The simulated responses are from the deterministic model defined by Equation (1) and the parameter matrices in the Appendix. Except for $C2$, which was 30% low, the initial states used for both the simulation runs and the Kalman filter derivation were the normal steady state operating conditions defined in the Notation section and by arrow heads on the vertical axis. Both process and measurement noise were added but the control and disturbance vectors were identically zero. The actual values of the state variables (which include the effect of the process noise $\mathbf{w}$) are shown by the solid curves in Figure 1. The state estimates from the Kalman filter (short dashes) are very close to the actual states and represent a considerable improvement over the unfiltered data. This improvement was expected since the simulation was based on the same model used in designing the Kalman filter (that is, no modeling errors) and the noise covariance matrices were set equal to the theoretically correct values (that is, $r/q = 1$). The gain matrix for this filter is also included in the Appendix.

**Effect of Using Incorrect Values for the Noise Covariance Matrices**

Given accurate information concerning noise statistics, the noise covariance matrices $\mathbf{Q}$ and $\mathbf{R}$ can be set to their

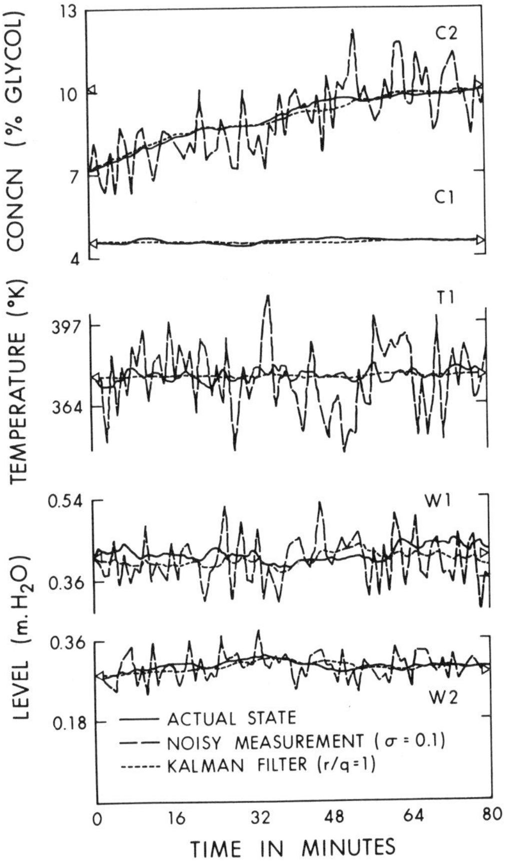


**Fig. 1. Simulated open-loop responses for all the state variables of the evaporator model starting with both the actual and estimated values of the product concentration 30% below the steady state value.**

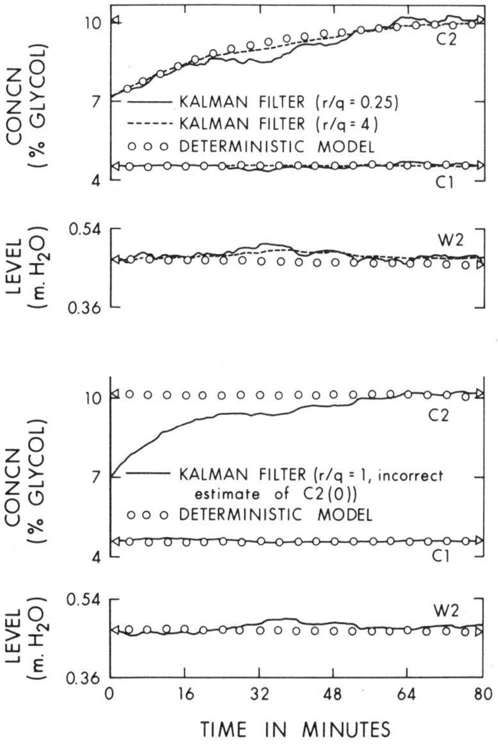


**Fig. 2. Simulated open-loop evaporator responses. Data in the top half shows that the filter output becomes smoother and closer to that of the deterministic model as $r/q$ increases. Data in the bottom half shows that errors in the filter estimate caused by incorrect estimates of the initial state are eliminated during a single transient response.**

theoretically optimal values. However, accurate statistical information is rarely available in practice and it is useful to know to what extent incorrect values of $\mathbf{R}$ and $\mathbf{Q}$ affect filter performance. The consequences of using an incorrect value of $r/q$ in designing a Kalman filter for the evaporator can be seen by comparing the filter estimates plotted in Figures 1 and the top half of Figure 2. As $r/q$ changes from 0.25 to 4.0, the estimated response becomes closer to the deterministic model response. This follows since a large value of $r/q$ implies relatively large measurement noise levels and small process noise levels; consequently, the resulting filter gain matrix $\mathbf{K}$ will have small elements and the estimated response will be close to the deterministic model response as is evident from Equation (18).

**Effect of Poor Initial State Estimates**

The lower half of Figure 2 shows the effect of an inaccurate initial state estimate on the performance of the Kalman filter. The deterministic model response was started at the correct initial state and since $\mathbf{u} = \mathbf{d} = \mathbf{0}$, the model responses remained constant. The Kalman filter provides satisfactory state estimates of the four state variables that were initialized to the correct values, and the $C2$ estimate gradually approaches the true value from an initial estimate of $C2$ which was 30% below the true value. Thus the filter estimates are reasonable after the transient due to the poor initial state estimate.

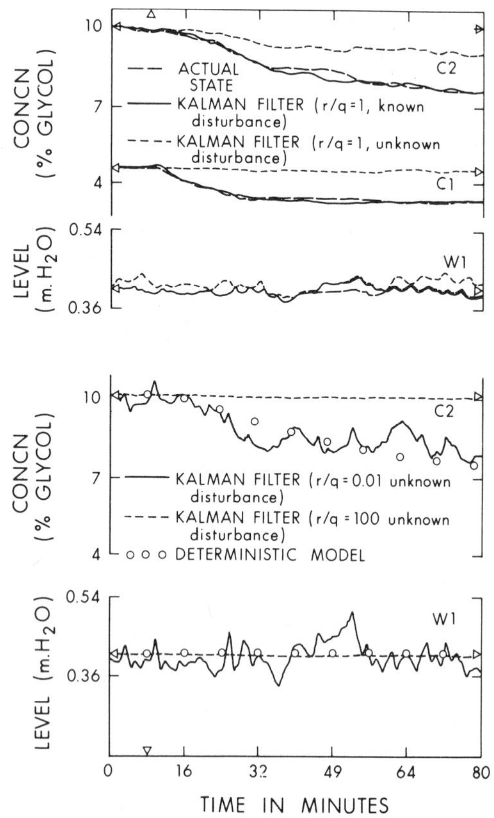


**Fig. 3. Simulated open-loop evaporator responses to a −30% step change in feed composition. The data in the top half show that unknown (unmeasured) disturbances produce significant errors in the filter estimates. The data in the bottom half show that this error can be reduced by decreasing the $r/q$ ratio (but the output is noisier).**

#### Effect of an Unmeasured Disturbance

Measured disturbances do not normally cause any difficulty because they can be introduced into the Kalman filter as part of the disturbance vector **d** in Equation (13). Unmeasured disturbances affect the measured process outputs y but have no effect on the state $\bar{x}$, calculated from Equation (13). If **K H** in Equation (18) is equal to the identity matrix **I**, then this causes no problems. For the other extreme where $\mathbf{K} = \mathbf{0}$ the filter outputs would never reflect the influence of the unmeasured disturbance. In practical applications the errors caused by unmeasured disturbances will lie between these two limits and will be determined by the magnitude of the elements of **K**, that is, by the choice of **Q** and **R**.

The effect of an unmeasured 30% step change in feed composition is shown in Figure 3. In the top set of curves in Figure 3, the theoretically appropriate value of $r/q = 1$ is used in the filter design. The state estimates are satisfactory when the step change in feed composition is known but are significantly in error when the filter is unaware of the disturbance. The bottom set of curves in Figure 3 indicates the effect of the same disturbance when different values of $r/q$ are used in designing the Kalman filter. For $r/q = 100$, the state estimates are very poor due to a **K** matrix with large elements. By contrast when $r/q = 0.01$, the elements of **K** are relatively small and although the state estimates are somewhat noisy, they correctly reflect trends in the state variables.

These results indicate that it may be advantageous to make the elements of **Q** artificially large (that is, $r/q$ small) to accommodate unmeasured disturbances (and/or process nonlinearities or modeling errors).

An alternative strategy for alleviating the effects of unmeasured disturbances is to estimate the disurbances online. This can be achieved by augmenting the state vector with the disturbance variables and designing a Kalman filter for the augmented system (Sage and Melsa, 1971). This approach, which can also be used to deal with measurement bias and drift, is the subject of a current study by the authors.

#### Effect of Errors in Model Parameters

A series of simulated runs were also made to determine the effect of model accuracy on the performance of the Kalman filter. This was done by arbitrarily specifying values for parameters (such as the holdups, $W1$ and $W2$) that were 25% above or below the true values and recalculating the coefficient matrices in the state space form of the model [Equations (1) and (2)]. These new (erroneous) models were then used as a basis for calculating new filter designs, that is, new values of **K** for Equation (18). Simulated responses were then obtained by applying these different Kalman filter designs to the same (accurate) process model. In general it was found that changes of $\pm$ 25% in $W1_{ss}$ and $W2_{ss}$ did not produce any noticeable changes in the filter estimates. Hence it was concluded that, in this application, the filter performance was not sensitive to changes in these model parameters.

#### Closed-Loop System

A series of closed-loop simulation runs was performed in order to evaluate the effectiveness of Kalman filters in multivariable feedback control systems. The state estimates produced by the Kalman filter were used in the multivariable control law defined by Equation (19). The optimal feedback matrix $\mathbf{K}_{FB}$ used in this work is given by Equation (21).

$$\mathbf{K}_{FB} = \begin{bmatrix} 10.78 & -1.61 & -4.82 & 0 & -19.57 \\ 5.35 & 0.36 & 0.55 & 0 & 12.49 \\ 7.52 & 1.27 & 0.18 & 24.61 & 32.69 \end{bmatrix} \qquad (21)$$

The performance of the Kalman filter in this optimal control scheme when process and measurement noise occur is illustrated in Figure 4. When the control calculations are based on unfiltered noisy measurements ($\sigma = 0.1$), the closed-loop response to a 20% step change in feed flow rate contains unacceptable oscillations (compare solid curves at top). By contrast, as shown by the curves in the center of Figure 4, when the Kalman filter is designed using the correct values of the noise statistics and when the disturbance is measured, then the system response is much smoother and closer to the deterministic model response. If the feed flow disturbance is not measured and the same filter is used, satisfactory control of $C2$ results but $W1$ drifts badly (data not plotted). Fortunately, the effects of unmeasured disturbances on the filter estimates can be alleviated as discussed previously by choosing smaller values of $r/q$. In Figures 4 to 7 the arrow heads

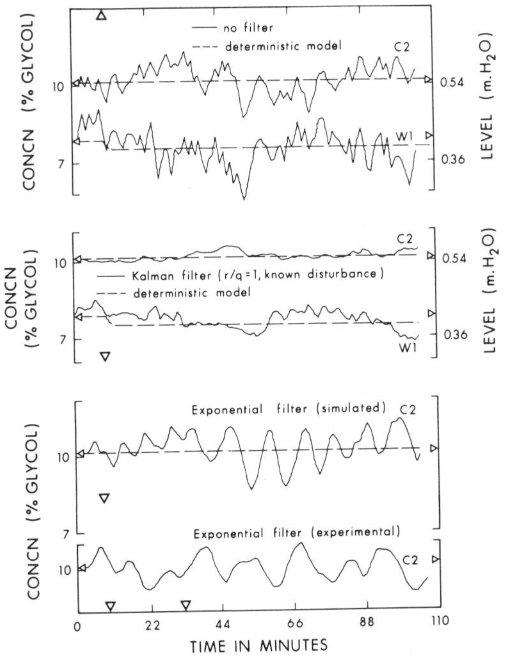


**Fig. 4. Simulated closed-loop evaporator responses to a —20% step in feed flow showing that the control system incorporating a Kalman filter (centre) is better than when no filter (top) or exponential filters (bottom) are used. (The bottom curve is an experimental response to a —20% followed by a +20% disturbance.)**

on the horizontal axis denote the start of a step disturbance.

## EXPERIMENTAL STUDY

The conclusions derived from the simulation studies were confirmed by experimental studies carried out on the pilot scale, double effect evaporator. Since the actual measurement and process noise levels in the evaporator are relatively low ($\sigma = 0.01$ and $0.02$ respectively), zero-mean Gaussian noise with $\sigma = 0.1$ was added to the output measurements in order to provide a more severe test of the Kalman filter. The filter for the experimental study was designed using a value of $r/q = 25$ which corresponds to assumed process and measurement noise levels of $\sigma =$ 0.02 and 0.10, respectively. The gain matrix for this Kalman filter is included in the Appendix. Details concerning the implementation of multivariable computer control techniques to the evaporator have been presented elsewhere (Newell et al., 1972b, c; Hamilton, 1972).

Experimental open-loop response data for a $-30\%$ step change in feed composition are shown, along with the corresponding filter estimates, in Figure 5. It should be noted that the experimental responses plotted in Figures 5 to 7 do not include the artificial Gaussian noise that was added to the measurements prior to the estimation and control calculations. When the Kalman filter is aware of the step disturbance, excellent state estimates result, as illustrated in the top half of Figure 5. However, when the disturbance is not measured a Kalman filter designed using $r/q = 25$ gives very poor estimates of the actual states. The estimates are improved, but noisier, when smaller $r/q$ values are used as illustrated in the bottom half of Figure 5.

Experimental closed-loop responses with and without artificially added measurement noise are shown in Figure 6 for optimal feedback control using the gain matrix of Equation (21). In the noise-free run (bottom half), noise

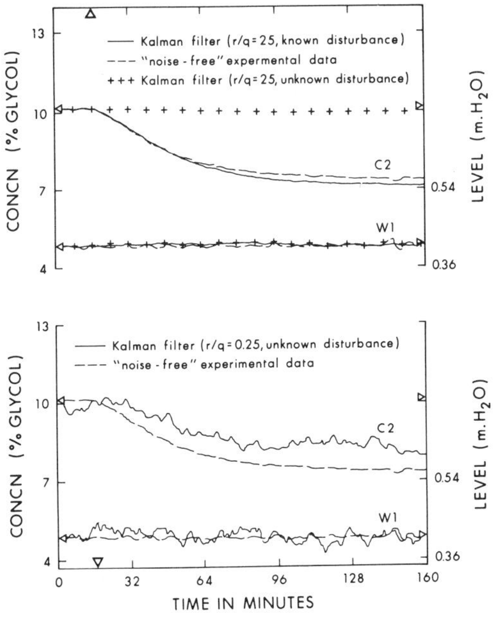


**Fig. 5. Experimental open-loop responses to a —30% step change in feed composition which show the detrimental effect of unknown disturbances and confirm the effect of varying r/q observed in the simulated data of Figure 3.**

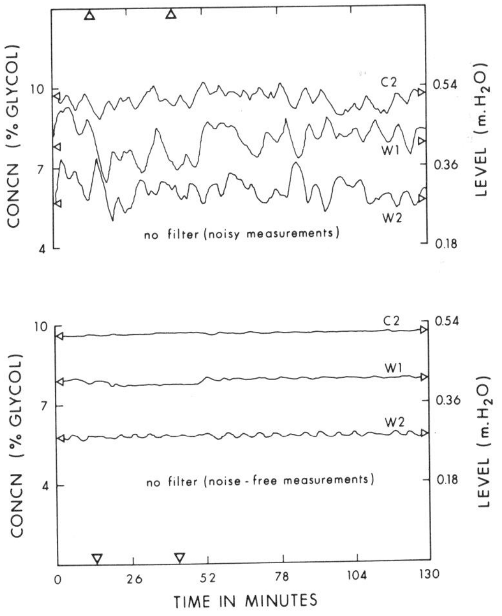


**Fig. 6. Experimental closed-loop evaporator responses to 20% step changes in feed flow demonstrating the detrimental effects of increased measurement noise on the multivariable computer control-schemes (compare Figure 4).**

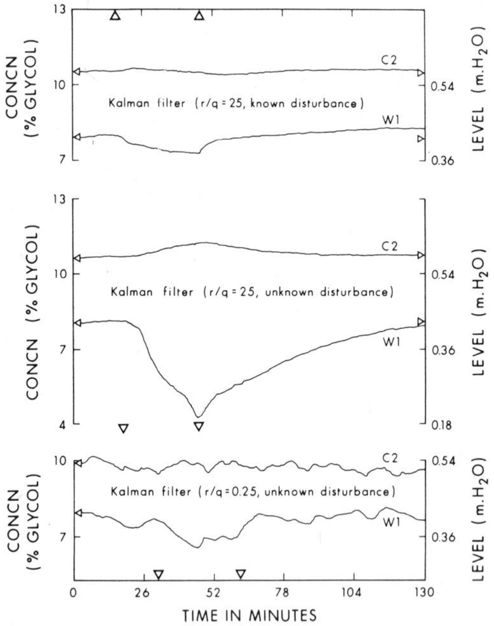


**Fig. 7. Experimental closed-loop evaporator responses to 20% step changes in feed flow showing the detrimental effect of unmeasured disturbances plus the effect of varying the *r/q* ratio (compare Figures 5 and 3).**

was not added to the measurements and excellent control was maintained in spite of two 20% feed flow disturbances. However, when unfiltered noisy measurements are used, the evaporator response exhibits excessive oscillations and is unsatisfactory as shown by the data in the top half of Figure 6. The control variables $u_i$ are not plotted but were very noisy as would be expected.

If a Kalman filter is used, the detrimental effects of noise can be greatly reduced as illustrated in the top part of Figure 7. Here the $C2$ and $W1$ responses represent a significant improvement over the noisy responses in Figure 6 where no filter was used. However, if the filter design is based on a $r/q$ value of 25 (which is the theoretically correct value) an unmeasured feed flow disturbance results in poor control (compare middle portion of Figure 7). A filter designed using a smaller value of $r/q$ is less seriously affected by unmeasured disturbances. For example, as illustrated in the bottom part of Figure 7, an $r/q$ ratio of 0.25 produces results that are a reasonable compromise between the oscillations in Figure 6 (top) and the sensitivity to disturbances in Figure 7 (center).

Thus the experimental results support the conclusions of the simulation studies concerning the effectiveness of the Kalman filter and demonstrate the utility of treating the **R** and **Q** matrices as design parameters that can be specified by the designer to tailor the process response to his specifications.

## EXPONENTIAL FILTER

For comparison with the Kalman filter, simulation and experimental studies were also carried out using an exponential filter of the type commonly available in commercial Direct Digital Control (DDC) programs.

An exponential or RC filter reduces noise in a set of measurements by combining the present measurement **y** and the previous estimate $\hat{\mathbf{y}}$ in a fixed proportion specified by the user. The scalar filter equation is

$$\hat{y}_i(k+1) = \hat{y}_i(k) + \alpha_i[y_i(k+1) - \hat{y}_i(k)] \quad (22)$$

Normally each filter constant $\alpha_i$ is assigned a value between zero and one since a value of one corresponds to no filtering and a value of zero corresponds to total filtering in which the measurements are not used. In selecting a filter constant for the exponential filter, a trade-off is involved since increasing the value of $\alpha_i$ increases the degree of filtering but also increases the dynamic lag introduced by the filter. It is important to realize that the exponential filter merely provides signal conditioning and does not furnish any information about unmeasured state variables.

The experimental and simulation results in Figure 4 provide a comparison of the closed-loop control resulting from the use of either a Kalman filter or an exponential filter in a multivariable control scheme. It is apparent that the Kalman filter provides better control of the primary controlled variable $C2$.

## ACKNOWLEDGMENT

Financial support from the National Research Council of Canada and the assistance of the staff in the department's Data Acquisition, Control and Simulation (DACS) Centre is gratefully acknowledged.

## NOTATION

$\mathbf{d}$ = disturbance vector, $p \times 1$
$E$ = expected value
$\mathbf{H}$ = constant coefficient matrix
$\mathbf{I}_l$ = identity matrix, $l \times l$
$\mathbf{I}_s$ = identity matrix, $s \times s$
$J$ = performance index for control problem
$j$ = time increment counter
$\mathbf{K}$ = gain matrix for Kalman filter
$\mathbf{K}_{FB}$ = feedback control matrix
$k$ = time increment counter
$l$ = dimension of output vector
$\mathbf{M}$ = variance matrix
$m$ = dimension of control vector
$N$ = indicator of final time period
$n$ = dimension of state vector
$\mathbf{P}$ = error covariance matrix
$p$ = dimension of disturbance vector
$\mathbf{Q}$ = process noise weighting matrix
$\mathbf{Q}_1$ = state weighting matrix in control problem
$q$ = process noise covariance, see Equation (20)
$\mathbf{R}$ = measurement noise weighting matrix
$\mathbf{R}_1$ = control weighting matrix in control problem
$r$ = measurement noise covariance, see Equation (20)
$\mathbf{S}$ = final state weighting matrix in control problem
$s$ = dimension of process noise vector
$T$ = discrete time interval
$\mathbf{u}$ = control vector, $m \times 1$
$\mathbf{v}$ = measurement noise vector, $l \times 1$
$\mathbf{w}$ = process noise vector, $s \times 1$
$\mathbf{x}$ = state vector, $n \times 1$
$\mathbf{y}$ = output measurement vector, $l \times 1$
cov = covariance

### Greek Letters

$\alpha$ = coefficient scalar in the exponential filter equation
$\mathbf{\Gamma}$ = coefficient matrix of the process noise

$\Delta$ = control coefficient matrix
$\Theta$ = disturbance coefficient matrix
$\sigma$ = standard deviation
$\Phi$ = transition matrix

**Superscripts**

$*$ = indicates optimal value of variable
$T$ = matrix (vector) transpose
$-1$ = matrix inverse
$-$ = indicates model calculation
$\wedge$ = indicates estimated variable

**Process Variables**

State vector, $\mathbf{x}$:
$W1$ = first-effect holdup, 20.8 kg
$C1$ = first-effect concentration, 4.59% glycol
$H1$ = first-effect enthalpy, 441 kJ/kg
$W2$ = second-effect holdup, 19.0 kg
$C2$ = second-effect concentration, 10.11% glycol
Control vector, $\mathbf{u}$:
$S$ = steam, 0.91 kg/min
$B1$ = first-effect bottoms, 1.58 kg/min
$B2$ = second-effect bottoms, 0.72 kg/min
Disturbance vector, $\mathbf{d}$:
$F$ = feed flow rate, 2.26 kg/min
$CF$ = feed concentration, 3.2% glycol
$HF$ = feed enthalpy, 365 kJ/kg
Output vector, $\mathbf{y}$:
$\mathbf{y}^T$ = $[W1, H1, W2, C2]$

## LITERATURE CITED


Åström, K. J., *Introduction to Stochastic Control Theory,* Academic Press, New York (1970).

Athans, M., "The Role and Use of the Stochastic Linear-Quadratic-Gaussian Problem in Control System Design," *IEEE Trans. Auto. Control, AC-16,* 529 (1971).

Bryson, A. E., Jr., and Y. C. Hö, *Applied Optimal Control,* Blaisdell, Waltham, Mass. (1969).

Bucy, R. S., "Linear and Nonlinear Filtering," *Proc. IEEE,* **58,** 854 (1970).

Choquette, P., A. R. M. Noton, and C. A. G. Watson, "Remote Computer Control of an Industrial Process," *ibid.,* 10.

Coggan, G. C., and A. R. M. Noton, "Discrete-Time Sequential State and Parameter Estimation in Chemical Engineering," *Trans. Instn. Chem. Engr.,* **48,** T255 (1970).

Coggan, G. C., and J. A. Wilson, "On-line State Estimation with a Small Computer," *Computer J.,* **14,** 61 (1971).

———, "Approximate Models and the Elimination of Bias in On-Line Estimation of Industrial Processes," *Proc. Joint Symp. on Online Computer Methods Relevant to Chem. Eng.,* Nottingham, British Computer Society, London (1971).

Gavalas, G. R., and J. H. Seinfeld, "Sequential Estimation of States and Kinetic Parameters in Tubular Reactors with Catalyst Decay," *Chem. Eng. Sci.,* **24,** 625 (1969).

Goldmann, S. E., and R. W. H. Sargent, "Applications of Linear Estimation Theory to Chemical Processes: A Feasibility Study," *Chem. Eng. Sci.,* **26,** 1535 (1971).

Hamilton, J. C., "An Experimental Investigation of State Estimation in Multivariable Control Systems," M.Sc. thesis, Univ. Alberta, Edmonton, Canada (1972).

Joffe, B. L., and R. W. H. Sargent, "The Design of On-line Control Scheme for a Tubular Reactor," *Trans. Inst. Chem. Engrs.,* **50,** T270 (1972).

Kalman, R. E., "A New Approach to Linear Filtering and Prediction Problems," *Trans. ASME, J. Basic Eng.,* **82,** 35 (1960).

———., and R. S. Bucy, "New Results in Linear Filtering and Prediction Theory," **83,** 95 (1961).

McGreavy, C., and A. Vago, "Application of Nonlinear Filtering Techniques to Adaptive Optimal Control," paper presented at the 65th Ann. meeting of Am. Inst. Chem. Engrs., New York (1972).

Mendel, J. M., and D. L. Gieseking, "Bibliography on the Linear-Quadratic-Gaussian Problem," *IEEE Trans. Auto. Control,* **AC-16,** 847 (1971).

Newell, R. B., and D. G. Fisher, "Model Development, Reduction and Experimental Evaluation for an Evaporator," *Ind. Eng. Chem. Process Design Develop.,* **11,** 213 (1972a).

———, "Experimental Evaluation of Optimal, Multivariable Regulatory Controllers with Model-Following Capabilities," *Automatica,* **8,** 247 (1972b).

———, and D. E. Seborg, "Computer Control Using Optimal, Multivariable, Feedforward-Feedback Algorithms", *AIChE J.,* **18,** 881 (1972c).

Noton, A. R. M., and P. Choquette, "The Application of Modern Control Theory to the Operation of an Industrial Process," *Proc. IFAC/IFIP Symp.,* Toronto, Ontario (1968).

Rhodes, I. B., "A Tutorial Introduction to Estimation and Filtering," *Proc. 1969 Nat. Conf. Auto. Control,* Edmonton,

Sage, A. P., and J. L. Melsa, *Estimation Theory with Applications to Communications and Control,* McGraw-Hill, N. Y. (1971).

———, and R. Caston, "The Application of Identification Methods to an Industrial Process," *1969 JACC Preprints,* p. 787, Boulder, Colorado (1969).

Sastry, V. A. and W. J. Vetter, "A Papermaking Wet-end Dynamics Model and Parameter Identification by Iterative Filtering," *Proc. 1969 Nat. Conf. Auto. Control,* Edmonton, Canada (1969).

Seinfeld, J. H., "Optimal Stochastic Control of Nonlinear Systems," *AIChE J.,* **16,** 1016 (1970).

———, G. R. Gavalas, and M. Hwang, "Control of Nonlinear Stochastic Systems," *Ind. Eng. Chem. Fundamentals,* **8,** 257 (1969).

Swerling, P., "Modern State Estimation Methods from the Viewpoint of the Method of Least Squares," *IEEE Trans. Auto. Control,* **AC-16,** 707 (1971).

Vakil, H., M. L. Michelsen, and A. S. Foss, "Fixed Bed Reactor Control with State Estimation," Paper presented at 65th Annual AIChE Meeting, New York (1972).

Wells, C. H., "Application of Modern Estimation and Identification Techniques to Chemical Processes," *AIChE J.,* **17,** 966 (1971).

———., "Optimum Estimation of Carbon and Temperature in a Simulated BOF," *1970 JACC Preprints,* p. 7, Atlanta, Georgia (1970).

———., and R. E. Larson, "Application of Combined Optimum Control and Estimation Theory to DDC," *Proc. IEEE,* **58,** 16 (1970).

Wells, C. H., and D. A. Wismer, "Advances in Process Control Applications," in *Advances in Control Systems,* Vol. 8, C. T. Leondes (ed.), Academic Press, N. Y. (1971).


## APPENDIX

### State Space Evaporator Model

The model is of the form of Equations (1) and (2) with the elements of **x, u, d** and **y** defined as normalized perturbation variables, for example,

$$x_1 = \frac{W1 - W1_{ss}}{W1_{ss}}$$

where $W1_{ss}$ is the normal steady state of $W1$. Previous evaporator studies by Newell and Fisher (1972b) have shown that a discrete time interval of $T = 64$ seconds gives satisfactory control. The coefficient matrices for $T = 64$ seconds are

$$\Phi = \begin{bmatrix} 1.0 & -0.0008 & -0.0912 & 0 & 0 \\ 0 & 0.9223 & 0.0871 & 0 & 0 \\ 0 & -0.0042 & 0.4377 & 0 & 0 \\ 0 & -0.0009 & -0.1052 & 1.0 & 0.0001 \\ 0 & 0.0391 & 0.1048 & 0 & 0.9603 \end{bmatrix}$$

$$\Delta = \begin{bmatrix} -0.0119 & -0.0817 & 0 \\ 0.0116 & 0 & 0 \\ 0.1569 & 0 & 0 \\ -0.0137 & 0.0847 & -0.0406 \\ 0.0137 & -0.0432 & 0 \end{bmatrix}$$

$$\boldsymbol{\Theta} = \begin{bmatrix} 0.1182 & 0 & -0.0050 \\ -0.0351 & 0.0785 & 0.0049 \\ -0.0135 & -0.0002 & 0.0662 \\ 0.0012 & 0 & -0.0058 \\ -0.0019 & 0.0016 & 0.0058 \end{bmatrix}$$

$$\mathbf{H} = \begin{bmatrix} 1 & 0 & 0 & 0 & 0 \\ 0 & 0 & 1 & 0 & 0 \\ 0 & 0 & 0 & 1 & 0 \\ 0 & 0 & 0 & 0 & 1 \end{bmatrix}$$

**Kalman Filter Gain Matrices**

for $r/q = 1$:

$$\mathbf{K} = \frac{1}{1000} \begin{bmatrix} 137. & -6.83 & -18.3 & -0.554 \\ -24.4 & 5.26 & -7.76 & 20.8 \\ -6.83 & 34.4 & -5.76 & 5.98 \\ -18.3 & -5.76 & 90.3 & -31.3 \\ -0.554 & 5.98 & -31.3 & 36.7 \end{bmatrix}$$

For $r/q = 25$:

$$\mathbf{K} = \frac{1}{1000} \begin{bmatrix} 29.0 & -0.339 & -1.67 & 0.422 \\ -2.17 & 0.248 & -0.639 & 1.41 \\ -0.339 & 1.44 & -0.284 & 0.284 \\ -1.67 & -0.284 & 20.4 & -2.48 \\ 0.422 & 0.284 & -2.48 & 3.87 \end{bmatrix}$$

*Manuscript received October 10, 1972; revision received March 20 and accepted April 5, 1973.*

# AN EXPERIMENTAL EVALUATION OF STATE ESTIMATION IN MULTIVARIABLE CONTROL SYSTEMS

by

Dale E. Seborg, D. Grant Fisher and James C. Hamilton
Department of Chemical Engineering
University of Alberta
Edmonton, Alberta, Canada

SUMMARY

This paper describes the application of two state estimation techniques, Kalman filters and Luenberger observers, to a computer-controlled, pilot plant evaporator. The estimation techniques are used to provide state estimates for an optimal feedback control system and are evaluated using both simulated and experimental response data.

## INTRODUCTION

Multivariable control systems designed using state-space techniques often require that all of the system state variables be available to the control system. However, in many practical problems it may not be economical or convenient to measure all of the state variables. An alternative approach is to use an estimation technique to provide estimates of the state variables which are not measured. Of the state estimation methods proposed in the literature, the Kalman filter[3,8,22] and the Luenberger observer[10-12] have received the most attention. Despite successful applications of the Kalman filter in the aerospace industry, there have been very few reported applications in the process industries[5,7]. Similarly, few if any experimental applications of Luenberger observers to process control problems have been reported.

In this paper Luenberger observers and Kalman filters are experimentally employed to provide state estimates to an optimal multivariable controller for a computer-controlled, pilot plant evaporator. Of particular interest was the performance of these techniques under non-ideal conditions including inaccurate initial conditions, unmeasured process disturbances and approximate process models.

This investigation is part of a continuing project at the University of Alberta whose concern is the design, evaluation and experimental application of modern multivariable control and estimation techniques. Previous studies have considered optimal regulators[15-17] and optimal servos[19,20], model reference adaptive control[21], reduced-order controllers[23] and time-delay compensation[1,2]. Eigenvalue assignment and frequency domain design techniques are the subjects of current investigations.

## FORMULATION

The mathematical formulation of both the Luenberger observer and the Kalman filter are well-known and consequently only a brief summary will be presented here. Consider the linear, stochastic, time-invariant, state-space model in equation (1):

$$\underline{x}(k+1) = \underline{\underline{\phi}}\,\underline{x}(k) + \underline{\underline{\Delta}}\,\underline{u}(k) + \underline{\underline{\theta}}\,\underline{d}(k) + \underline{\underline{\Gamma}}\,\underline{w}(k) \qquad (1)$$

$$\underline{y}(k) = \underline{\underline{C}}\,\underline{x}(k) + \underline{v}(k)$$

where $\underline{x}$ is the nx1 state vector, $\underline{u}$ is the mx1 control vector, $\underline{d}$ is the px1 disturbance vector, $\underline{y}$ is the ℓx1 output vector and $\underline{\underline{\phi}}$, $\underline{\underline{\Delta}}$, $\underline{\underline{\theta}}$, $\underline{\underline{\Gamma}}$ and $\underline{\underline{C}}$ are constant matrices. (Note: vectors are denoted by a single underline and matrices by a double underline). It is assumed that the measurement noise vector $\underline{v}$ and the s-dimensional process noise vector $\underline{w}$ are zero mean, uncorrelated and have covariance matrices of $\underline{\underline{Q}}$ and $\underline{\underline{R}}$, respectively. Suppose that the control law of interest is a proportional state feedback controller with a constant gain matrix, $\underline{\underline{K}}_{FB}$:

$$\underline{u}(k) = \underline{\underline{K}}_{FB}\,\underline{x}(k) \qquad (2)$$

Proc. of the 4th IFAC/IFIP Conference on Digital Computer Applications to Process Control, pp. 144-155, Zurich, March, 1974.

Newell and Fisher[16] have demonstrated that the multivariable controller in (2) can be easily extended to include integral, feedforward and setpoint control modes.

In order to implement the control law in (2), measurements of all n state variables are required, but in many practical control problems, only measurements of the $\ell$ output variables are available where $\ell<n$. One possible strategy is to use a state estimation technique to generate an estimate of the state vector, $\hat{\underline{x}}$, and then substitute the estimate in the control law of (2). This approach results in the following controller:

$$\underline{u}(k) = \underline{\underline{K}}_{FB}\hat{\underline{x}}(k) \tag{3}$$

Two of the most popular state estimation techniques, the Luenberger observer and the Kalman filter, are briefly summarized below.

Luenberger Observer

The Luenberger observer was originally proposed as a deterministic state estimation technique by Luenberger in 1964[10]. In the last decade the approach has been extended to a broad class of deterministic and stochastic systems as described recently by Luenberger[12] and Yuksel and Bongiorno[24]. Although observers have been the subject of numerous theoretical and simulation studies, relatively few experimental applications have been reported. The derivation below of a minimal-order observer for a discrete-time system closely follows the derivation of MacFarlane[13]. The deterministic version of (1) is employed in this derivation (i.e. $\underline{w}=\underline{o}$ and $\underline{v}=\underline{o}$).

Let $\underline{z}$ be a $(n-\ell)$-dimensional vector defined by

$$\underline{z} = \underline{\underline{T}}\underline{x} \tag{4}$$

If the estimate of $\underline{z}$ is denoted by $\hat{\underline{z}}$, then the associated error in the estimate, $\Delta\underline{z}$, is defined as

$$\Delta\underline{z} \equiv \hat{\underline{z}}-\underline{z} \tag{5}$$

The minimal-order observer is a linear, time-invariant dynamic system which generates $\hat{\underline{z}}$ while being driven by the plant outputs $\underline{y}$ and the plant inputs, $\underline{u}$ and $\underline{d}$. Suppose that the observer is given by,

$$\hat{\underline{z}}(k+1) = \underline{\underline{E}}\,\hat{\underline{z}}(k) + \underline{\underline{H}}\,\underline{y}(k) + \underline{\underline{F}}\,\underline{d}(k) + \underline{\underline{G}}\,\underline{u}(k) \tag{6}$$

where $\underline{\underline{E}}$, $\underline{\underline{H}}$, $\underline{\underline{F}}$ and $\underline{\underline{G}}$ are constant matrices. Combining (1) and (4)-(6) gives

$$\Delta\underline{z}(k+1) = \underline{\underline{E}}\,\Delta\underline{z}(k) + (\underline{\underline{E}}\,\underline{\underline{T}} + \underline{\underline{H}}\,\underline{\underline{C}} - \underline{\underline{T}}\,\underline{\underline{\phi}})\,\underline{x}(k) + (\underline{\underline{G}} - \underline{\underline{T}}\underline{\underline{\Delta}})\,\underline{u}(k) + (\underline{\underline{F}} - \underline{\underline{T}}\,\underline{\underline{\theta}})\,\underline{d}(k) \tag{7}$$

If matrices $\underline{\underline{E}}$, $\underline{\underline{H}}$, and $\underline{\underline{F}}$ and $\underline{\underline{G}}$ are selected so that

$$\underline{\underline{T}}\,\underline{\underline{\phi}} - \underline{\underline{E}}\,\underline{\underline{T}} = \underline{\underline{H}}\,\underline{\underline{C}} \tag{8}$$

$$\underline{\underline{F}} = \underline{\underline{T}}\,\underline{\underline{\theta}} \tag{9}$$

$$\underline{\underline{G}} = \underline{\underline{T}}\,\underline{\underline{\Delta}} \tag{10}$$

then (7) reduces to

$$\Delta\underline{z}(k+1) = \underline{\underline{E}}\Delta\underline{z}(k) \tag{11}$$

Equation (11) implies that if $\underline{\underline{E}}$ is a stable matrix, then the error in the estimate, $\Delta\underline{z}$, will asymptotically approach zero i.e. $\hat{\underline{z}} \to \underline{\underline{T}}\,\underline{x}$ as $k \to \infty$. Furthermore, if there is zero error in the initial estimate (i.e. $\hat{\underline{z}}(o) = \underline{\underline{T}}\,\underline{x}(o)$, then $\hat{\underline{z}}(k) = \underline{\underline{T}}\,\underline{x}(k)$ for all $k \geq o$. An $(n-\ell)$ - dimensional observer of the form of (6) will always exist providing that the $\ell$ output variables are linearly independent and the system in (1) is completely observable [12].

The observer can also be used to generate a state estimate, $\hat{\underline{x}}$. Rearrangement of (1) and (4) gives

$$\underline{x}(k) = \begin{bmatrix} \underline{\underline{T}} \\ \underline{\underline{C}} \end{bmatrix}^{-1} \begin{bmatrix} \underline{z}(k) \\ \underline{y}(k) \end{bmatrix} \tag{12}$$

or since in general, only an estimate of $\underline{z}$ is available,

$$\hat{\underline{x}}(k) = \begin{bmatrix} \underline{\underline{T}} \\ \underline{\underline{C}} \end{bmatrix}^{-1} \begin{bmatrix} \hat{\underline{z}}(k) \\ \underline{y}(k) \end{bmatrix} \tag{13}$$

It is instructive to consider the use of the observer in conjunction with a state feedback control system. If the state estimate is used in the control law of (3), then the resulting closed-loop system can be written as:

$$\begin{bmatrix} \underline{x}(k+1) \\ \Delta\underline{z}(k+1) \end{bmatrix} = \begin{bmatrix} \underline{\underline{\phi}} + \underline{\underline{\Delta}}\,\underline{\underline{K}}_{FB} & \underline{\underline{\Delta}}\,\underline{\underline{K}}_2 \\ \underline{\underline{0}} & \underline{\underline{E}} \end{bmatrix} \begin{bmatrix} \underline{x}(k) \\ \Delta\underline{z}(k) \end{bmatrix} \tag{14}$$

Equation (14) yields the important result that the eigenvalues of the composite system are those of the observer matrix $\underline{\underline{E}}$ plus those of the closed-loop system, $\underline{\underline{\phi}} + \underline{\underline{\Delta}}\,\underline{\underline{K}}_{FB}$, that would result if the state feedback control law in (2) were implemented directly. Thus the observer merely adds its eigenvalues to those of the state feedback control system[9,12,13].

## Kalman Filter

The Kalman filter[3,7,8] generates a state estimate $\hat{\underline{x}}$ which is the solution to the following optimal estimation problem. Given the stochastic system of (1), the known (or assumed) noise covariance matrices, $\underline{\underline{Q}}$ and $\underline{\underline{R}}$, and the initial state covariance matrix $\underline{\underline{P}}(o)$, generate an estimate which is linear with respect to the measured outputs $\underline{y}$ and which provides a minimal variance estimate. The resulting optimal estimate is generated from the Kalman filter:

$$\hat{\underline{x}}(k) = \bar{\underline{x}}(k) + \underline{\underline{K}}(k)\,[\underline{y}(k) - \underline{\underline{C}}\,\bar{\underline{x}}(k)] \tag{15}$$

where $\bar{\underline{x}}(k)$ is calculated from the deterministic model using the state estimate plus the measured process inputs from the previous time interval, i.e.

$$\bar{\underline{x}}(k) = \underline{\underline{\phi}}\,\hat{\underline{x}}(k-1) + \underline{\underline{\Delta}}\,\underline{u}(k-1) + \underline{\underline{\theta}}\,\underline{d}(k-1) \tag{16}$$

The filter gain matrix, $\underline{\underline{K}}(k)$, is generated from a recursive relation[3,8,22]. If it is assumed that the observation period is long compared to the dominant time constants of the process, then the gain matrix in (15) becomes a time-invariant matrix, $\underline{\underline{K}}$. There is a significant practical advantage in using the stationary form of the Kalman filter since only a single matrix, $\underline{\underline{K}}$, rather than a sequence of time-varying matrices $\{\underline{\underline{K}}(k)\}$, need be stored.

It should be noted that according to the Separation Theorem[3,9], the optimal control policy for the stochastic system of (1) and a quadratic performance index consists of the optimal feedback controller for the deterministic system acting on the optimal estimate from the Kalman filter. The Separation Theorem is valid if the process noise, the measurement noise and the initial state are Gaussian, an assumption which has not been necessary up to this point.

## DESCRIPTION OF THE PROCESS AND PROCESS MODEL

The pilot plant evaporator which served as the subject of the experimental and simulation studies is shown schematically in Fig. 1. The primary control objective is regulation of the product concentration C2 at the setpoint value of approximately 10% despite disturbances in feed flowrate F, feed concentration CF, and feed temperature TF. The liquid holdups in the first and second effect, W1 and W2, must also be maintained within acceptable operating limits. The primary control variables are the inlet steam flowrate S, the bottoms flowrate from the first effect B1, and the product flowrate from the second effect, B2. The pilot plant unit is fully instrumented and can be controlled by either conventional electronic controllers or by an IBM 1800 digital computer.

Several state-space models of the evaporator have been derived by Newell and Fisher[15] using linearized mass and energy balances. In this investigation a theoretical fifth-order model of the form of (1) was employed with the states, controls etc. defined as normalized, perturbation variables (see Appendix). The definitions of these variables and their steady-state values are presented in the Notation section. Of the five state variables in this model, only the first effect concentration, C1, is not measured.

The process noice vector $\underline{w}$ in (1) was assumed to consist of six elements corresponding to the three disturbance variables and the three control variables. Thus matrix $\underline{\underline{\Gamma}}$ in (1) was specified to be $\underline{\underline{\Gamma}} = [\underline{\underline{\Delta}} \vdots \underline{\underline{\theta}}]$.

## APPLICATION OF THE LUENBERGER OBSERVER

The fifth order evaporator model was used to design several first and second order observers which provided estimates of two state variables, the first effect concentration C1 and the first effect temperature T1 (which is a linear function of state variable H1). These observers were then employed in a multivariable feedback controller and were evaluated using simulated and experimental response data.

### Design Considerations

The basic decisions that must be made in designing an observer are the choices of $\underline{\underline{E}}$, $\underline{\underline{H}}$, $\underline{\underline{F}}$ and $\underline{\underline{G}}$ subject to the constraints imposed by (8)-(10) and by the invertibility of the matrix in (12). Several design methods are available in the literature[4,14,18, 24]including the original approach of Luenberger[10]which was used in this investigation.

In Luenberger's design method, $\underline{\underline{E}}$ and $\underline{\underline{H}}$ are initially specified which then allows the transformation matrix $\underline{\underline{T}}$ to be calculated from (8). Matrices $\underline{\underline{F}}$ and $\underline{\underline{G}}$ are then calculated from (9) and (10) and the reconstruction matrix, $\underline{\underline{L}}$, defined by

$$\underline{\underline{L}} \equiv \begin{bmatrix} \underline{\underline{T}} \\ \underline{\underline{C}} \end{bmatrix}^{-1} \tag{17}$$

is calculated from (17). Luenberger[10]has shown that if $\underline{\underline{\phi}}$ and $\underline{\underline{E}}$ have no common eigenvalues and if the system is observable, then $\underline{\underline{T}}$ exists and $\underline{\underline{L}}$ is invertible. The major advantage of this design method is that the designer can arbitrarily specify $\underline{\underline{E}}$ and hence the dynamics of the observer. The disadvantages of Luenberger's approach are the difficulty of calculating $\underline{\underline{T}}$ and the possible ill-conditioning of $\underline{\underline{L}}$. Also, it is not obvious how matrix $\underline{\underline{H}}$ should be assigned.

The specification of the eigenvalues of $\underline{\underline{E}}$ is an important design decision. As indicated in the previous section, the eigenvalues of the composite system consist of the eigenvalues of the observer plus those of the closed-loop system that exists when the state feedback controller acts on $\underline{x}$ rather than $\hat{\underline{x}}$. Consequently, it is desirable that the dynamics of the observer be somewhat faster than the system dynamics[12,13]. However, if the eigenvalues of $\underline{\underline{E}}$ are made too small, then the observer becomes overly sensitive to noise. Consequently, a compromise is required. In designing observers for the evaporator, it was arbitrarily assumed that the $\underline{\underline{E}}$ matrix was diagonal.

### Simulation Results

Under ideal conditions where the process model, initial conditions and disturbances are known, an observer will reconstruct the state vector exactly. In the simulation study it was of interest to evaluate the performance of observers under more realistic conditions where the initial state was not known exactly and/or where process disturbances occurred of which the observer was unaware. Such disturbances will be cited in this study as "unknown" or "unmeasured" disturbances.

The effectiveness of two second-order observers in providing estimates of C1 and T1 under open-loop conditions is shown in Figs. 2 and 3. The "fast observer" has an $\underline{\underline{E}}$ matrix with small eigenvalues, $\underline{\underline{E}}$ = diag. [0.001, 0.002], while the "slow observer" consists of an $\underline{\underline{E}}$ matrix with large eigenvalues, $\underline{\underline{E}}$ = diag. [0.945, 0.940]. The fast observer provides much better estimates of T1 and recovers more rapidly from the incorrect initial estimate of C1 in Fig. 2.

Steady-state estimation errors due to unknown disturbances during open-loop operation are shown in Table 1. These results suggest that there is no single $\underline{\underline{E}}$

TABLE 1. STEADY-STATE ESTIMATION ERRORS FOR SECOND ORDER LUENBERGER OBSERVERS

| Eigenvalues of $\underline{\underline{E}}$ | UNMEASURED STEP DISTURBANCE | | | | | |
|---|---|---|---|---|---|---|
| | +20% feed flow | | -30% feed conc. | | +20% feed enthalpy | |
| | error (%) $\hat{C}1$ | $\hat{H}1$ | error (%) $\hat{C}1$ | $\hat{H}1$ | error (%) $\hat{C}1$ | $\hat{H}1$ |
| 0.945, 0.940 | 2 | 21 | 4 | 168 | 7 | -62 |
| 0.7, 0.8 | 29 | 8 | 14 | -4 | -4 | -3 |
| 0.5, 0.6 | 46 | 0.4 | 5 | -0.1 | -6 | -2 |
| 0.3, 0.4 | 54 | 0.1 | 2 | -0.02 | -7 | -2 |
| 0.1, 0.2 | 59 | 0.6 | 0.1 | -0.2 | -8 | -2 |
| 0.001, 0.002 | 62 | 1 | -0.8 | -0.4 | -8 | -2 |

which is superior for all three disturbances. A similar conclusion was drawn concerning $\underline{\underline{C}}$ with (18) providing a reasonable compromise. (It was used for Table 1 and Figs. 2-6).

$$\underline{\underline{C}} = \begin{bmatrix} 1 & 1 & 1 \\ 1 & 1 & 1 \end{bmatrix} \tag{18}$$

Several first order observers were also designed to estimate C1[6] and the resulting steady-state estimation errors are shown in Table 2 for various values of the scalar, E. Here the larger values of E result in smaller steady-state estimation errors. (For Table 2, $\underline{\underline{C}}$ = [1 1 1 1]).

TABLE 2. STEADY-STATE ESTIMATION ERRORS FOR FIRST ORDER LUENBERGER OBSERVERS

| Value of E | UNMEASURED STEP DISTURBANCE | | |
|---|---|---|---|
| | +20% feed flow | -30% feed concentration | +20% feed enthalpy |
| | error in $\hat{C}1$(%) | error in $\hat{C}1$(%) | error in $\hat{C}1$(%) |
| 0.9 | 16 | 23 | -5 |
| 0.8 | 35 | 11 | -15 |
| 0.7 | 43 | 7 | -31 |
| 0.6 | 49 | 5 | -61 |
| 0.5 | 53 | 4 | -134 |
| 0.4 | 64 | 3 | -588 |
| 0.3 | 56 | 2 | 420 |
| 0.2 | 51 | 2 | 184 |
| 0.1 | 53 | 2 | 128 |
| 0.01 | 55 | 1 | 105 |

In Fig. 4 the second-order observers supply estimates of T1 and C1 to the optimal multivariable feedback controller used by Hamilton et al[7] and Oliver et al[21]. As in the open-loop runs of Figs. 2 and 3, the fast observer recovers more rapidly and generates significantly better estimates of T1. (The triangles along the vertical axis denote the normal steady state while those along the horizontal axis denote the start of a step disturbance).

Experimental Results

The experimental open-loop response data in the bottom half of Fig. 5 indicates that neither observer performs satisfactorily when an unknown step disturbance in CF occurs. The fast observer yields very noisy estimates while the T1 estimate of the slow observer is very poor. Since both of these observers were unsatisfactory, a third observer with $\underline{\underline{E}}$ = diag [0.7, 0.8] and the $\underline{\underline{C}}$ in (18) was considered. Its estimates of T1 and C1 are shown in the top half of Fig. 5 and represent a significant improvement over those in the bottom half of Fig. 5. This second order observer also performed well under experimental closed-loop conditions as illustrated in Fig. 6. When the optimal feedback controller [7,21] is used, the control is satisfactory although the response to the feed flow disturbance is oscillatory. The first order observer (E=0.9) also resulted in good closed-loop control (cf. bottom of Fig. 6).

## APPLICATION OF THE KALMAN FILTER

In both the simulation and experimental studies it was assumed that all state variables except C1 are measured and that $\underline{\underline{Q}}$ and $\underline{\underline{R}}$ are diagonal matrices with equal diagonal elements, i.e.

$$\underline{\underline{Q}} = q\underline{\underline{I}}_s \quad \text{and} \quad \underline{\underline{R}} = r\underline{\underline{I}}_\ell \tag{19}$$

where $\underline{\underline{I}}_s$ and $\underline{\underline{I}}_\ell$ are identity matrices of dimension s and $\ell$. The scalar ratio, r/q, uniquely determines the time-invariant gain matrix, $\underline{\underline{K}}$.

This application has been presented in greater detail elsewhere [6,7]. The following sections of this paper have two purposes: i) to summarize the pertinent results for comparison with the observer and ii) to present some new results not reported in [7].

Simulation Results

In evaluating the Kalman filters, Gaussian noise elements with zero means and standard deviations of 0.1 were employed. Hence the theoretically correct value of r/q was 1.0.

The effective filtering of noisy measurements ($\sigma$=0.1) is shown in Fig. 7. When no filter is used, the closed-loop response is excessively oscillatory due to the high noise levels and large gains in the optimal controller [6,7]. By contrast the Kalman filter provides smoother control which is closer to the "deterministic response" for the ideal state feedback controller.

Other simulation results [6,7] demonstrate that when the theoretically correct value of r/q=1 is used, the closed-loop system can be quite sensitive to unknown process disturbances. These results and those shown in Table 3 demonstrate that the sensitivity can be reduced by using smaller r/q values than are justified based on the actual (or assumed) noise statistics.

Comparing Tables 2 and 3 suggests that in this application, the Kalman filter is much less sensitive to unknown step disturbances than is the Luenberger observer. However the advantage of the observer is that only unmeasured state variables are affected.

TABLE 3. EXPECTED STEADY-STATE ESTIMATION ERRORS FOR THE KALMAN FILTER

| r/q | Error (%) $\hat{W}1$ | $\hat{C}1$ | $\hat{H}1$ | $\hat{W}2$ | $\hat{C}2$ | Unmeasured Disturbance |
|---|---|---|---|---|---|---|
| 1:1 | -30 | 4 | ≈0 | -10 | -1 | |
| 1:100 | ~0 | -23 | ~0 | ~0 | ~0 | +20% feed flow |
| 100:1 | -248 | 9 | ~0 | -388 | 11 | |
| 1:1 | 2 | 26 | ~0 | 6 | 16 | |
| 1:100 | ~0 | 5 | ~0 | ~0 | ~0 | -30% feed concentration |
| 100:1 | ~0 | 30 | ~0 | 8 | 30 | |
| 1:1 | 5 | -1 | -2 | 4 | -3 | |
| 1:100 | ~0 | ~0 | ~0 | ~0 | ~0 | +20% feed enthalpy |
| 100:1 | 35 | -4 | -2 | 1826 | -12 | |

Experimental Results

The conclusions drawn from the simulation studies were confirmed by the experimental results[6,7]. Since the actual measurement and process noise levels are relatively low ($\sigma$=0.01 and 0.02, respectively), zero-mean Gaussian noise with $\sigma$=0.1 was added to the measurements (via computer software) in order to provide a more severe test of the Kalman filter. Consequently, the filter for the experimental investigation was designed using a value of r/q = 25 corresponding to assumed process and measurement noise levels of 0.02 and 0.10, respectively. Details concerning the implementation of multivariable computer control techniques have been reported elsewhere[6,16,17].

The effect of using theoretically incorrect r/q values and/or having unmeasured process disturbances are illustrated by the closed-loop experimental results of Figs. 8 and 9. When the theoretically correct r/q value of 25 is used and the filter is aware of the step disturbances, the optimal controller maintains excellent control of the primary controlled variable, C2 (cf. top part of Fig. 8). However, when this filter is unaware of the disturbance, control of C2 and the liquid holdups is very poor. (cf middle portion of Fig. 8).

If the filter is designed using smaller r/q values, the closed-loop system is less sensitive to unknown disturbances (cf. Figs. 8 and 9). As r/q decreases the offset in W1 is reduced but the filter is less successful in eliminating the effects of noise as illustrated by the oscillatory responses for r/q=0.25.

## CONCLUSIONS

Several Kalman filters and Luenberger observers have been successfully applied to a computer-controlled, pilot scale evaporator. In both experimental and simulation studies, these state estimation techniques provided satisfactory estimates to an optimal multivariable controller.

In the evaporator application, the Luenberger observer performed well under normal conditions but was quite sensitive to process noise and unmeasured process disturbances. Even though there is considerable freedom in designing observers, it was difficult to specify an observer which gave reasonable state estimates for all three types of process disturbances.

The stationary form of the discrete Kalman filter provided satisfactory state estimates even in the presence of significant noise levels, uncertain noise statistics and modelling errors. Its sensitivity to unmeasured process disturbances could be reduced by treating the process noise covariance matrix as a design parameter and increasing the numerical values of its elements.

ACKNOWLEDGEMENT

Financial support from the National Research Council of Canada and the University of Alberta is gratefully acknowledged.

REFERENCES

[1] ALEVISAKIS, G. and D.E. SEBORG: An extension of the Smith predictor method to multivariable linear systems containing time delays, Int. J. Control, 17, 541-551 (1973).

[2] ALEVISAKIS, G. and D.E. SEBORG: Control of multivariable systems containing time delays using a multivariable Smith predictor, Chem. Eng. Science (in press).

[3] ATHANS, M.: The role and use of the stochastic-linear-quadratic-Gaussian problem in control system design, IEEE Trans. Auto. Control, AC-16, 529-552 (1971).

[4] FORTMANN, T.E. and D. WILLIAMSON: Design of low-order observers for linear feedback control laws, IEEE Trans. Auto. Control, AC-17, 301-308 (1972).

[5] GOLDMANN, S.E. and R.W.H. SARGENT: Applications of linear estimation theory to chemical processes: a feasibility study, Chem. Eng. Science, 26, 1535-1553 (1971).

[6] HAMILTON, J.C.: An experimental investigation of state estimation in multivariable control systems, M.Sc. Thesis, Univ. of Alberta (1972).

[7] HAMILTON, J.C., D.E. SEBORG and D.G. FISHER: An experimental evaluation of Kalman filtering, AIChE Journal (in press).

[8] KALMAN, R.E.: A new approach to linear filtering and prediction problems, Trans ASME, J. Basic Eng., 82, 35-45 (1960).

[9] KWAKERNAAK, H. and R. SIVAN: Linear optimal control systems, Wiley-Interscience, New York (1972).

[10] LUENBERGER, D.G.: Observing the state of a linear system, IEEE Trans. Mil. Elect., MIL-8, 74-80 (1964).

[11] LUENBERGER, D.G.: Observers for multivariable systems, IEEE Trans. Auto. Control, AC-11, 190-197 (1966).

[12] LUENBERGER, D.G.: An introduction to observers, IEEE Trans. Auto. Control, AC-16, 596-602 (1971).

[13] MACFARLANE, A.G.J.: A survey of some recent results in linear multivariable feedback theory, Automatica, 8, 455-492 (1972).

[14] MUNRO, N.: Computer-aided-design procedure for reduced-order observers, Proc. IEE, 120, 319-324 (1973).

[15] NEWELL, R.B. and D.G. FISHER: Model development, reduction and experimental evaluation for a double effect evaporator, Ind. Eng. Chem. Process Design & Develop., 18, 976-984 (1972).

[16] NEWELL, R.B. and D.G. FISHER: Experimental evaluation of optimal, multivariable regulatory controllers with model-following capabilities, Automatica,8,247-262(1972).

[17] NEWELL, R.B., D.G. FISHER and D.E. SEBORG: Computer control using optimal multivariable feedforward-feedback algorithms, AIChE Journal, 18, 976-984 (1972).

[18] NEWMANN, M.M.: Design algorithms for minimal-order Luenberger observers, Elect. Letters, 5, 391-393 (1969).

[19] NIEMAN, R.E. and D.G. FISHER: Experimental evaluation of time-optimal, open-loop control, Trans. Instn. Chem. Engrs., 51, 132-140 (1973).

[20] NIEMAN, R.E. and D.G. FISHER: Experimental evaluation of optimal, multivariable servo control in conjunction with conventional regulatory control, Chem. Eng. Communications (in press).

[21] OLIVER, W.K., D.E. SEBORG and D.G. FISHER: Model reference adaptive control based on Liapunov's direct method, Chem. Eng. Communications (in press).

[22] SAGE, A.P. and J.L. MELSA: Estimation theory with applications to communications and control, McGraw-Hill (1971).

[23] WILSON, R.G., D.E. SEBORG and D.G. FISHER: Modal approach to control law reduction, Proc. 1973 JACC., 554-565 (1973).

[24] YUKSEL, Y.O. and J.H. BONGIORNO: Observers for linear multivariable systems with applications, IEEE Trans. Auto. Control, AC-16, 603-613 (1971).

## NOTATION

| | State Vector, $\underline{x}$: | Normal Steady State Value |
|---|---|---|
| W1 | First effect holdup | 20.8 kg |
| C1 | First effect concentration | 4.59% glycol |
| H1 | First effect enthalpy | 441 kJ/kg |
| W2 | Second effect holdup | 19.0 kg |
| C2 | Second effect concentration | 10.11% glycol |
| | Control Vector, $\underline{u}$: | |
| S | Steam flowrate | 0.91 kg/min |
| B1 | First effect bottoms flowrate | 1.58 kg/min |
| B2 | Second effect bottoms flowrate | 0.72 kg/min |
| | Disturbance Vector, $\underline{d}$: | |
| F | Feed flowrate | 2.26 kg/min |
| CF | Feed concentration | 3.2% glycol |
| HF | Feed enthalpy | 365 kJ/kg |
| | Output Vector, $\underline{y}$: | |

$\underline{y}^T$ = [W1, H1, W2, C2] or [W1, W2, C2]

## APPENDIX

The state-space evaporator model is of the form of (1) with the elements of $\underline{x}$, $\underline{u}$, $\underline{y}$ etc. defined as normalized perturbation variables, e.g.

$$x_1 = \frac{W1 - W1_{ss}}{W1_{ss}}$$

where $W1_{ss}$ is the steady-state value of W1. The coefficient matrices for a sampling period of 64 seconds are:

$$\underline{\underline{\phi}} = \begin{bmatrix} 1 & -0.0008 & -0.0912 & 0 & 0 \\ 0 & 0.9223 & 0.0871 & 0 & 0 \\ 0 & -0.0042 & 0.4376 & 0 & 0 \\ 0 & -0.0009 & -0.1052 & 1 & 0.0001 \\ 0 & 0.0391 & 0.1048 & 0 & 0.9603 \end{bmatrix}$$

$$\underline{\underline{\Delta}} = \begin{bmatrix} -0.0119 & -0.0817 & 0 \\ 0.0116 & 0 & 0 \\ 0.1569 & 0 & 0 \\ -0.0138 & 0.0848 & -0.0406 \\ 0.0137 & -0.0432 & 0 \end{bmatrix} \quad \underline{\underline{\theta}} = \begin{bmatrix} 0.1182 & 0 & -0.0050 \\ -0.0351 & 0.0785 & 0.0049 \\ -0.0136 & -0.0002 & 0.0662 \\ 0.0012 & 0 & -0.0058 \\ -0.0019 & 0.0016 & 0.0058 \end{bmatrix}$$

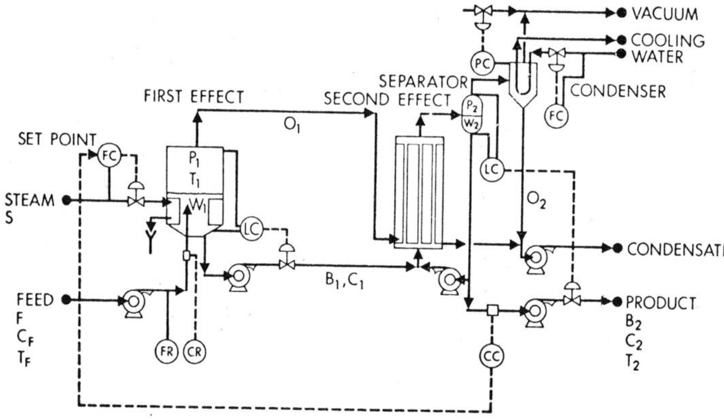


Fig 1. Schematic diagram of the pilot plant evaporator and a conventional multiloop controller.

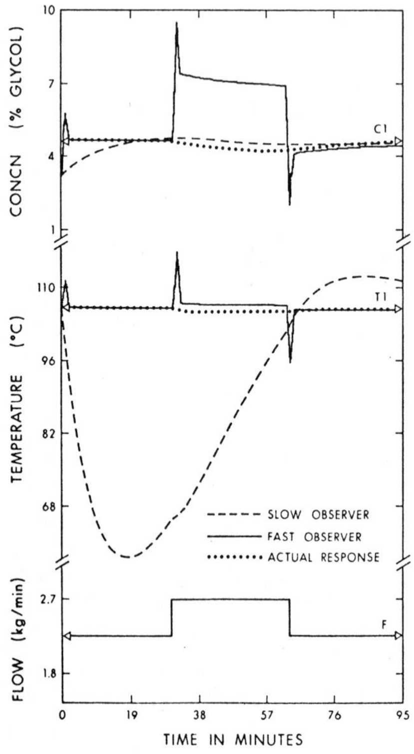


Fig 2. Simulated open-loop responses for an incorrect initial estimate of C1 and two "unknown", 20% step disturbances in feed flowrate, F.

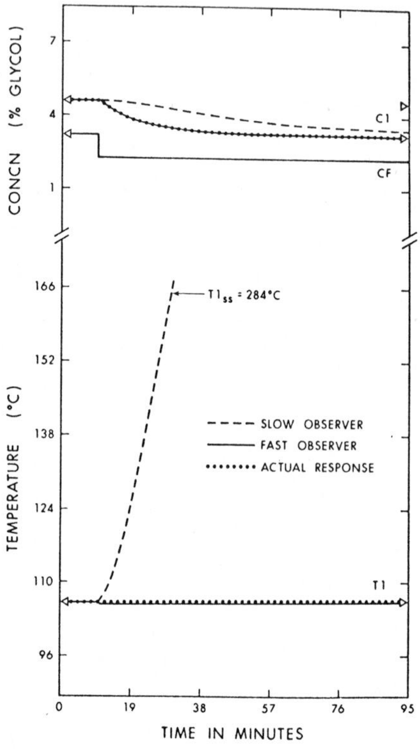


Fig 3. Simulated open-loop responses for an "unknown", -30% step disturbance in feed concentration, CF.

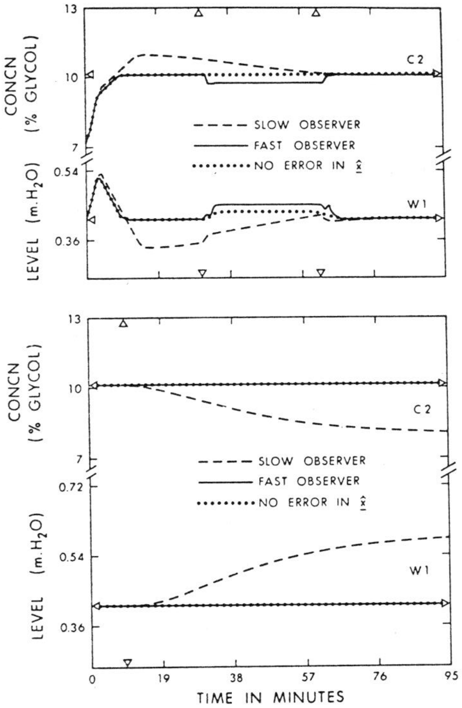


Fig 4. Simulated closed-loop responses. Top: incorrect initial C1 estimate and two "unknown", 20% step disturbances in F. Bottom: "unknown", -30% step disturbance in CF.

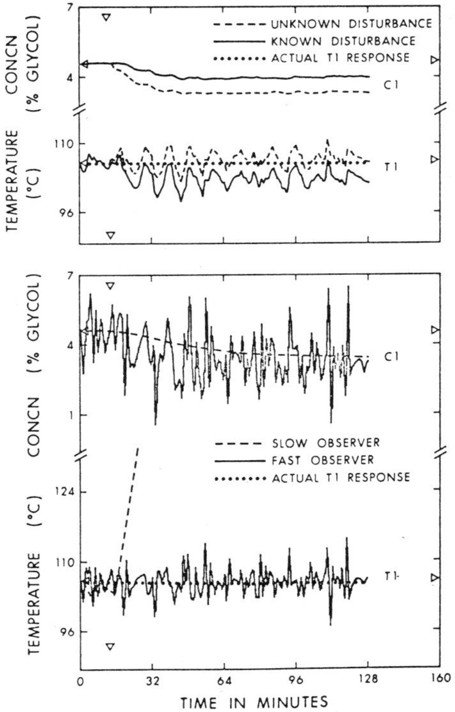


Fig 5. Observer estimates based on experimental open-loop response data for a -30% step disturbance in CF.

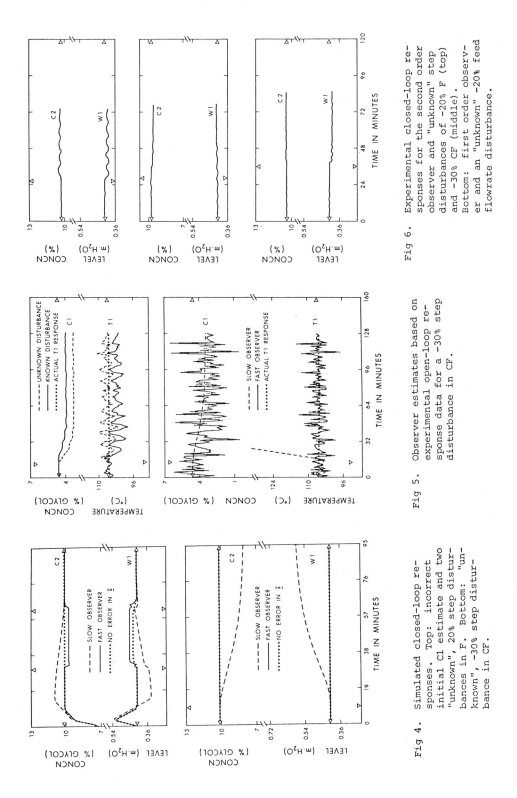

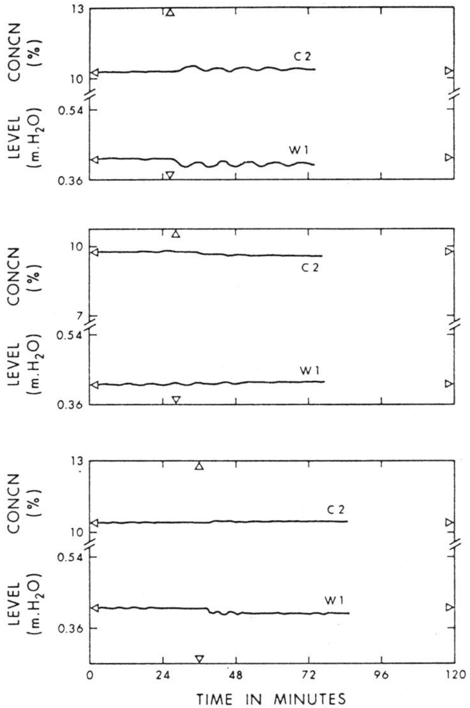


Fig 6. Experimental closed-loop responses for the second order observer and "unknown" step disturbances of -20% F (top) and -30% CF (middle). Bottom: first order observer and an "unknown" -20% feed flowrate disturbance.

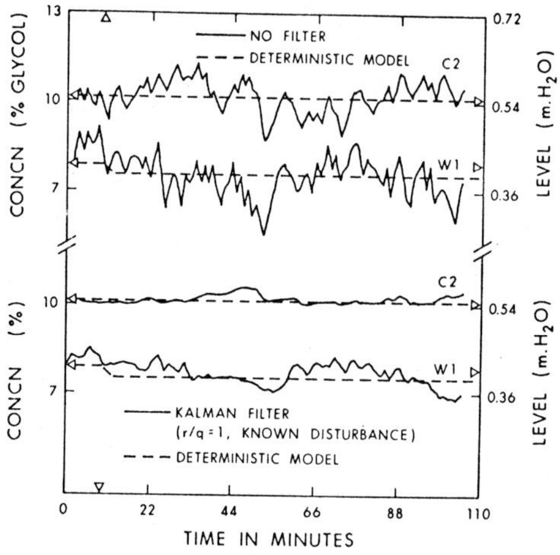


Fig 7. Simulated closed-loop responses to an "unknown", -20% step disturbance in F.

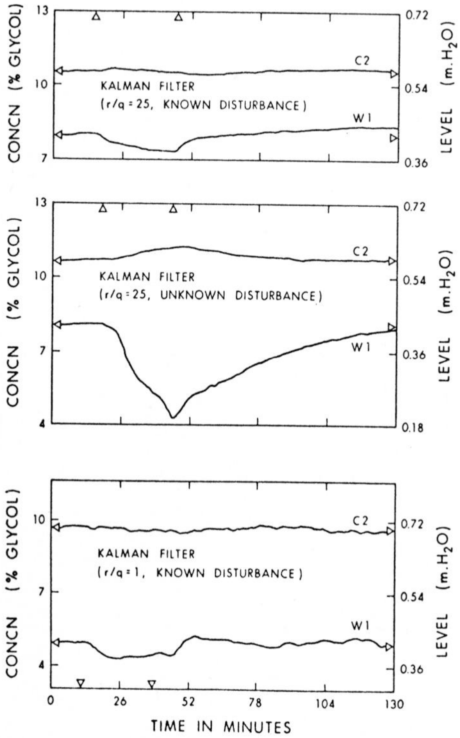


Fig 8. Experimental closed-loop responses to two, "unknown", 20% step changes in F.

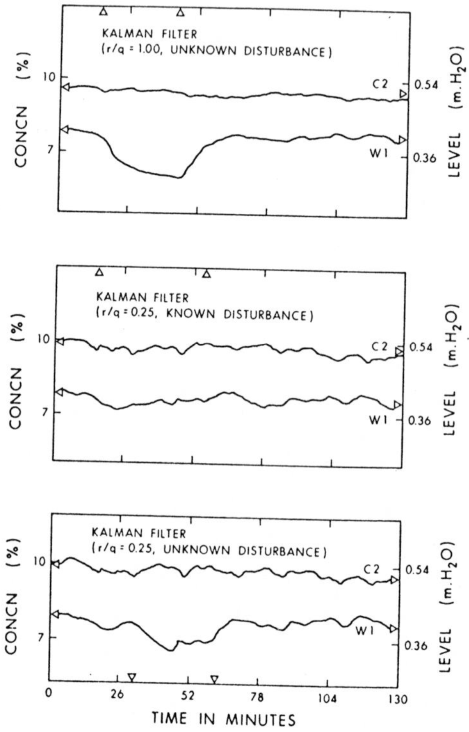


Fig 9. Experimental closed-loop responses to two, "unknown", 20% step changes in F.

# Section 8: EDUCATIONAL ASPECTS OF COMPUTER CONTROL

CONTENTS:



COMMENTS:

Since this work was performed at a university, one of the prime considerations was the education of students at both the graduate and undergraduate levels. The paper in this section describes some of the student laboratories that were associated with these computer control studies and summarizes some of the educational aspects.

REFERENCES:

[1] Seborg, D.E. and D.G. Fisher, "Computer-Aided Student Experiments at the University of Alberta", Proc. of the Workshop on the Undergraduate Chemical Engineering Laboratory, ASEE Summer School, Boulder, Colorado, August, 1972.

[2] Interim Report of the Computer Aids for Chemical Engineering Education (CAChE) Real-Time Task Force, October, 1973. (This report is available from the United States National Academy of Engineering's Commission on Education, Washington, D.C.)

**EDUCATION:**

# Real-Time Computing in Engineering Education

**D. Grant Fisher**
Department of Chemical and Petroleum Engineering
University of Alberta
Edmonton, Alberta, Canada

This paper describes the experience, over the last three years, of the Department of Chemical and Petroleum Engineering at the University of Alberta with real-time computers applied to graduate, undergraduate, and research programs. The Department's DACS Centre includes a staff of five full-time people and operates an IBM 1800 digital computer, a DEC PDP8I–based communication system, an AD32 analog computer, and an EAI590 hybrid computing system. Time-shared applications include direct digital control, multivariable control, gas chromatograph and IR spectrophotometer monitoring, on-line optimization, control system analysis, computer-aided instruction, optimal control of pilot-plant units, and hybrid computer simulations. Instructional programs include computer-oriented fourth-year control courses, a graduate course in real-time computer applications, "hands-on" computer control labs, and extension courses and workshops for people in industry. Real-time interactive computing has been an effective tool in student education and has increased the quantity, quality, and scope in many areas of department research.

THIS PAPER DESCRIBES THE REAL-TIME COMPUTER facilities in the Department of Chemical and Petroleum Engineering at the University of Alberta, and reports on our experience over the last two and one-half years. The objective is to present a broad overview of real-time computing in engineering education and research and to illustrate by example some factors that arise in the design, operation, and use of such facilities.

In the following sections on computing hardware and software and student education and research projects, an attempt is made to list general observations and conclusions. Unfortunately, due to the breadth of the field and a shortage of space, many of the conclusions will be left largely unsupported and some of the points will be expressed in computer-oriented jargon. However, further information is available in a 70-page booklet published by the Department (*1*)* and in the reports and theses associated with projects summarized in the booklet.

***Copies of reference 1 may be obtained free of charge by writing the author.***

## DACS CENTRE

Planning for the Department's Data Acquisition, Control and Simulation (DACS) Centre began in 1965–66, motivated primarily by the growing need to familiarize engineers with real-time computer systems and to permit research and development in the various application areas. Planning and program development started immediately, but equipment delivery was delayed until December 1967, so that it could be installed in the Department's new facilities in Phase I of the new Engineering Centre. The department has 14 full-time academic staff members working in a wide variety of research areas. A typical graduating class would be 40 students and there are normally about 50 graduate students in residence. Therefore the computing facilities had to be designed to meet a wide variety of continuously changing applications.

Present facilities of the DACS Centre include an IBM 1800 digital computer, a Digital Equipment Corporation 680 (PDP8I) communications system, an Electronics Associates, Inc. 590 hybrid computing system, an Applied Dynamics 32 PB analog computer, and support equipment such as patchable digital logic units, recorders, etc. The design and operation of the Centre has been determined primarily by our objective of providing a powerful, flexible system that would be readily accessible and convenient for relatively large numbers of students and research personnel to use.

Figure 1 summarizes the facilities in block diagram form. The IBM 1800, 2 $\mu$ sec. central processing unit, 48K of 16-bit word length core storage, and three 500K disk cartridge storage units form the heart of the digital system. The conventional data-processing peripherals are shown at the top of Figure 1 and include a card reader/punch, a line printer, a digital plotter, typewriter units, and provision for linkage to the University's IBM 360/67 computer. These facilities, under the control of the "nonprocess" operating system, comprise a conventional digital computer such as would be found in any small data-processing center.

Chemical Engineering Computing Workshop, Vol. I, 52-58 (1972).

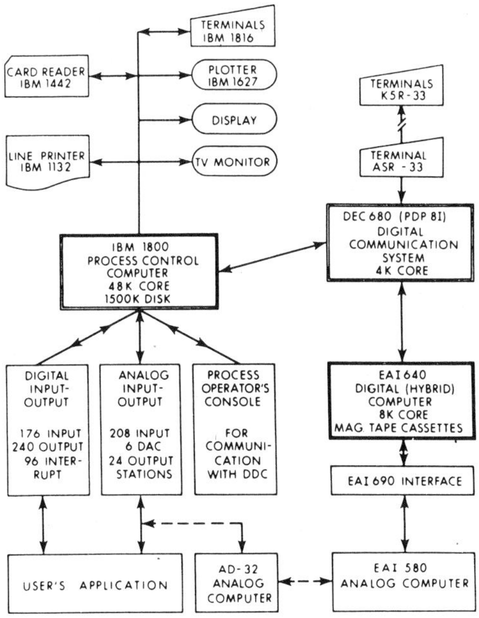


**Figure 1. Data Acquisition, Control and Simulation Centre computing facilities.**

It has the ability, for example, to compile and execute student FORTRAN programs.

The principal features that distinguish a real-time computer from a data-processing unit are: (1) process input/output equipment; (2) real-time clocks; and (3) priority interrupt system.

As an overgeneralization, the process I/O equipment shown at the bottom of Figure 1 permits the digital computer to gather, or send, information directly to a process or experimental installation in a manner analogous to use of a card reader or line printer. For example, thermocouples can be connected directly to the computer. Under program control the particular thermocouple of interest is "addressed" (that is, connected to the computer by closing the appropriate multiplexer switch) and the voltage is amplified, sent to the analog-to-digital converter (ADC) and the equivalent digital value is stored in core. Once in core the value is functionally indistinguishable from data obtained from more familiar sources, such as punched cards, and can be processed further by other computer programs. Digital values can also be converted into an equivalent analog voltage, by using digital-to-analog (DAC) converters, and sent directly to conventional process instruments such as control valves. Digital input/output hardware has the obvious use of determining whether a voltage is present or not and for opening or closing electrical switches. However, it is also used at the University of Alberta to interface to digital instruments (voltmeters, thermometers) and other digital computers. In general, through the provision of proper interface equipment, any electrical instrument or device can be read or actuated under control of statements in the user's computer program. Thus the process input/output hardware can be used to eliminate the need for a user to read a value from his experiment, punch it onto a card, and then feed the card to the computer. However, with the capability of handling thousands of readings per second it has much more sophisticated uses.

Real-time clocks and timers permit the user to have his program executed once, at a specific time of day, or repeatedly, at any specified interval. For example, our basic DDC program is executed every second and can be used to acquire a series of data points from an experiment.

Hardware "interrupt" capability means that when a process event occurs, to which the user has assigned a higher priority than the program the computer is currently executing, the computer will transfer control to a program associated with the new event. For example, if the computer is doing a routine job and an emergency occurs in a process it is monitoring, then control of the computer will be transferred to a program that will deal with the emergency. For example, in the case of a high-temperature alarm from a chemical reactor the computer could generate an analog signal that would increase the cooling water flow to the reactor. When the high priority program is finished the computer returns to where it left off in the original job. An interrupt may be communicated to the computer by a digital signal (for example, a switch connected to the computer) or by a statement in another program. Hardware interrupts accessible to the user are essential for using the computer to deal with several jobs "simultaneously" and to operate asynchronously with slower peripherals and process applications.

## Selection of Equipment

The facilities summarized in Figure 1 include most of the features common to real-time computer installations but with a configuration designed to meet the special needs of a university. It is perhaps instructive to examine some of the factors that led to selection of the present DACS Centre facilities. Our contemplated applications had no critical requirements that seriously restricted our choice of computers. We wanted a complete hardware/software system that would allow us to get a quick start on use of the system for applications and student instruction. No one associated with the project at that time had any previous experience in operating a computer, and the University Centre could not provide extensive assistance because of other commitments, so reliable equipment and local systems support were essential. The reasons for purchasing a medium size system such as the IBM 1800 (in 1966) included:

1. Full range of desired hardware features
2. Reliable, proven data-processing peripherals
3. "Complete" time-sharing operating system, including master interrupt servicing program, clock management program, program sequence control, nonprocess monitor, FORTRAN IV compiler, FORTRAN callable programs for process I/O, disk and file management utilities
4. Local systems support and maintenance
5. Training courses and extensive documentation
6. Stability and continuity of product line
7. Availability of application programs (for example, DDC)
8. Large users' group and program library
9. Potential support for link to University's 360/67

Also implicit in the selection was the desire, for educational purposes, to have a system typical of what the students might encounter in industry and to provide a broad range of services to a large number of users. It was decided that rather than allocating the system among users by time-slicing or by giving each user a virtual computer of his own, the DACS Centre would develop monitor programs to perform the most common data acquisition, control, and service functions. The success of this approach is demonstrated by experience that shows that, not counting nonprocess users, about 25% of the DACS Centre users (mainly student labs) do essentially no programming; 40% (mainly graduate student projects) write relatively small programs to augment the standard options or implement specific data-processing requirements; 25% do relatively extensive FORTRAN programming that requires some knowledge of how the computer system operates; and about 10% get intimately involved with the computer system and/or assembly language programming. Provision of such a high level of system support requires a full-time staff of at least five people. In our experience three people are required to maintain and operate (and make the inevitable changes and extensions to) the basic computer system and two people to maintain the application program library and assist users. The DACS Centre operates on an open-shop, 24-hr. day basis with no staff operator.

## Terminal System

Another basic design decision was to orient the DACS Centre around a system of terminals located in the user's own lab or research area. The selection of the DEC 680 system was made on the basis of the number of terminals envisaged; the desire to have relatively inexpensive, general-purpose terminals that users could purchase; and provision for buffered I/O from the main computer system which would always be "live" from the point of view of the user. A live terminal enables the user to enter information at any time and have it routed by the monitor system to the correct system program or to a file for later use by the user's program. The ability to "queue" (initiate execution of) a program through the terminal keyboard and to change parameters in system programs, such as DDC, gives the necessary control to the users.

Unfortunately, system support is not provided to permit program development, compilation, execution, and modification via the terminal system. However, the availability of the card reader and line printer makes these functions much more convenient when done in the DACS Centre. Our experience while developing the programming for the DEC 680 system (PDP8I, 4K core, and teletype I/O) illustrated the inconvenience of relying on teletypes and/or slow-speed paper tape systems for I/O. Basically after about six weeks of working with the PDP8 our systems people wrote a compiler for the PDP8 that would execute on the IBM 1800. The increased speed of development due to the better peripherals, increased core size, etc., of the 1800 quickly paid back the one month of development time. The DEC 680 system is interfaced to the 1800 (and will permit communication at up to 50,000 words per second) so the object code could be loaded directly into the PDP8I for execution.

## Hybrid Computing

The hybrid computing facilities were motivated primarily by research interests and were partially supported by a grant from the National Research Council. However the EAI 590 system will be interfaced to the DEC 680 system so that information can be passed among all three computers. Although it will not operate in a time-sharing mode, the EAI 590 provides an independent system for single user program development and real-time applications in addition to its primary purpose of hybrid computation. As discussed in the next section, hybrid computation is also used by graduate and undergraduate students in process dynamics, control, and optimization studies.

## Cost of Facilities

An important consideration in any computer installation is cost. The DEC 680 communications system and the hybrid computing facilities were purchased, but the IBM 1800 system is rented for approximately $8000/month. The equivalent capital costs are shown in Table 1. Since the payout time for computing equipment is about 40 months, it is normally financially advantageous to purchase the equipment. On the other hand, rental gives the flexibility to reconfigurate the system to meet changing applications and/or incorporate new

**Table 1. Approximate capital costs.**

| | |
|---|---|
| IBM 1800 | $325,000 |
| DEC 680 | $ 30,000 |
| EAI 590 | $135,000 |

equipment. However rental puts the computer in a distinctly different position than most research equipment and means that it must be rejustified annually. With the exception of the $37,000 grant from the National Research Council for hybrid facilities, all the finances are obtained from the regular Faculty of Engineering budget.

## EDUCATION

Other than instruction in computers and computing techniques the most extensive educational application of the DACS Centre has been in the six graduate and undergraduate process dynamics and control courses offered by the Department. The main advantages concern: (1) assistance with design and analysis; (2) flexible implementation of labs and demonstrations; and (3) comparison of methods and identification of system structure.

### Design and Analysis

The Department has developed two digital simulation programs for control system design and analysis (CSAP and CSDAP) that, in addition to calculating the time domain response, can also provide Bode diagrams, Nyquist diagrams, root locus plots, the equivalent state-space matrix representation of a control system, estimations of suitable control constants, and evaluation of different criteria for comparing the performance of control systems. Both programs are "conversational" in operation and can display output on typewriters, display scopes, TV monitors, and/or the digital plotter. Thus the user can define his own problem, apply different design and/or analysis techniques, and then evaluate his result by analog, digital, hybrid, or physical implementation. The opportunity to formulate his own problem, the immediate feedback of results, and the direct comparison of alternative methods are all powerful educational advantages. In general the computer programs eliminate the busy work and make it feasible for the student to evaluate alternatives before proceeding with implementation.

We have also noted in labs dealing with optimization that a good program library and rapid turn around of computer jobs make it possible for a class to get first-hand experience on the relative advantages and pitfalls of different techniques.

### Laboratory Implementation

The demonstration units for computer control are a pilot-plant size, double-effect evaporator, an eight-tray distillation column, and other smaller pieces of equipment such as heat exchangers. These units can be operated with conventional industrial instruments and/or under computer supervisory, or direct digital, control. The principal educational advantage is that the control configurations, the control modes, and the control constants are all implemented by computer programs and are easily changed by the student. Thus instead of "cookbook" procedures to make the lab equipment do what it was designed to do the student has relatively unlimited opportunity to innovate and apply knowledge learned in the courses. (In some cases it has been observed that students are bewildered by the large choice of alternatives and would rather have a cookbook assignment.) The data acquisition and control monitor programs are powerful enough that students do not have to do any programming to implement data acquisition, limit checking, input data processing such as square root calculations or digital filtering, standard proportional-integral-derivative or nonlinear control algorithms, cascade control, etc. Service programs are also available to plot or list the experimental data. The most effective procedure seems to be: (1) familiarize the student with the physical application and equipment; (2) let him operate the equipment under manual and/or conventional control; and (3) introduce computer control.

In many cases a fourth step, analog or hybrid simulation, is extremely effective. The student can program the analog computer to solve the mathematical model he has developed based on a theoretical analysis or experimental testing. The same digital computer control programs and techniques as used on the physical experiment can then be used to control the simulated process, and the results compared. All that is required in the most direct case is to change the input/output addresses and possibly change the timing routines. This permits a very direct evaluation of the suitability of the model, the need for experimental tuning of parameters, and the significance of physical assumptions. It also illustrates directly the relationship between terms in the model and components of the physical system. The convenience of hybrid computing can also be used to screen a number of alternatives so that only the most promising need be implemented on the physical system. Finally, the user can turn to purely digital simulation and evaluate its pros and cons. Several of the student projects have encompassed all these steps, albeit with varying degrees of success.

### System Structure

The principal advantage claimed for computer-assisted instruction of this type is that it allows the user to "see the forest in spite of the trees." That is, when each step in the design and analysis procedure is computerized, the student, even though he might have some doubts about "how" some steps are implemented, can learn "what" each step does and see its relation to the overall procedure. Thus he can experiment with and evaluate system design techniques—something which is generally too time-consuming to do by hand. With proper design of the course material, the various examples and assignments done throughout the term can be combined at the end of the term into an effective demonstration of the total design process and a review of the individual steps.

Many of the points brought out in the next section on research applications also apply to student labs. Other practical points with respect to students labs are that the amount of busy work (that is, data processing) can be reduced; it is difficult for the student to "fake" data; and since the computer will not usually accept "vague" instructions his understanding of the application is well-documented by the computer log. (We have noticed, however, that many students are very reluctant to try new options for fear of making errors and careful guidance must be given so they can learn without being embarrassed by their mistakes.)

If the system is sufficiently protected against user errors and is programmed to assist the uninformed user (that is, upon demand, types out step by step instructions), then one of the most profitable experiences for many students is to simply "experiment" on his own. The only supervision required is to answer questions and give general guidance. (However, it is a lot harder for the instructor to "debug" and grade individual problems as compared with a single class assignment.)

## RESEARCH

It is impossible to deal with individual research projects in detail because most of them require separate reports to describe the work adequately. However, it is possible to generalize about the types of research applications and some of the principal benefits of using real-time computers. Work within our Department can be categorized as: (1) research projects in which the computer is an essential part; (2) research projects that use the computer as an effective tool to assist with applications that could be, or were, run without a computer; (3) research and development of computer hardware and software systems; and (4) hybrid computation and digital simulation.

Most of the process dynamics and control studies fall into the first category. The computer has been used successfully to implement multiloop (*4*), feed forward (*4–6*), inferential (*4, 5*), multivariable-optimal-regulatory (*7*), optimal state-driving (*8*), noninteracting, and adaptive control techniques on the pilot-plant equipment. Other studies include computer-implemented process identification of pilot-plant units with on-line display of results (*9, 10*); computer control of an 8-in., eight-tray distillation column (*11*); a real-time checking and adjustment of process data so that it is consistent with material and energy balance constraints (*12*); and the design/analysis of control systems (*13–15*). A more complete list of research projects making use of the DACS Centre is found in Appendices C and F of reference *1*. Most of the research papers describing the results of these studies are just in the process of being submitted for publication, but it is hoped that these studies will help to bridge the gap between theoretical developments and the practical applications of interest to industry. Most of these applications and parallel studies of parameter and process identification are simply not practical without a real-time computer.

Other research projects in chemical and petroleum engineering make use of the computer for data acquisition, process monitoring, logging, data reduction, and experimental documentation. Kinetic studies use a computer-controlled PE 621 infrared spectrophotometer and use the computer for acquisition of data from gas chromatograph analyzers and other process instruments. The computer is also used to monitor operation of the evaporator pilot plant so that it can be run 24 hr./day without any operator supervision.

Projects in category 3 are concerned primarily with application programs rather than developing alternatives to the computer manufacturer's operating systems. Typical examples include the CSAP and CSDAP simulation programs mentioned earlier, and a generalized monitor system to supervise the execution of series of discrete events such as are found in plant startups and shutdowns, batch operations, and system checkout. In general, these projects require a considerable degree of familiarity with the computer system as well as the application area and usually take longer than "typical" graduate student projects.

Hybrid computing is an area that has been widely reported in the literature. The interests of our Department are not to work extensively on the development of purely hybrid techniques or to get involved in that class of problems that are only practical if solved by hybrid techniques. Rather, the interests are directed toward student education and assisting research in other areas.

Specific advantages of the use of real-time computers include:

1. Increased quantity of research data due to faster operation or extended periods of operation. Some of our M.Sc. thesis projects now involve several times the amount of data collected in earlier Ph.D. studies.

2. Increased quality, precision, and reproducibility of data due to precisely implemented procedures, replicated runs, automatic recalibration, continuous monitoring of data during each experiment, and elimination of random human errors and bias between different operators.

3. Broader experimental studies because the incremental effort required to extend the data acquisition and/or processing is often minimal.

4. Cooperation between different people because the computer acts as a standard "interface" so the experimentalist can implement the work of the theoretician and the data-processing specialist can work with "real" data. Projects tend to become more interdisciplinary.

5. More continuity and carryover from one research

student to the next because of computerized procedures.

6. Precise documentation (the program itself!) and standardized, tested, and approved methods for data reduction, presentation, and interchange between groups.

7. Reduction of busy work and more challenge to the researcher to examine critically and improve both his techniques and his results. (From a student point of view the use of a computer makes many traditional areas of research more attractive as thesis projects.)

8. Permits design or evolution of projects into areas that are not possible without a computer due to the speed of operation, the degree of control required, or the amount of data that must be processed.

9. A much more "flexible" investment for the research dollar than special purpose instruments such as multichannel recorders, etc. They can be pooled for large applications or reallocated to meet changing needs of smaller projects.

The literature abounds with reports of research work that made use of computers and hence the reader is referred to the literature in his own area of interest for specific applications and details.

## OTHER APPLICATIONS

### Laboratory Automation

The replacement of laboratory technicians by laboratory automation is a justification for many industrial computer installations. However, it is not as important a factor in universities because of the rapidly changing requirements, specialized techniques, and relatively small number of any specific test. However, a computer can be of tremendous assistance since many of the advantages discussed earlier are applicable. In our Department the best examples are control of an IR spectrophotometer (*2*) and computer control of gas chromatographs (*3*). Several man-years of work have gone into developing and supporting GC applications. As a starting point the IBM "stand-alone" GC application package was modified to run in conjunction with the DDC and other application programs on the IBM 1800. The system was then modified to handle both research and process chromatographs, and results from process GCs can be transferred at regular intervals to the DDC monitor program for use in closed-loop control. The program is currently undergoing extensive modifications to make it better suited for our relatively small, time-shared applications.

### Batch Processing

As discussed earlier, the time-sharing capabilities of a computer such as the IBM 1800 can be used for standard data-processing functions such as handling student FORTRAN programs. However, the reader is advised to evaluate carefully the implications of small word size (round off errors), absence of "floating-point hardware" (long execution time), relatively small amounts of available core, low-speed peripherals, FORTRAN compilers designed primarily for real-time applications, and such factors as interactions between the "nonprocess" applications and the real-time computer applications. Our DACS Centre is used extensively for program development and reduction of data but we do not encourage or support use of the system for programs not related to real-time applications.

### Continuing Education

Finally, it is worthwhile to consider use of real-time computer systems in the areas of continuing education for engineers, demonstration of advanced techniques of interest to industry, and as a basis for cooperative industry-university programs. Our Department has cosponsored, with the Federal Department of Industry, seminars on computer applications, presented workshops and tours in conjunction with national meetings of technical societies, and offered extension courses in the evening for the benefit of local engineers. These activities can be regarded as "public relations" or "professional service" but in many universities they are beginning to be regarded as an important basic function.

## SYSTEM DESIGN CONSIDERATIONS

No discussion of real-time computer applications is complete without consideration of the role of minicomputers.

Minicomputers are appearing on the market in increasing numbers and there has been a corresponding improvement in the availability of low-cost, high-performance peripherals. Minicomputers represent an excellent means of implementing specific applications such as data acquisition, control, and other dedicated or single-user applications. However, they do not appear, as yet, to meet the need for a powerful, multi-user facility for program development and checkout. Perhaps the solution in the university environment will be an interconnected hierarchy of computers consisting of a large central computer, several satellite supervisory computers, and minicomputers in individual laboratories. The large central computers could be used, either through simulation or emulation, to develop and debug application programs. The satellite computers would act as concentrators and buffers between the large central computer and the minicomputers, and perhaps supervise large, shared, bulk storage facilities and handle communications with the on-line users. The minicomputers could then be dedicated to implementation of specific applications such as process control, data acquisition from research experiments, etc. This arrangement has several advantages:

1. The central computer can provide support for powerful high level languages and could maintain a common library of service programs and subroutines.

2. The minicomputers would in most cases be dedicated to a single application and problems of interaction between users, and conflicting demands for system resources would be largely eliminated.

3. Direct communication could be handled between the minicomputer and the user but more powerful communication systems, possibly involving CRT display units, would probably be more advantageously handled by the satellite computers.

4. Finally, blocks of data that have been collected in the minicomputers could be checked, converted, condensed, formatted, etc., and then sent to the larger computer for further processing.

However, while the hierarchy of computers might be desirable to meet the needs of an entire university, it is obviously not suitable for smaller installations such as that presently met by our DACS Centre. The trend will be toward several interconnected computers, but the division of functions between the computers and the best way of providing the flexibility required for university applications is not yet clear.

## REFERENCES

1. DACS Centre Booklet, Dept. Chem. Petrol. Eng., Univ. Alberta, Edmonton, Canada (1970).
2. Chuang, T., G. Misko, I. G. Dalla Lana, and D. G. Fisher, in *Computers in Analytical Chemistry*, Plenum Press, New York (1969).
3. Coxhead, P., M.Sc. thesis, Univ. Alberta, Edmonton, Canada (1969).
4. Jacobson, B. A., M.Sc. thesis, Univ. Alberta, Edmonton, Canada (1970).
5. Fehr, M., M.Sc. thesis, Univ. Alberta, Edmonton, Canada (1969).
6. Wilson, A. H., M.Sc. thesis, Univ. Alberta, Edmonton, Canada (1967).
7. Newell, R. B., and D. G. Fisher, paper presented at IFAC meeting, Helsinki, Finland (June 1971).
8. Nieman, R. E., and D. G. Fisher, *Proc. Can. Natl. Conf. Automatic Control* (Aug. 1970).
9. Lees, R. S., M.Sc. thesis, Univ. Alberta, Edmonton, Canada (1969).
10. Wood, R. K., and T. A. Wildman, *Proc. Can. Natl. Conf. Automatic Control* (Aug. 1970).
11. Wood, R. K., and W. C. Pacey, paper presented at 20th Can. Chem. Eng. Conf., Sarnia, Ont. (Oct. 1970).
12. Nieman, R. E., and D. G. Fisher, *Dept. Res. Rept.* 700401, Univ. Alberta, Edmonton, Canada (1970).
13. Lofkrantz, E., M.Sc. thesis, Univ. Alberta, Edmonton, Canada (1967).
14. Agostinis, W., M.Sc. thesis, Univ. Alberta, Edmonton, Canada (1969).
15. Farwell, R. A., M.Sc. thesis, Univ. Alberta, Edmonton, Canada (1970).

# APPENDIX A

## STATE SPACE MODELS OF THE PILOT PLANT EVAPORATOR

A variety of state space models have been developed for the pilot plant evaporator based on linearized material and energy balances and model reduction techniques. Andre [1,2] built the evaporator and proposed the first mathematical model, a fifth order state space model. The most rigorous modelling study of the evaporator to date is due to Newell [3,4] who derived a nonlinear tenth order model based on material and energy balances. Linearization of this model resulted in a continuous-time, linear state space model of the form of equations (A.1) and (A.2):

$$\dot{\underline{x}} = \underline{\underline{A}}\underline{x}(t) + \underline{\underline{B}}\underline{u}(t) + \underline{\underline{D}}\underline{d}(t) \qquad (A.1)$$

$$\underline{y}(t) = \underline{\underline{C}}\underline{x}(t) \qquad (A.2)$$

The equivalent discrete-time state space model can be written as [3]:

$$\underline{x}(j+1) = \underline{\underline{\Phi}}\underline{x}(j) + \underline{\underline{\Delta}}\underline{u}(j) + \underline{\underline{\Theta}}\underline{d}(j) \qquad (A.3)$$

$$\underline{y}(j) = \underline{\underline{C}}\underline{x}(j) \qquad (A.4)$$

The elements of vectors $\underline{x}$, $\underline{u}$, $\underline{d}$, and $\underline{y}$ are defined as normalized perturbation variables, e.g.

$$x_3 = \frac{W1 - W1_{ss}}{W1_{ss}}$$

where $W1_{ss}$ is the normal steady-state value of W1. These vectors are defined in terms of the process variables in Figure A.1 as follows:

| State Vector, $\underline{x}$: | | Normal Steady State Value |
|---|---|---|
| TS | Steam temperature ($x_1$) | 118°C |
| TW1 | First effect tube wall temperature ($x_2$) | 108°C |
| W1 | First effect holdup ($x_3$) | 20.8 kg. |
| C1 | First effect concentration ($x_4$) | 4.59% glycol |
| H1 | First effect enthalpy ($x_5$) | 441 kJ/kg. |
| TW2 | Second effect tube wall temperature ($x_6$) | 83°C |
| W2 | Second effect holdup ($x_7$) | 19 kg. |
| C2 | Second effect concentration ($x_8$) | 10.1% glycol |
| H2 | Second effect enthalpy ($x_9$) | 312 kJ/kg. |
| TW3 | Condenser tube wall temperature ($x_{10}$) | 42°C |

| Control Vector, $\underline{u}$: | | Normal Steady State Value |
|---|---|---|
| S | Steam flow ($u_1$) | 0.91 kg/min. |
| B1 | First effect bottoms flow ($u_2$) | 1.58 kg/min. |
| B2 | Second effect bottoms flow ($u_3$) | 0.72 kg/min. |

Disturbance Vector, $\underline{d}$:

| | | |
|---|---|---|
| F | Feed flow ($d_1$) | 2.26 kg/min. |
| CF | Feed concentration ($d_2$) | 3.2% glycol |
| HF | Feed enthalpy ($d_3$) | 365 kJ/kg. |

Output Vector, $\underline{y}$:

$\underline{y} = (W1, W2, C2)^T$

The coefficient matrices of the continuous-time model (for these steady state conditions) are shown in Table A.1 and those of the corresponding discrete-time model and a 64 second sampling interval are shown in Table A.2. The system modal matrix, $\underline{\underline{M}}$, and the corresponding eigenvalues of $\underline{\underline{A}}$ and $\underline{\underline{\phi}}$ are shown in Table A.3.

A fifth-order state space model derived by Newell [3,4] has also been widely used in the evaporator studies. The state vector in the fifth-order model is $\underline{x}^T = [W1, C1, H1, W2, C2]$ and the coefficient matrices for the continuous-time and discrete-time models are shown in Tables A.4 and A.5, respectively. These coefficient matrices are for the normal steady-state conditions cited above.

Wilson [5,6] used several model reduction techniques to derive third-order models from the continuous and discrete tenth-order models. When Marshall's modal approach [6,7] was used with $\underline{x}^T = [W1, W2, C2]$, the third-order models in Tables A.6 and A.7 were obtained.

The state space models in Tables A.1 through A.7 have been used in the evaporator research projects since 1973. The earlier studies (prior to 1973) used similar models but the coefficient matrices were different because the models were linearized about different steady state conditions.

REFERENCES:

[1] Andre, H., Ph.D. Thesis, University of Alberta, Edmonton, Canada (1966).

[2] Andre, H. and R.A. Ritter, "Dynamic Response of a Double Effect Evaporator", Can. J. Chem. Eng., 46, 259 (1968).

[3] Newell, R.B., Ph.D. Thesis, University of Alberta, Edmonton, Canada (1971).

[4] Newell, R.B. and D.G. Fisher, "Model Development, Reduction and Experimental Evaluation for a Double Effect Evaporator", Ind. Eng. Chem. Design and Develop., 11, 212 (1972).

[5] Wilson, R.G., Ph.D. Thesis, University of Alberta, Edmonton, Alberta (1974).

[6] Wilson, R.G., D.G. Fisher and D.E. Seborg, "Model Reduction for Discrete Time Dynamic Systems", Internat. J. Control, 16, 549 (1972).

[7] Marshall, S.A., "An Experimental Method for Reducing the Order of a Linear System", Control, 10, 642 (1966).

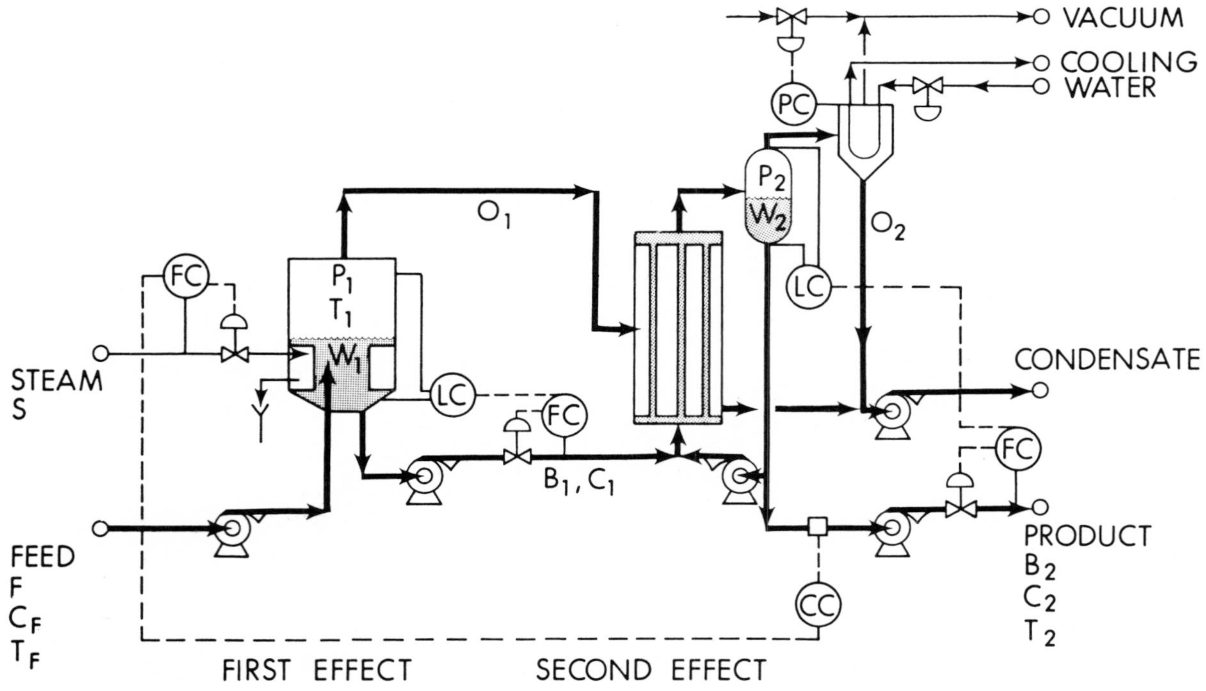


Fig. A.1: Schematic diagram of the pilot scale evaporator and a conventional multiloop control system.

TABLE A.1

CONTINUOUS TENTH ORDER EVAPORATOR MODEL

A =

| | | | | | | | | | |
|---|---|---|---|---|---|---|---|---|---|
| -0.2978E 03 | 0.2724E 03 | 0.0000E 00 | 0.0000E 00 | 0.0000E 00 | 0.0000E 00 | 0.0000E 00 | 0.0000E 00 | 0.0000E 00 | 0.0000E 00 |
| 0.1043E 02 | -0.4862E 02 | 0.0000E 00 | 0.2820E 00 | 0.3262E 02 | 0.0000E 00 | 0.0000E 00 | 0.0000E 00 | 0.0000E 00 | 0.0000E 00 |
| 0.0000E 00 | -0.1348E 00 | -0.5238E-06 | -0.2818E-03 | -0.3265E-01 | 0.1360E 00 | 0.0000E 00 | 0.0000E 00 | 0.0000E 00 | 0.0000E 00 |
| 0.0000E 00 | 0.1348E 00 | -0.1790E-04 | -0.7630E-01 | 0.3265E-01 | -0.1360E 00 | 0.0000E 00 | 0.0000E 00 | 0.0000E 00 | 0.0000E 00 |
| 0.0000E 00 | 0.1126E 02 | -0.3660E-06 | -0.8823E-01 | -0.1028E 02 | 0.6947E 00 | 0.0000E 00 | 0.0000E 00 | 0.0000E 00 | 0.0000E 00 |
| 0.0000E 00 | 0.0000E 00 | 0.0000E 00 | 0.8278E-01 | 0.9574E 01 | -0.4552E 02 | 0.0000E 00 | 0.5587E 00 | 0.2738E 02 | 0.0000E 00 |
| 0.0000E 00 | 0.0000E 00 | 0.0000E 00 | -0.1148E-05 | -0.1670E-03 | -0.6412E-02 | 0.0000E 00 | -0.2596E-02 | -0.1273E 00 | 0.1021E 00 |
| 0.0000E 00 | 0.0000E 00 | 0.0000E 00 | 0.3811E-01 | 0.1670E-03 | 0.6412E-02 | -0.2915E-04 | -0.3548E-01 | 0.1273E 00 | -0.1021E 00 |
| 0.0000E 00 | 0.0000E 00 | 0.0000E 00 | -0.8138E-05 | 0.1171E 00 | 0.4505E 01 | -0.1539E-05 | -0.8959E-01 | -0.4429E 01 | 0.7620E 00 |
| 0.0000E 00 | 0.0000E 00 | 0.0000E 00 | 0.0000E 00 | 0.0000E 00 | 0.0000E 00 | 0.0000E 00 | 0.4205E-01 | 0.2062E 01 | -0.8204E 01 |

B =

| | | |
|---|---|---|
| 0.0000E 00 | 0.0000E 00 | 0.0000E 00 |
| 0.0000E 00 | 0.0000E 00 | 0.0000E 00 |
| 0.1100E 00 | -0.7334E-05 | -0.1028E-02 |
| -0.3348E-01 | 0.7660E-01 | 0.1028E-02 |
| -0.1765E-01 | -0.3819E-04 | 0.8586E-01 |
| 0.0000E 00 | 0.0000E 00 | 0.0000E 00 |
| 0.0000E 00 | 0.0000E 00 | 0.0000E 00 |
| 0.0000E 00 | 0.0000E 00 | 0.0000E 00 |
| 0.0000E 00 | 0.0000E 00 | 0.0000E 00 |
| 0.0000E 00 | 0.0000E 00 | 0.0000E 00 |

D =

| | | |
|---|---|---|
| 0.2053E 02 | 0.0000E 00 | 0.0000E 00 |
| 0.0000E 00 | 0.0000E 00 | 0.0000E 00 |
| 0.0000E 00 | -0.7658E-01 | 0.0000E 00 |
| 0.0000E 00 | -0.3482E-09 | 0.0000E 00 |
| 0.0000E 00 | 0.0000E 00 | 0.0000E 00 |
| 0.0000E 00 | 0.0000E 00 | 0.0000E 00 |
| 0.0000E 00 | 0.8391E-01 | -0.3808E-01 |
| 0.0000E 00 | -0.4579E-01 | -0.8545E-08 |
| 0.0000E 00 | 0.3414E-01 | 0.0000E 00 |
| 0.0000E 00 | 0.0000E 00 | 0.0000E 00 |

## TABLE A.2

### DISCRETE TENTH ORDER EVAPORATOR MODEL

$\underline{\underline{\phi}}$ =

| | | | | | | | | | |
|---|---|---|---|---|---|---|---|---|---|
| 0.3836E-02 | 0.1092E 00 | -0.2568E-06 | 0.3219E-02 | 0.3757E 00 | 0.1501E-01 | -0.1414E-06 | 0.1911E-02 | 0.9376E-01 | 0.8400E-02 |
| 0.4188E-02 | 0.1193E 00 | -0.2814E-06 | 0.3515E-02 | 0.4102E 00 | 0.1641E-01 | -0.1553E-06 | 0.2090E-02 | 0.1024E 00 | 0.9192E-02 |
| -0.6923E-03 | -0.1978E-01 | 0.9998E 00 | -0.5016E-03 | -0.5753E-01 | 0.5370E-02 | -0.4962E-07 | 0.6722E-03 | 0.3296E-01 | 0.2933E-02 |
| 0.6573E-03 | 0.1878E-01 | -0.1835E-04 | 0.9217E 00 | 0.5487E-01 | -0.5013E-02 | 0.4812E-07 | -0.6369E-03 | -0.3123E-01 | -0.2799E-02 |
| 0.4961E-02 | 0.1413E 00 | -0.1851E-06 | -0.3800E-02 | 0.4860E 00 | 0.1967E-01 | -0.1934E-06 | 0.2518E-02 | 0.1234E 00 | 0.1110E-01 |
| 0.2950E-02 | 0.8418E-01 | -0.1406E-06 | 0.2438E-02 | 0.2919E 00 | 0.2297E-01 | -0.7174E-06 | 0.3745E-02 | 0.1837E 00 | 0.1900E-01 |
| -0.2556E-03 | -0.7329E-02 | 0.5755E-08 | -0.2180E-03 | -0.2633E-01 | -0.6145E-02 | 0.9998E 00 | -0.1168E-02 | -0.5731E-01 | 0.7482E-02 |
| 0.2687E-03 | 0.7704E-02 | -0.3787E-06 | 0.3846E-01 | 0.2721E-01 | 0.5823E-02 | -0.3054E-04 | 0.9610E 00 | 0.5503E-01 | -0.7216E-02 |
| 0.3158E-02 | 0.9017E-01 | -0.1113E-06 | 0.1806E-02 | 0.3140E 00 | 0.3064E-01 | -0.5159E-06 | -0.1436E-01 | 0.2560E 00 | 0.2725E-01 |
| 0.7696E-03 | 0.2197E-01 | -0.2511E-07 | 0.6338E-03 | 0.7682E-01 | 0.8611E-02 | -0.2671E-06 | 0.1498E-02 | 0.7357E-01 | 0.7963E-02 |

$\underline{\underline{\Delta}}$ =

| | | |
|---|---|---|
| 0.1857E 00 | 0.2200E-02 | 0.2128E-08 |
| 0.1280E 00 | 0.2417E-02 | 0.2346E-08 |
| -0.9644E-02 | -0.8090E-01 | 0.7565E-09 |
| 0.9347E-02 | -0.7480E-03 | -0.7402E-09 |
| 0.1338E 00 | 0.3009E-02 | 0.3003E-08 |
| 0.5457E-01 | 0.1116E-01 | 0.1688E-07 |
| -0.2161E-02 | 0.8823E-01 | -0.4063E-01 |
| 0.2283E-02 | -0.4662E-01 | 0.6155E-06 |
| 0.4601E-01 | 0.1867E-01 | 0.1546E-07 |
| 0.9651E-02 | 0.4158E-02 | 0.5803E-08 |

$\underline{\underline{\theta}}$ =

| | | |
|---|---|---|
| -0.8858E-02 | 0.3069E-03 | 0.4238E-01 |
| -0.9710E-02 | 0.3364E-03 | 0.4645E-01 |
| 0.1180E 00 | -0.3028E-04 | -0.4170E-02 |
| -0.3491E-01 | 0.7847E-01 | 0.4038E-02 |
| -0.1155E-01 | -0.2673E-03 | 0.5671E-01 |
| -0.4878E-02 | 0.1656E-03 | 0.2334E-01 |
| 0.2005E-03 | -0.6683E-05 | -0.9603E-03 |
| -0.9045E-03 | 0.1601E-02 | 0.1013E-02 |
| -0.4134E-02 | 0.1053E-03 | 0.1985E-01 |
| -0.8757E-03 | 0.2913E-04 | 0.4192E-02 |

TABLE A.3

MODAL MATRIX AND EIGENVALUES OF $\underline{\underline{A}}$ AND $\underline{\underline{\phi}}$ FOR TENTH ORDER EVAPORATOR MODEL

$\underline{\underline{M}}$ =

| | | | | | | | | | |
|---|---|---|---|---|---|---|---|---|---|
| -0.3410E-07 | -0.1097E-06 | 0.2825E-05 | -0.1559E-03 | -0.3852E 00 | -0.1901E 00 | -0.2514E-01 | -0.8693E 00 | 0.8483E 00 | 0.9985E 00 |
| -0.3253E-07 | -0.1154E-06 | 0.6100E-05 | -0.1711E-03 | -0.4207E 00 | -0.2070E 00 | 0.3171E-01 | 0.4336E 00 | -0.1778E 00 | 0.1907E-01 |
| -0.7428E 00 | -0.7034E 00 | -0.1962E-02 | -0.3791E-04 | -0.7805E-01 | -0.5549E-01 | 0.1348E-02 | 0.5399E-03 | -0.1810E-02 | 0.2844E-03 |
| 0.1741E-03 | 0.1650E-03 | -0.9696E-05 | 0.7121E 00 | 0.1056E 00 | 0.5815E-01 | -0.1383E-02 | 0.5311E-03 | 0.1826E-02 | -0.2860E-03 |
| -0.1552E-05 | -0.1571E-05 | 0.9881E-05 | -0.6361E-02 | -0.5009E 00 | -0.2372E 00 | -0.2371E-02 | -0.1209E 00 | 0.3609E-01 | -0.1512E-01 |
| -0.6031E-06 | -0.6979E-06 | 0.1005E-04 | -0.2021E-03 | -0.3732E 00 | 0.4339E 00 | -0.6787E-01 | 0.2029E 00 | 0.4948E 00 | -0.4950E-01 |
| 0.6696E 00 | 0.7108E 00 | 0.3921E-03 | -0.4630E-03 | -0.1628E 00 | 0.4483E-01 | -0.1363E-01 | -0.2309E-03 | 0.7180E-04 | -0.1180E-04 |
| -0.3404E-03 | -0.3814E-03 | -0.9998E 00 | -0.7019E 00 | 0.1717E 00 | -0.4719E-01 | 0.1372E-01 | 0.1496E-03 | -0.4512E-04 | 0.7214E-05 |
| 0.5949E-05 | 0.6661E-05 | 0.2041E-01 | 0.1406E-01 | -0.4450E 00 | 0.7779E 00 | -0.1046E 00 | -0.1905E-01 | -0.5072E-01 | 0.5352E-02 |
| -0.2670E-06 | -0.2985E-06 | 0.3026E-05 | -0.6663E-04 | -0.1150E 00 | 0.2465E 00 | 0.9912E 00 | 0.1018E-01 | -0.5961E-02 | 0.8351E-03 |

Eigenvalues of $\underline{\underline{A}}$

| | | | | | | | | | |
|---|---|---|---|---|---|---|---|---|---|
| 0.1905E-05 | -0.1566E-04 | -0.3807E-01 | -0.7659E-01 | -0.2876 | -0.1705E 01 | -0.8420E 01 | -0.4656E 02 | -0.4911E 02 | -0.3089E 03 |

Eigenvalues of $\underline{\underline{\phi}}$

| | | | | | | | | | |
|---|---|---|---|---|---|---|---|---|---|
| 0.9999E 00 | 0.9999E 00 | 0.9600E 00 | 0.9212E 00 | 0.7354E 00 | 0.1622E 00 | 0.1271E-03 | 0.2166E-04 | 0.6938E-05 | -0.3728E-06 |

TABLE A.4

CONTINUOUS FIFTH ORDER EVAPORATOR MODEL

$$\mathbf{A} = \begin{bmatrix} 0.0000E\ 00 & -0.1085E{-}02 & -0.1254E\ 00 & 0.0000E\ 00 & 0.0000E\ 00 \\ 0.0000E\ 00 & -0.7549E{-}01 & 0.1254E\ 00 & 0.0000E\ 00 & 0.0000E\ 00 \\ 0.0000E\ 00 & -0.6036E{-}02 & -0.7740E\ 00 & 0.0000E\ 00 & 0.0000E\ 00 \\ 0.0000E\ 00 & -0.1219E{-}02 & -0.1447E\ 00 & 0.0000E\ 00 & 0.1258E{-}03 \\ 0.0000E\ 00 & 0.3930E{-}01 & 0.1447E\ 00 & 0.0000E\ 00 & -0.3796E{-}01 \end{bmatrix}$$

$$\mathbf{B} = \begin{bmatrix} 0.0000E\ 00 & -0.7658E{-}01 & 0.0000E\ 00 \\ 0.0000E\ 00 & 0.0000E\ 00 & 0.0000E\ 00 \\ 0.2159E\ 00 & 0.0000E\ 00 & 0.0000E\ 00 \\ 0.0000E\ 00 & 0.7945E{-}01 & -0.3808E{-}01 \\ 0.0000E\ 00 & -0.4135E{-}01 & 0.0000E\ 00 \end{bmatrix} \qquad \mathbf{D} = \begin{bmatrix} 0.1097E\ 00 & 0.0000E\ 00 & 0.0000E\ 00 \\ -0.3328E{-}01 & 0.7658E{-}01 & 0.0000E\ 00 \\ -0.1876E{-}01 & 0.0000E\ 00 & 0.9110E{-}01 \\ 0.0000E\ 00 & 0.0000E\ 00 & 0.0000E\ 00 \\ 0.0000E\ 00 & 0.0000E\ 00 & 0.0000E\ 00 \end{bmatrix}$$

TABLE A.5

DISCRETE FIFTH ORDER EVAPORATOR MODEL

$$\boldsymbol{\phi} = \begin{bmatrix} 0.9999E\ 00 & -0.7888E{-}03 & -0.9116E{-}01 & 0.0000E\ 00 & 0.0000E\ 00 \\ 0.0000E\ 00 & 0.9223E\ 00 & 0.8705E{-}01 & 0.0000E\ 00 & 0.0000E\ 00 \\ 0.0000E\ 00 & -0.4187E{-}02 & 0.4377E\ 00 & 0.0000E\ 00 & 0.0000E\ 00 \\ 0.0000E\ 00 & -0.8736E{-}03 & -0.1051E\ 00 & 0.9999E\ 00 & 0.1316E{-}03 \\ 0.0000E\ 00 & 0.3909E{-}01 & 0.1048E\ 00 & 0.0000E\ 00 & 0.9603E\ 00 \end{bmatrix}$$

$$\boldsymbol{\Delta} = \begin{bmatrix} -0.1191E{-}01 & -0.8169E{-}01 & 0.0000E\ 00 \\ 0.1157E{-}01 & 0.0000E\ 00 & 0.0000E\ 00 \\ 0.1568E\ 00 & 0.0000E\ 00 & 0.0000E\ 00 \\ -0.1374E{-}01 & 0.8474E{-}01 & -0.4062E{-}01 \\ 0.1371E{-}01 & -0.4323E{-}01 & 0.0000E\ 00 \end{bmatrix} \qquad \boldsymbol{\theta} = \begin{bmatrix} 0.1181E\ 00 & -0.3655E{-}04 & -0.5025E{-}02 \\ -0.3511E{-}01 & 0.7847E{-}01 & 0.4882E{-}02 \\ -0.1354E{-}01 & -0.1974E{-}03 & 0.6618E{-}01 \\ 0.1212E{-}02 & -0.4070E{-}04 & -0.5798E{-}02 \\ -0.1902E{-}02 & 0.1633E{-}02 & 0.5786E{-}02 \end{bmatrix}$$

TABLE A.6

CONTINUOUS THIRD ORDER EVAPORATOR MODEL

$$\underline{\underline{A}} = \begin{bmatrix} -0.4534E-06 & -0.3554E-07 & -0.1532E-06 \\ 0.8207E-07 & 0.2477E-06 & 0.1369E-05 \\ -0.9030E-05 & -0.2938E-04 & -0.3808E-01 \end{bmatrix}$$

$$\underline{\underline{D}} = \begin{bmatrix} 0.1124E\ 00 & -0.9173E-04 & -0.1263E-01 \\ 0.3052E-02 & -0.7374E-04 & -0.1466E-01 \\ -0.2091E-01 & 0.3823E-01 & 0.2095E-01 \end{bmatrix} \quad \underline{\underline{B}} = \begin{bmatrix} -0.3050E-01 & -0.7603E-01 & 0.7019E-08 \\ -0.3541E-01 & 0.8007E-01 & -0.3808E-01 \\ 0.5060E-01 & -0.4222E-01 & -0.2524E-07 \end{bmatrix}$$

TABLE A.7

DISCRETE THIRD ORDER EVAPORATOR MODEL

$$\underline{\underline{\phi}} = \begin{bmatrix} 0.9998E\ 00 & -0.3795E-07 & -0.2348E-06 \\ 0.8633E-07 & 0.9998E\ 00 & 0.1662E-05 \\ -0.9408E-05 & -0.3070E-04 & 0.9600E\ 00 \end{bmatrix}$$

$$\underline{\underline{\theta}} = \begin{bmatrix} 0.1200E\ 00 & -0.9780E-04 & -0.1345E-01 \\ 0.3249E-02 & -0.7866E-04 & -0.1561E-01 \\ -0.2177E-01 & 0.3979E-01 & 0.2183E-01 \end{bmatrix} \quad \underline{\underline{\Delta}} = \begin{bmatrix} -0.3250E-01 & -0.8108E-01 & 0.8226E-08 \\ -0.3770E-01 & 0.8539E-01 & -0.4063E-01 \\ 0.5272E-01 & -0.4413E-01 & 0.6023E-06 \end{bmatrix}$$